YOU JUST GAINED ACCESS TO

ATLAS™

YOUR ONLINE LEARNING PLATFORM!

Atlas will take you through this book and give you access to a host of additional resources, including:

• A free, full-length adaptive practice exam that simulates the real test, plus in-depth performance analytics and thorough explanations

• Additional official-format problems to hone your skills and test-wide strategies for maximizing your performance

HOW TO ACCESS ATLAS

If you purchased this guide from Manhattan Prep's online store:
Your resources are already in your account! Go to manhattanprep.com/gmat/studentcenter and log in with the same username and password you used to create your account.

If you purchased this guide at a retail location:
Go to manhattanprep.com/gmat/access and follow the instructions on the screen.

Online resource access can only be granted for a single account; the guide cannot be registered a second time for a different account.

You have one year from the date of purchase to complete your registration.

MANHATTAN
PREP

GMAT®
Foundations of Math

This supplemental guide provides in-depth and easy-to-follow explanations of the fundamental math skills necessary for a strong performance on the GMAT.

Acknowledgments

A great number of people were involved in the creation of the book you are holding.

Our Manhattan Prep resources are based on the continuing experiences of our instructors and students. The overall vision for this edition was developed by Stacey Koprince and Emily Madan, who determined what strategies to cover and how to weave them into a cohesive whole.

Stacey Koprince was the primary author of this edition and Emily Madan was the primary editor. Avi Gutman played a key role in developing and testing reasoning-based approaches that make math easier to do. Helen Tan provided project management and Mario Gambino managed production for all images.

Matthew Callan coordinated the production work for this guide. Once the manuscript was done, Naomi Beesen and Emily Meredith Sledge edited and Cheryl Duckler proofread the entire guide from start to finish. Carly Schnur designed the covers.

GMAT® Foundations of Math

Retail ISBN: 978-1-5062-4923-0
Course ISBN: 978-1-5062-4925-4
Retail eISBN: 978-1-5062-4924-7
Course eISBN: 978-1-5062-4926-1

GMAT® Strategy Guides

GMAT All the Quant

GMAT All the Verbal

GMAT Integrated Reasoning and Essay

Strategy Guide Supplements

Math

GMAT Foundations of Math

GMAT Advanced Quant

Verbal

GMAT Foundations of Verbal

January 7, 2020

Dear Student,

Thank you for picking up a copy of *Foundations of Math*. I hope this book provides just the guidance you need to get the most out of your GMAT studies.

At Manhattan Prep, we continually aspire to provide the best instructors and resources possible. If you have any questions or feedback, please do not hesitate to contact us.

Email our Student Services team at gmat@manhattanprep.com or give us a shout at 212-721-7400 (or 800-576-4628 in the United States or Canada). We try to keep all our books free of errors, but if you think we've goofed, please visit manhattanprep.com/GMAT/errata.

Our Manhattan Prep Strategy Guides are based on the continuing experiences of both our instructors and our students. The primary author of this edition of the Foundations of Math guide was Stacey Koprince and the primary editor was Emily Madan. Project management and design were led by Matthew Callan, Mario Gambino, and Helen Tan. I'd like to send particular thanks to instructors Avi Gutman and Emily Meredith Sledge for their content contributions.

Finally, we are indebted to all of the Manhattan Prep students who have given us excellent feedback over the years. This book wouldn't be half of what it is without their voice.

And now that *you* are one of our students too, please chime in! I look forward to hearing from you. Thanks again and best of luck preparing for the GMAT!

Sincerely,

Chris Ryan
Executive Director,
Product Strategy

TABLE OF CONTENTS

The GMAT Mindset

The GMAT is a complex exam. It feels like an academic test—math, grammar, logical reasoning—but it's really not! At heart, the GMAT is a test of your executive reasoning skills.

Executive reasoning is the official term for your ability to make all kinds of decisions in the face of complex and changing information. It makes sense, then, that graduate management programs would want to test these skills. It's crucial for you to understand *how* they do so because that understanding will impact both how you study for the GMAT and how you take the test.

You do need to know various math and grammar facts, rules, and concepts in order to do well on the GMAT—and this makes the test feel similar to tests that you took in school. There's one critical difference though: When your teachers gave you tests in school, they tested you on material they expected you to know how to handle. Your teachers wouldn't put something on the test that they *expected* you to get wrong. That would be cruel!

Well, it would be cruel if the main point of the exam was to test your mastery of those facts, rules, and concepts. But that isn't the main point of the GMAT. Rather, the GMAT wants to know how well you make decisions regarding when to invest your limited time and mental energy—and when *not* to.

In other words, the GMAT wants to know how you make business decisions. And no good businessperson invests in every single opportunity placed in front of them, just because it's there. A good businessperson evaluates each opportunity, saying yes to some and no to others. That's what you're going to do on the GMAT, too. You'll invest in a majority of the problems presented to you, but you *will* say no to some—the ones that look too hard or seem like they'll take too long to solve. These are literally bad investments.

So the GMAT will offer you questions that it thinks you will not be able to do. How does it accomplish this? The GMAT is an adaptive test; that is, it adapts to you as you take it, offering easier or harder questions based on how you're doing on the test. Ideally, you'll do well on the material that you know how to answer in a reasonable amount of time. Your reward? You'll earn questions that are too hard for you to do—either they'll take too long to answer or they'll be so hard that you wouldn't be able to do them even if you had unlimited time.

Then what? If you try to use a "school mindset" on the test, you'll keep trying to answer the questions even though you really can't do them. You'll waste a bunch of time and then, later, you'll have to rush on other questions. As a result, you'll start to miss questions that you actually do know how to answer and your score will go down. This is the business equivalent of spending most of your annual budget by August…and then not having enough money left to run the business well from September through December.

Instead, use your "business mindset" to carry you through the exam. When the test finds your limit, acknowledge that! Call it a bad investment and let that problem go (ideally before you've spent very much time on it). Choose an answer, any answer, and move on.

Extend the business mindset to your studies as well. If there are certain topics that you really hate, decide that you're not going to study them in the first place. You're just going to bail (guess quickly and move on) when one of those "opportunities" comes up. (One caveat: You can't bail on huge swaths of content. For example, don't bail on all of algebra; that represents too great a portion of the Quant section. You can, though, bail on a subset—say, absolute values and sequences.)

Start orienting yourself around your business mindset today. You aren't going to do it all. You're going to choose the best opportunities as you see them throughout the test. When you decide not to pursue a particular "investment," you're going to say no as quickly as you can and forget about it—don't waste precious resources on a poor investment opportunity! Move on to the next opportunity, feeling good about the fact that you're doing what you're supposed to do on the GMAT: making sound investment decisions about what to do and what *not* to do.

Get the Most out of FoM

FoM is short for Foundations of Math, the fundamental quant material that you need to know to do well on the GMAT. But FoM is going to help you with so much more—it's going to help you work with quant concepts in grad school and even during your career.

That might seem odd since, in the real world, you're always going to have access to a calculator, a computer, or Excel. Your boss is never going to make you do math on paper. So how will FoM help you on more than just this test?

What you're really going to learn in FoM is the ability to *think logically about quantitative concepts*. In fact, that's what standardized tests such as the GMAT are really testing—how well you understand what math means.

Since the exam is not *really* a math test, it's built to allow you to minimize or outright avoid a significant number of actual calculations. Throughout this guide and other guides in the Manhattan Prep series, we'll teach you exactly how to think about math in a way that gets you to the answer you need while avoiding tedious calculations whenever possible. And when you know how to do that, you'll also find it easier to learn how to analyze an earnings report, or understand the significance of a balance sheet, or do any number of other necessary things at work or in grad school.

When studying from FoM, keep two principles in mind. First, look for opportunities to minimize or avoid computation whenever possible. Eyeball, estimate, and think logically about the quant concepts whenever you can.

Second, don't do every last problem you see in this guide or in the associated online materials. Instead, do what *you* need. Try a couple problems of each type to see how it goes. If the work feels smooth and you get the correct answer, move on to something else in this guide or to higher-level material. If you make a mistake or the material just feels clunky, practice more now *and* set a reminder on your calendar to try this material again in a week or two.

FoM will give you a strong foundation to learn how to think logically about quant. As you progress through the guide, keep this overall principle in mind.

Arithmetic

In This Chapter

- Quick-Start Definitions
  - Basic Numbers
  - Greater Than and Less Than
  - Adding and Subtracting Positives and Negatives
  - Multiplying and Dividing
  - Distributing and Factoring
  - Multiplying Positives and Negatives
  - Fractions and Decimals
  - Divisibility and Even and Odd Integers
  - Exponents and Roots (and π)
  - Variable Expressions and Equations
- PEMDAS
- Combining Like Terms
- Distribution
- Pulling Out a Common Factor
- Long Addition and Subtraction
- Long Multiplication
- Long Division

1

In this chapter, you will learn to recognize math vocabulary so that you can set problems up correctly. You'll also learn how to apply PEMDAS, simplify math, and solve in ways that avoid tedious calculations.

CHAPTER 1 Arithmetic

This book has three goals for you. First, you will review fundamental math skills. Second, you will learn how to approach the math with a business focus: how to get to the correct answer quickly and effectively while avoiding drawn-out textbook math calculations that aren't necessary on the GMAT. Third, you will practice applying these skills.

To practice applying these skills, there are a number of "Check Your Skills" questions throughout each chapter. After learning a topic, try these problems, checking your answers at the end of the chapter as you go.

If you find these skill checks challenging, reread the section you just finished. Then try the questions again. Whenever needed, use the solution to help you work through the math step-by-step. When you get stuck, don't read the entire solution immediately. Read as much as you need to get yourself unstuck, then continue to try to do the work on your own. Return to the solution each time you get stuck and use only the portion of the solution you need to help unstick yourself.

Quick-Start Definitions

Whether you work with numbers every day or only do math when calculating a tip on your restaurant bill, carefully read this first section, which provides definitions for certain core concepts in order to give your studies a quick start. You'll come back to many of these concepts during your studies. You can also find a glossary of common math terms at the back of this guide.

Basic Numbers

All the numbers that come into play on the GMAT can be shown as a point somewhere on the **number line**:

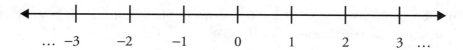

Another word for *number* is *value*.

Positive integers are the counting numbers 1, 2, 3, and so on. These are the first numbers that you ever learned—the numbers that you use to count individual items. They are **whole numbers**; they do not include any fractions or decimals:

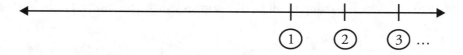

Digits are 10 symbols (0, 1, 2, 3, 4, 5, 6, 7, 8, and 9) used to represent numbers. If the GMAT asks you specifically for a digit, it wants one of these 10 symbols.

Counting numbers above 9 are represented by two or more digits. The number *four hundred twelve* is represented by three digits in this order: 412.

Place value tells you how much a digit in a specific position is worth. The 4 in 412 is worth 4 hundreds (400), so 4 is the *hundreds digit* of 412. Meanwhile, 1 is the *tens digit* and is worth 1 ten (10). Finally, 2 is the *units digit* and is worth 2 units, or just plain old 2:

412	=	400	+	10	+	2
Four hundred twelve	equals	four hundreds	plus	one ten	plus	two units (or two)

The GMAT always separates the thousands digit from the hundreds digit by a comma. For readability, big numbers are broken up by commas placed three digits apart:

1,298,023 equals one million two hundred ninety-eight thousand twenty-three.

Addition (+, or *plus*) is the most basic operation in arithmetic. If you add one positive number to another, you get a third positive number farther to the right on the number line. For example:

7	+	5	=	12
Seven	plus	five	equals	twelve

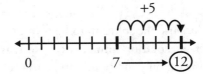

Therefore, 12 is the **sum** of 7 and 5. By the way, the statement $7 + 5 = 12$ is called an **equation**.

When you add two numbers in either order, you get the same result:

5	+	7	=	12
Five	plus	seven	equals	twelve

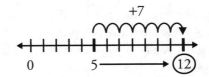

Subtraction (−, or *minus*) is the opposite of addition. Subtraction undoes addition. For example:

7	+	5	−	5	=	7
Seven	plus	five	minus	five	equals	seven

Order matters in subtraction: $6 − 2 = 4$, but $2 − 6 =$ something else (more on this in a minute). By the way, since $6 − 2 = 4$, the **difference** between 6 and 2 is 4.

Any number minus itself equals zero (0):

7	−	7	=	0
Seven	minus	seven	equals	zero

Any number plus zero equals that starting number. The same is true if you subtract zero. For example:

$$8 + 0 = 8 \qquad 9 - 0 = 9$$

In either case, you're moving *zero* units away from the original number on the number line, or staying right where you started.

Numbers can be positive, negative, or neither. **Positive** numbers lie to the right of zero on a number line. **Negative** numbers lie to the left of zero. Zero itself is neither positive nor negative—it's the only number that is neither:

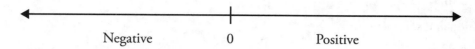

The **sign** of a number indicates whether the number is positive or negative.

Negative integers are −1, −2, −3, and so on:

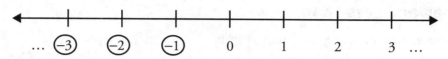

When you subtract a greater number from a smaller number, you will end up with a negative number. For example:

2	–	6	=	−4
Two	minus	six	equals	negative four

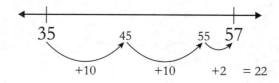

Negative numbers can be used to represent deficits. If you have $2 but you owe $6, your net worth is −$4.

Most people tend to make more mistakes when dealing with negative numbers than with positive ones. If you're asked to find *small minus big*, you may instead want to figure out *big minus small*, then make the result negative. For example:

What is 35 − 57 ?

$$57 - 35 = 22$$

$$35 - 57 = -22$$

Since *big minus small* is equal to 22, the reverse—*small minus big*—is equal to −22.

The work above represents what is called textbook math. There are other, often better, ways to think your way through the math that needs to be done on the GMAT—and these methods are more useful in the real world as well.

For example, when you want to find 57 minus 35, you are really looking for the distance between the two numbers. Visualize 35 and 57 on a number line, and then count the distance between them using whatever increments seem reasonable to you:

As the diagram above shows, first count by 10 up to 45, then by another 10 up to 55, and finally by 2 to get to 57. The distance between the two numbers is 10 + 10 + 2 = 22.

If the problem asks for $57 - 35$, then the answer is 22. If the problem asks for $35 - 57$, then add a negative sign: The answer is -22.

Check Your Skills

Perform addition and subtraction.

1. What is $37 + 141$?
2. What is $23 - 136$?

Answers can be found on page 37.

Greater Than and Less Than

Greater than ($>$) means *to the right of* on a number line:

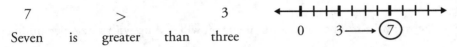

The statement $7 > 3$ is an example of an **inequality**.

In the real world, people also say *bigger than* or *larger than*, but there is one drawback to using this terminology. Take a look at this example:

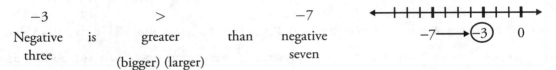

A lot of people will intuitively call -7 the "bigger" number because, when the numbers are positive, 7 is greater than 3. But that thinking doesn't work the same way for negative numbers! The value of -7 is less than the value of -3.

So stick to *greater than* and *less than*. Greater numbers are *to the right of* smaller numbers on the number line.

Any positive number is greater than any negative number. For example:

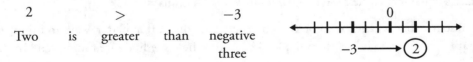

Likewise, zero is greater than every negative number:

0   >   −3
Zero   is   greater   than   negative   three

Less than ($<$) means *to the left of* on a number line. You can always reexpress a *greater than* relationship as a *less than* relationship—just flip it around. For example:

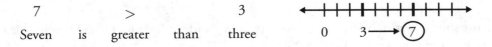

3		<		7
Three	is	less	than	seven

If 7 is greater than 3, then 3 is less than 7.

If you find yourself making mistakes with negatives, look for opportunities to think about negatives in terms of the number line. Test out the following true statements, or **inequalities**, on a number line:

−7 is less than −3	$-7 < -3$
−3 is less than 2	$-3 < 2$
−3 is less than 0	$-3 < 0$

Inequalities are very similar to equations, but equations always use equals signs ($=$), while inequalities use greater than ($>$) or less than ($<$) signs. Later in this guide, you'll learn more about how to work with both equations and inequalities.

Check Your Skills

3. What is the sum of the greatest negative integer and the smallest positive integer?

For questions 4 and 5, plug in $>$ and $<$ symbols and say the resulting statement aloud.

4. 5 ___ 16

5. −5 ___ −16

Answers can be found on page 37.

Adding and Subtracting Positives and Negatives

Positive plus positive gives you a positive result. For example:

7	+	5	=	12
Seven	plus	five	equals	twelve

When adding two positives, you move farther to the right of zero, so the result is always greater than either starting number.

Positive minus positive, on the other hand, could give you either a positive or a negative:

Big positive	−	*small positive*	=	*positive*
8	−	3	=	5
Eight	minus	three	equals	five

1

$$Small\ positive\ -\ big\ positive\ =\ negative$$
$$3\ -\ 8\ =\ -5$$
Three   minus   eight   equals   negative five

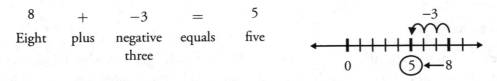

Either way, the result is less than where you started because you move left when you subtract a positive.

Adding a negative is the same as subtracting a positive—you move left. For example:

$$8\ +\ -3\ =\ 5$$
Eight   plus   negative three   equals   five

In fact, you can write $8 + (-3)$ as $8 - 3$. They both mean the exact same thing.

Negative plus negative always gives you a negative because you move even farther to the left of zero:

$$-3\ +\ -5\ =\ -8$$
Negative three   plus   negative five   equals   negative eight

Subtracting a negative is the same as adding a positive—you move right. Add in parentheses to keep the two minus signs straight:

$$7\ -\ (-5)\ =\ 12$$
Seven   minus   negative five   equals   twelve

The equation above is the same as the equation below:

$$7\ +\ 5\ =\ 12$$
Seven   plus   five   equals   twelve

In both cases, you end up with the following:

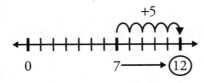

In general, any subtraction can be rewritten as an addition. If you're subtracting a positive, you can instead add a negative. If you're subtracting a negative, you can instead add a positive.

Check Your Skills

6. Which is greater, a positive minus a negative or a negative minus a positive?

Answer can be found on page 37.

Multiplying and Dividing

Multiplication ($\times$, or *times*) is the same as adding the same number multiple times. Imagine that there are three tennis balls per can and you have four cans. How many tennis balls do you have?

$$4 \qquad \times \qquad 3 \qquad = \qquad 3 + 3 + 3 + 3 \qquad = \qquad 12$$

Four   times   three   equals   four threes added up,   and this equals   twelve

Therefore, 12 is the **product** of 4 and 3. Also, 4 and 3 are called **factors** of 12.

Parentheses can be used to indicate multiplication. Parentheses are usually written with ( ), but brackets [ ] can also be used, especially if you have parentheses within parentheses. If a set of parentheses bumps up right against something else, multiply that something by whatever is in the parentheses:

$$4(3) = (4)3 = (4)(3) = 4 \times 3 = 12$$

When writing multiplication, if you use a $\times$ symbol, make sure to write it in such a way that you won't confuse it for the variable *x*. Also, you may have learned to use a dot to mean multiplication:

$$4 \bullet 3 = 4 \times 3 = 12$$

The potential danger in using a dot is that you might mistake the dot for a decimal point. If you do want to write a dot, make that dot big and high, as shown above, so it doesn't look like a decimal point. In general, it's a good idea to get in the habit of using parentheses rather than $\bullet$ or $\times$ so that you minimize your chances of making a careless mistake on the real test.

You can always multiply in either order; you will get the same result, as shown below. (Pop quiz: What other operation that you learned earlier in this section can also be done in either order?)

$$4 \qquad \times \qquad 3 \qquad = \qquad 3 + 3 + 3 + 3 \qquad = \qquad 12$$

Four   times   three   equals   four threes added up,   and this equals   twelve

$$3 \qquad \times \qquad 4 \qquad = \qquad 4 + 4 + 4 \qquad = \qquad 12$$

Three   times   four   equals   three fours added up,   and this equals   twelve

Pop quiz answer: Addition can also be performed in either order.

Since you are allowed to multiply in any order, don't necessarily multiply in the given order. Rather, invest a few seconds to think about the easiest order in which to multiply. Try this:

$$2 \times 7 \times 5$$

Most people work left to right, so the second step would be 14×5. That's not terrible, but it's also not the easiest calculation. Go back to the first step and play around with the numbers. Can you find a way that's easier?

In general, it's easy to multiply by multiples of 10, so whenever you see 2's and 5's in the mix, rearrange to group them together:

$$(2 \times 5) \times 7 =$$
$$10 \quad \times 7 = 70$$

Try this one:

$$6 \times 5 \times 7 =$$

This one has a 5 but no 2. Now what? Actually, there is a 2, lurking inside the 6. Split up the 6 into 2×3, and rearrange to pair the 2 with the 5:

$$2 \times 3 \times 5 \times 7 =$$
$$(2 \times 5) \times (3 \times 7) =$$
$$10 \quad \times \quad 21 \quad = 210$$

In general, think before you multiply. Do the math in whatever order looks best to you—and keep an eye out for 2's and 5's.

Division (÷, or *divided by*) is the opposite of multiplication. Division undoes multiplication:

2	×	3	÷	3	=	2
Two	times	three	divided by	three	equals	two

As with subtraction, order matters in division. $12 \div 3 = 4$, but $3 \div 12 =$ something else (more on this soon).

Multiplying any number by 1 leaves the number the same. One times anything is equal to whatever you started with:

1	×	5	=	5	=	5
One	times	five	equals	one five by itself,	and this equals	five

5	×	1	=	$1 + 1 + 1 + 1 + 1$	=	5
Five	times	one	equals	five ones added up,	and this equals	five

Multiplying any number by zero (0) gives you zero. Anything times zero is zero:

5	×	0	=	$0 + 0 + 0 + 0 + 0$	=	0
Five	times	zero	equals	five zeros added up,	and this equals	zero

Since order doesn't matter in multiplication, this means that zero times anything is zero, too:

0	×	5	=	5×0	=	0
Zero	times	five	equals	five times zero,	and this equals	zero

Multiplying a number by zero destroys it permanently, in a sense. So you're not allowed to undo that destruction by dividing by zero.

Never divide by zero: $13 \div 0 =$ undefined. Stop right there—don't do this! The GMAT avoids getting into undefined-number territory, thankfully.

You *are* allowed to divide zero by any nonzero number. The answer is—surprise!—zero:

0	÷	13	=	0
Zero	divided by	thirteen	equals	zero

Check Your Skills

Complete the operations.

7.   What is 7×6 ?

8.   What is $52 \div 13$?

Answers can be found on page 37.

Distributing and Factoring

What is $4 \times (3 + 2)$? Here's one way to solve it. Turn $(3 + 2)$ into 5, then multiply 4 by that 5:

4	×	(3 + 2)	=	4 × 5	=	20
Four	times	the quantity three plus two	equals	four times five,	and this equals	twenty

If you're working with real numbers, the process above will usually be the fastest way to solve. But you can solve another way—and that other way is usually necessary when working with algebra or other more advanced math. The other way to solve this problem is to **distribute** the 4 to both the 3 and the 2:

4	×	(3 + 2)	=	4 × 3	+	4 × 2
Four	times	the quantity three plus two	equals	four times three	plus	four times two,
			=	12	+	8
			and this equals	twelve	plus	eight,
			=	20		
			and that equals	twenty		

Notice that you multiply the 4 by *both* the 3 and the 2. Distributing is extra work in this case, but the technique will come in handy later when you're doing algebra.

Another way to see how distributing works is to put the sum in front:

$$\overbrace{(3+2) \times 4}^{20} = \overbrace{3 \times 4 + 2 \times 4}^{20}$$

$\underbrace{(3+2)}_{5}$

Five	times	four	equals	three times four	plus	two times four
Five fours added together			equals	three fours added together	plus	two fours added together
Twenty			equals		twenty	

In a sense, you're splitting up the sum $3 + 2$. Just be sure to multiply both the 3 and the 2 by 4.

You distribute when the terms inside the parentheses are connected by addition or subtraction signs. Do *not* distribute if the terms inside the parentheses are connected by multiplication or division; in that case, drop the parentheses and multiply or divide straight across. For example:

DO distribute: $\qquad$ $3(4 + y) = (3)(4) + (3)(y) = 12 + 3y$

Do NOT distribute: $\qquad$ $3(4 \times y) = 3 \times 4 \times y = 12y$

Distributing works similarly for subtraction. Just keep track of the minus sign:

6	$\times$	(5 − 2)		=	6 × 5	−	6 × 2
Six	times	the quantity five minus two	equals		six times five	minus	six times two,

	=	30	−	12
	and this equals	thirty	minus	twelve,

	=	18
	and that equals	eighteen

Again, for that problem, distribution is more work than it's worth. You can do the subtraction first, $5 - 2 = 3$, and then multiply 6×3 to get 18. Just learn how distribution works because you'll need to do it later with algebra.

You can also go in reverse. You can **factor** the sum of two products if the products contain the same factor. Pull out the **common factor** of 4 from each of the products 4×3 and 4×2. Next, put the sum of 3 and 2 into parentheses:

4 × 3	+	4 × 2	=	4	×	(3 + 2)
Four times three	plus	four times two	equals	four	times	the quantity three plus two

Most of the time, you'll see the result written in this form: $4(3 + 2)$. By the way, *common* here doesn't mean *frequent* or *typical*. Rather, it means *belonging to both products*. A common factor is a factor in common (like a friend in common).

You can also write the common factor at the back of each product, if you like:

3 × 4	+	2 × 4	=	(3 + 2)	×	4
Three times four	plus	two times four	equals	the quantity three plus two,	times	four or five,

This is the equivalent of writing $(3 + 2)4$. Since the order of multiplication doesn't matter, $4(3 + 2) = (3 + 2)4$. Reminder: You can distribute when the math in the parentheses is added or subtracted. Don't distribute if the math in the parentheses is multiplied or divided.

Example	Can I distribute?	
$3(4 + y)$	Yes! The math in the parentheses is added.	$3(4 + y) =$ $(3)(4) + (3)(y) = 12 + 3y$
$3(4 \times 5)$	No. When the math in the parentheses is multiplied (or divided), do the math straight out.	$3(4 \times 5) =$ $3 \times 4 \times 5 = 60$

You will use both distributing and factoring in more advanced ways later in this guide. Keep in mind that using simple numbers to understand the concepts now will help you to apply these same concepts to more complex problems in the future. As you move further in your studies, you will find yourself confused about a problem at times. Try some real numbers to help understand what's going on, and you may find that you can solve that problem after all!

Check Your Skills

9. Use distribution to solve. What is $5 \times (3 + 4)$?

10. Factor a 6 out of the following expression: $36 - 12$.

Answers can be found on page 38.

Multiplying Positives and Negatives

Positive × *positive* is always *positive*:

3	×	4	=	$4 + 4 + 4$	=	12
Three	times	four	equals	three fours added up,	and this equals	twelve

Positive × *negative* is always *negative*:

3	×	-4	=	$-4 + (-4) + (-4)$	=	-12
Three	times	negative four	equals	three negative fours, all added up,	and this equals	negative twelve

Since order doesn't matter in multiplication, the same outcome happens when you have *negative times positive*. You again get a *negative*:

-4	×	3	=	$3 \times (-4)$	=	-12
Negative four	times	three	equals	three times negative four,	and this equals	negative twelve

What is *negative* × *negative*? *Positive*. This fact may seem weird, but it's consistent with the rules developed so far. In the same way that something minus a negative turns into something plus a positive ($7 - (-3) = 7 + 3$), a negative times a negative also turns positive. In either case, two negatives make a positive.

All the same rules hold true for division:

Positive Result					**Negative Result**				
Positive	÷	Positive	=	Positive	Positive	÷	Negative	=	Negative
Negative	÷	Negative	=	Positive	Negative	÷	Positive	=	Negative
Positive	×	Positive	=	Positive	Positive	×	Negative	=	Negative
Negative	×	Negative	=	Positive	Negative	×	Positive	=	Negative

1

Notice any pattern there? (Really—examine the information in that table before you keep reading!)

Regardless of whether you're doing multiplication or division, if the two starting numbers have the *same* sign (both positive or both negative), then you end up with a *positive*.

But if the two starting numbers have *opposite* signs (one positive and one negative), then you end up with a *negative*.

Check Your Skills

11. What is $(2)(-5)$?

12. Use distribution to solve. What is $-6 \times (-3 + (-5))$?

Answers can be found on page 38.

Fractions and Decimals

Adding, subtracting, or multiplying integers always results in an integer, whether positive or negative:

$$\text{Integer} \quad + \quad \text{Integer} \quad = \quad \text{Integer}$$

$$\text{Integer} \quad - \quad \text{Integer} \quad = \quad \text{Integer}$$

$$\text{Integer} \quad \times \quad \text{Integer} \quad = \quad \text{Integer}$$

However, dividing an integer by another integer does *not* always give you an integer:

$$\text{Integer} \quad \div \quad \text{Integer} \quad = \quad \text{Sometimes an integer, sometimes not!}$$

When you don't get an integer, you get a **fraction** or a **decimal**—a number that falls between the integers on the number line:

$$7 \div 2 \quad = \quad \frac{7}{2} \quad = \quad 3.5$$

Seven divided by two	equals	seven halves,	and this equals	three point five
		Fraction		Decimal

A horizontal **fraction line**, or bar, expresses the division of the **numerator** (above the fraction line) by the **denominator** (below the fraction line):

Numerator

Fraction line $\longrightarrow \dfrac{7}{2} = 7 \div 2$

Denominator

In fact, the division symbol $\div$ is just a miniature fraction. People often say things such as "seven *over* two" rather than "seven halves" to express a fraction.

You can express division in three ways: with a fraction line, with the division symbol ($\div$), or with a slash (/), as shown here:

$$\frac{7}{2} = 7 \div 2 = 7/2$$

A **decimal point** is used to extend place value to the right for decimals. Each place to the right of the decimal point is worth a tenth $\left(\frac{1}{10}\right)$, a hundredth $\left(\frac{1}{100}\right)$, and so on. For example:

3.5	=	3	+	$\frac{5}{10}$		
Three point five	equals	three	plus	five-tenths		

1.25	=	1	+	$\frac{2}{10}$	+	$\frac{5}{100}$
One point two five	equals	one	plus	two-tenths	plus	five-hundredths

A decimal such as 3.5 has an *integer part* (3) and a *fractional part* or *decimal part* (0.5). In fact, an integer is just a number with no fractional or decimal part.

Every fraction can be written as a decimal, although you might need an unending string of digits in the decimal to properly express the fraction:

$4 \div 3$	=	$\frac{4}{3}$	=	1.333…	=	$1.\overline{3}$
Four divided by three	equals	four-thirds (or four over three),	and this equals	one point three three three dot dot dot, forever and ever,	and that equals	one point three, repeating

Fractions and decimals obey all the rules you've seen so far about how to add, subtract, multiply, and divide. Everything you've learned for integers applies to fractions and decimals as well: how positives and negatives work, how to distribute, etc.

Check Your Skills

13. Which arithmetic operation involving integers does NOT always result in an integer?

14. Write $2 \div 7$ in fraction form, then multiply that fraction by 0. What is the result?

Answers can be found on page 38.

Divisibility and Even and Odd Integers

Sometimes you do get an integer out of integer division:

$15 \div 3$	=	$\frac{15}{3}$	=	5	=	int
Fifteen divided by three	equals	fifteen-thirds (or fifteen over three),	and this equals	five,	which is	an integer

In this case, 15 and 3 have a special relationship. You can express this relationship in any of the following ways:

15 is **divisible** by 3.

15 divided by 3 equals an integer: $15 \div 3 = \text{int}$

3 is a **divisor** of 15.

15 divided by 3 equals an integer: $15 \div 3 = \text{int}$

15 is a **multiple** of 3.

15 equals 3 times an integer: $15 = 3 \times \text{int}$

1

3 is a **factor** of 15.

3 times an integer equals 15: $3 \times \text{int} = 15$

People also say the following, though you won't see this language on the test:

3 goes into 15 evenly.

3 divides 15 evenly.

Even integers are divisible by 2:

14 is even because $14 \div 2 = 7 =$ an integer.

All even integers have 0, 2, 4, 6, or 8 as their units digit.

Zero is even because it is divisible by 2:

$0 \div 2 = 0$, which is an integer.

Odd integers are not divisible by 2:

15 is odd because $15 \div 2 = 7.5 =$ not an integer.

All odd integers have 1, 3, 5, 7, or 9 as their units digit.

Even and odd integers alternate on the number line:

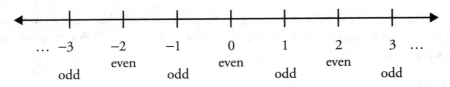

...   −3   −2   −1   0   1   2   3   ...

odd   even   odd   even   odd   even   odd

Only integers can be said to be even or odd. Fractions or decimals are not considered even or odd.

Check Your Skills

15. Fill in the blank. If 7 is a factor of 21, then 21 is a _____ of 7.

16. Is 2,284,623 divisible by 2 ?

Answers can be found on page 38.

Exponents and Roots (and π)

Earlier, you learned that multiplication is really just repeated addition. Exponents are one level "up" in the food chain: They represent repeated *multiplication*.

In 5^2, the **exponent** is 2 and the **base** is 5. The exponent tells you how many of the bases to multiply together to get your answer. When the exponent is 2, you usually say *squared*:

5^2	=	5×5	=	25
Five squared	equals	two fives multiplied together, or five times itself,	and this equals	twenty-five

When the exponent is 3, you usually say *cubed*:

4^3	=	$4 \times 4 \times 4$	=	64
Four cubed	equals	three fours multiplied together, or four times four times four,	and this equals	sixty-four

For other exponents, you say *to the _____ power* or *raised to the _____ power*. You could also say *to the power of _____*. For example:

2^5	=	$2 \times 2 \times 2 \times 2 \times 2$	=	32
Two to the fifth power	equals	five twos multiplied together,	and this equals	thirty-two

Be careful when writing exponents on your own paper. Make them much tinier than regular numbers and put them *clearly* up to the right. You don't want to mistake 5^2 for 52, or vice versa.

By the way, a number raised to the first power equals the original number:

7^1	=	7	=	7
Seven to the first power	equals	just one seven by itself,	and this equals	seven

A **perfect square** is the square of an integer:

25 is a perfect square because $25 = 5^2 = \text{int}^2$.

A **perfect cube** is the cube of an integer:

64 is a perfect cube because $64 = 4^3 = \text{int}^3$.

Roots undo exponents. The simplest and most common root is the **square root**, which undoes squaring. The square root is written with the **radical sign** ($\sqrt{}$); if a problem refers to the **radical**, it's talking about that symbol. Here's how exponents and roots connect:

5^2	=	5×5	=	25,	so	$\sqrt{25}$	=	5
Five squared	equals	five times five,	and this equals	twenty-five,	so	the square root of twenty-five	equals	five

As a shortcut, *the square root of twenty-five* can just be called *root twenty-five*.

Asking for the square root of 49 is the same as asking what number, times itself, gives you 49:

$\sqrt{49}$	=	7	because	7×7	=	7^2	=	49
Root forty-nine	equals	seven,	because	seven times seven	equals	seven squared,	and this equals	forty-nine

The square root of a perfect square is an integer, because a perfect square is an integer squared:

$\sqrt{36}$	=	int	because	36	=	int^2
The square root of thirty-six	is	an integer,	because	thirty-six	equals	an integer squared

1

The square root of any *non*-perfect square is an unending decimal that never even repeats, as it turns out:

$\sqrt{2}$	$\approx$	$1.414213562\ldots$	because	$(1.414213562\ldots)^2$	$\approx$	2
Root two	is about	one point four one four two blah blah,	because	that thing squared	is about	two

The square root of 2 can't be expressed as a simple fraction, either. So you can leave it as is $\left(\sqrt{2}\right)$, or you can approximate it $\left(\sqrt{2} \approx 1.4\right)$.

It's useful on this test to know that root 2 is approximately 1.4 and root 3 is approximately 1.7. Here's an easy way to remember that:

Root	Value	Date
$\sqrt{2}$	~1.4	2/14 is Valentine's Day
$\sqrt{3}$	~1.7	3/17 is St. Patrick's Day

Speaking of dates, have you ever heard of Pi Day? (Feel free to google it.) You'll encounter one other common number with an ugly decimal in geometry: **pi** (π).

Pi is the ratio of a circle's circumference to its diameter. It's about $3.14159265\ldots$ without ever repeating.

Since pi can't be expressed as a simple fraction, it is typically represented by the Greek letter π, or you can approximate it ($\pi \approx 3.14$, or a little more than 3). And—you guessed it—Pi Day is 3/14!

Cube roots undo cubing. The cube root has a little 3 tucked into its notch ($\sqrt[3]{}$):

$\sqrt[3]{8}$	$=$	2	because	2^3	$=$	8
The cube root of eight	equals	two,	because	two cubed	equals	eight

Other roots occasionally show up. For example, the **fourth root** undoes the process of taking a base to the fourth power:

$\sqrt[4]{81}$	$=$	3	because	3^4	$=$	81
The fourth root of eighty-one	equals	three,	because	three to the fourth power	equals	eighty-one

Check Your Skills

17. $2^6 =$

18. $\sqrt[3]{27} =$

Answers can be found on page 38.

Variable Expressions and Equations

Up to now, every number you've dealt with has been an actual, known number. Algebra is the art of dealing with *unknown* numbers.

A **variable** is an unknown number, or an **unknown** for short. You represent a variable with a single letter, such as x or y.

When you see y, imagine that it represents a number that you don't happen to know. At the start of a problem, the value of y is hidden from you. It could be anywhere on the number line, in theory, unless you're told something about y.

The letter x is the stereotypical letter used for an unknown. Since x looks so much like the multiplication symbol $\times$, you can prevent mistakes by not using the $\times$ symbol when writing out algebra. To represent multiplication, do other things.

To multiply variables, put them next to each other:

What You See	What You Say	What It Means
xy	"x y"	x times y
abc	"a b c"	The product of a, b, and c
		a times b times c

To multiply a known number by a variable, write the known number in front of the variable:

What You See	What You Say	What It Means
$3x$	"three x"	3 times x

Here, 3 is called the **coefficient** of x. If you want to multiply x by 3, write $3x$, not $x3$, which could look too much like x^3 (*x cubed*).

All the operations besides multiplication look the same for variables as they do for known numbers:

What You See	What You Say	What It Means
$x + y$	"x and y" "the sum of x and y"	x plus y
$x - y$	"subtract y from x" "the difference between x and y" "the difference of x and y"	x minus y
$\dfrac{x}{y}$	"x over y" "x divided by y"	x divided by y
x^2	"x squared" "x raised to the power of 2" "x to the second"	x squared, which is x times x
$\sqrt{x}$	"root x" "root of x" "x rooted"	the square root of x

By the way, be careful when you have variables in exponents:

What You See	What You Say	What It Means
$3x$	"three x"	3 times x
3^x	"three to the x"	3 multiplied by itself x times

Never refer to 3^x as "three x." It's called "three *to the x*." Calling it "three x" is likely to lead to a careless mistake, since three x typically means three times x.

An **expression** is something that ultimately represents a number; for example, $x + 3$ is an expression. You might not know that number, but you *express* it using variables, numbers you know, and operations such as adding, subtracting, etc.

An expression is like a recipe. If you follow the recipe, you will get the correct number, as shown in these examples:

Expression	What You Say	The Number Represented by the Expression
$x + y$	"x plus y"	Add x and y.
$3xz - y^2$	"3 x z minus y squared"	First, multiply 3, x, and z, then subtract the square of y.
$\dfrac{\sqrt{2w}}{3}$	"The square root of 2 w, all over 3"	First, multiply 2 and w together. Next, take the square root. Finally, divide by 3.

Within an expression, you have one or more **terms**. A single term involves no addition or subtraction (typically). Often, a term is just a product of variables and known numbers. Here are the terms in the previous expressions, with the separate terms separated by commas:

Expression	Terms	Number of Terms
$x + y$	x, y	Two
$3xz - y^2$	$3xz, y^2$	Two
$\dfrac{\sqrt{2w}}{3}$	$\dfrac{\sqrt{2w}}{3}$	One

It's useful to notice terms so that you can **simplify** expressions, or reduce the number of terms in those expressions. For example, how can you simplify the expression below?

$x + 2x$

The expression currently has two terms, but they both use the same variable, x, so you can combine them to get a new expression with a single term: $3x$.

An **equation** sets one expression equal to another using the **equals sign** ($=$), which you've seen plenty of times in your life—and in this book already.

What you might not have thought about, though, is that the equals sign is a *verb*. In other words, an equation is a complete, grammatical sentence or statement:

Something	*equals*	something else.
Some expression	*equals*	some other expression.

Here's an example:

$3 + 2x$	$=$	11
Three plus two x	equals	eleven

Each equation has a *left side* (the subject of the sentence) and a *right side* (the object of the verb *equals*). You can say *is equal to* rather than *equals* if you want:

$3 + 2x$	$=$	11
Three plus two x	is equal to	eleven

Solving an equation is solving this mystery:

What is x?

Or, more precisely:

What is the value or values of x that make the equation *true*?

Since an equation is a sentence, it can be true or false, at least in theory. You always want to focus on how to make the equation *true*, or keep it so, by finding the correct values of any variables in that equation.

The process of solving an equation usually involves rearranging the equation and performing identical operations on each side until the equation tells you what the variable equals. For example:

$3 + 2x$	$=$	11	Three plus two x	equals	eleven
-3		-3	Subtract 3		Subtract 3
$2x$	$=$	8	Two x	equals	eight
$\div 2$		$\div 2$	Divide by 2		Divide by 2
x	$=$	4	x	equals	four

The solution to the original equation is $x = 4$. If you replace x with 4 in the equation $3 + 2x = 11$, then you get $11 = 11$, which is always true. Any other value of x would make the equation false.

If the GMAT gives you the equation $3 + 2x = 11$, it's telling you something very specific about x. For this particular equation, in fact, just one value of x makes the equation work (namely, 4).

Check Your Skills

19. If $x = 4$ and $y = -1$, what is the value of the expression $2x - 3y$?

Answer can be found on page 38.

PEMDAS

Consider the expression $3 + 2 \times 4$.

Should you add 3 and 2 first, then multiply by 4? If so, you get 20.

Or, should you multiply 2 and 4 first, then add 3? If so, you get 11.

There's no ambiguity—mathematicians have decided on the second option. **PEMDAS** is an acronym to help you remember the proper **order of operations**.

When you simplify an expression, don't automatically perform operations from left to right, even though that's how you read English. Instead, follow PEMDAS:

Parentheses	Do P first
Exponents	Then E
Multiplication	Then either M or D
Division	
Addition	Then either A or S
Subtraction	

1

Try these two problems. How are they different?

$$3 + 2 \times 4 = ? \qquad (3 + 2) \times 4 = ?$$

In the first example, do the M (multiplication) first, then the A (addition):

$$3 + 2 \times 4 = \qquad \text{Multiply first.}$$
$$3 + \ 8 \ = 11 \qquad \text{Then add.}$$

But in the second example, the math inside the parentheses must be done first:

$$(3 + 2) \times 4 = \qquad \text{Do the parentheses first.}$$
$$5 \ \times 4 = 20 \qquad \text{Then multiply.}$$

Parentheses always come first, then exponents, then multiplication and division, and finally addition and subtraction.

Multiplication and division are at the same level of importance in PEMDAS, because any multiplication can be expressed as division, and vice versa:

$$7 \div 2 = 7 \times \frac{1}{2}$$

In a sense, multiplication and division are two sides of the same coin.

Likewise, addition and subtraction are at the same level of importance. Any addition can be expressed as subtraction, and vice versa:

$$3 - 4 = 3 + (-4)$$

So you can think of PEMDAS this way:

$$\underrightarrow{\text{PE} \ {}^M\!/_D \ {}^A\!/_S}$$

Earlier in this chapter, you learned that you can do multiplication in any order. The same is true when you have a mix of multiplication and division:

$$4 \times 5 \div 2 = \qquad\qquad 4 \times 5 \div 2 =$$
$$20 \ \div 2 = 10 \qquad\qquad 4 \div 2 \times 5 =$$
$$\qquad\qquad\qquad\qquad 2 \ \times 5 = 10$$

You can multiply first, but you can also rearrange and divide first. When the numbers are greater, multiplying first can make the math much more cumbersome; dividing first usually makes things easier. As always, invest a few seconds to think about the best order in which to do the math.

For addition and subtraction, you can technically do the same thing—but there's a greater chance of making a careless mistake. Therefore, if you have multiple steps of addition and subtraction, do them *left to right*:

$$3 - 2 + 4 =$$
$$1 \ + 4 = 5 \qquad \text{Correct}$$

$$3 - 2 + 4 =$$
$$3 - \ 6 \ = -3 \qquad \text{Incorrect}$$

In the incorrect example, a mistake was made: The given operation is not $2 + 4$. It's actually $-2 + 4$. It's really easy to make that mistake if you work out of order, but you're much less likely to make that mistake if you work left to right.

Note, though, that parentheses always come first. If you have parentheses, always do them first:

$$3 - (2 + 3) =$$
$$3 - \quad 5 \quad = -2$$

Try this more complicated problem:

What is $3 + 4(5 - 1) - 3^2 \times 2$?

Here is the correct order of steps to simplify:

$3 + 4(5 - 1) - 3^2 \times 2 =$	Parentheses
$3 + \quad 4(4) \quad - 3^2 \times 2 =$	Exponents
$3 + \quad 4(4) \quad - 9 \times 2 =$	Multiplication/Division
$3 + \quad 16 \quad - 18 =$	Addition/Subtraction
$19 \quad - \quad 18 = 1$	

Try another problem:

What is $6 - 3 \times 2^3 \div (7 - 1)$?

Here's the solution:

$6 - 3 \times 2^3 \div (7 - 1) =$	Parentheses
$6 - 3 \times 2^3 \div \quad 6 \quad =$	Exponents
$6 - 3 \times 8 \div \quad 6 \quad =$	Multiplication/Division
$6 - \quad 24 \quad \div \quad 6 \quad =$	
$6 - \quad 4 \quad = 2$	Addition/Subtraction

Try one more:

What is $32 \div 2^4 \times \left(5 - 3^2\right)$?

Here's the solution:

$32 \div 2^4 \times \left(5 - 3^2\right) =$	Parentheses
$32 \div 2^4 \times \quad (5 - 9) =$	
$32 \div 2^4 \times \quad (-4) =$	Exponents
$32 \div 16 \times \quad (-4) =$	Multiplication/Division
$2 \quad \times \quad (-4) = -8$	

Check Your Skills

Evaluate the following expressions.

20. What is $-4 + \dfrac{12}{3}$?

21. What is $(5 - 8) \times 10 - 7$?

22. What is $-3 \times 12 \div 4 \times 8 + (4 - 6)$?

23. What is $\dfrac{2^4 \times (8 \div 2 - 1)}{(9 - 3)}$?

Answers can be found on pages 38–39.

Combining Like Terms

How can you simplify this expression?

$3x^2 + 7x + 2x^2 - x$

Here's the expression in words:

$3x^2$	$+$	$7x$	$+$	$2x^2$	$-$	x
The square of x multiplied by 3	plus	x multiplied by 7	plus	the square of x multiplied by 2	minus	x

There are four terms in the expression above. Reminder: A term is an expression that doesn't contain addition or subtraction. Quite often, a term is just various things multiplied together.

Some of the terms are very similar to one another—they are **like terms**. Like terms differ only by a numerical coefficient (the number in front of the variable), but the form of the variables is the same. For example, $7x$ and $-x$ are like terms. The form of the variable is the same in both, but the coefficients are different.

There are two pairs of like terms in the expression:

Pair one:	$3x^2$	and	$2x^2$
Pair two:	$7x$	and	$-x$

To be like terms, the variables must be identical, including exponents. Otherwise, the terms aren't *like*.

You can combine any like terms into one term by adding or subtracting the coefficients. Keep track of $+$ and $-$ signs. For example:

$3x^2$	$+$	$2x^2$	$=$	$5x^2$
Three x squared	plus	two x squared	equals	five x squared

$7x$	$-$	x	$=$	$6x$
Seven x	minus	x	equals	six x

Whenever a variable does not have a number in front, the coefficient is 1. In the second pair, x can be rewritten as $1x$:

$7x$	$-$	$1x$	$=$	$6x$
Seven x	minus	one x	equals	six x

Or you could say that you're adding $-1x$:

$$7x \qquad + \qquad -1x \qquad = \qquad 6x$$

Seven x plus negative one x equals six x

Either way is fine. A negative sign in front of a term on its own is really a -1 coefficient. For instance, $-xy^2$ has a coefficient of -1.

Combining like terms is allowed because, for like terms, everything *but* the coefficient is a **common factor**. So you can *pull out* that common factor and group the coefficients into a sum (or difference). This is when factoring starts to become really useful.

For a review of factoring, see the section *Distributing and Factoring* earlier in this chapter. When you read something and think, "That rings a bell. I'm pretty sure I learned about this earlier," flip back and remind yourself about that topic. Spending two minutes recalling an earlier lesson that hasn't fully imprinted in your brain yet will help you to better recall that same material in future—including during the test.

In the first pair, the coefficients are 3 and 2, and the common factor is x^2:

$$3x^2 \qquad + \qquad 2x^2 \qquad = \qquad (3+2)x^2$$

Three x squared plus two x squared equals the quantity three plus two, times x squared

The right side then reduces by PEMDAS to $5x^2$. You don't have to write out $(3+2)$ first (unless you want to); you can go straight to $3x^2 + 2x^2 = 5x^2$.

By the way, when you *pronounce* $(3+2)x^2$, you should technically say "the quantity three plus two..." The word *quantity* indicates parentheses. If you just say "three plus two x squared," someone could (and should) interpret what you said as $3 + 2x^2$, with no parentheses.

In the case of $7x - x$, the common factor is x:

$$7x \qquad - \qquad 1x \qquad = \qquad (7-1)x$$

Seven x minus one x equals the quantity seven minus one, times x

The right side reduces by PEMDAS to $6x$.

So if you combine like terms, you can simplify the original expression this way:

$$3x^2 + \qquad 7x \qquad + \qquad 2x^2 \quad - \quad x$$
$$\left(3x^2 + 2x^2\right) + (7x - x)$$
$$5x^2 \qquad + \qquad 6x$$

The common factor in like terms does not have to be a single variable expression such as x^2 or x. It could involve more than one variable:

$$-xy^2 + 4xy^2 = (-1+4)xy^2 = 3xy^2 \qquad \text{Common factor: } xy^2$$

Reminder: The coefficient on the first term is -1.

Be careful when you see multiple variables in a single term. To be like terms, the exponents have to match for every variable.

In $-xy^2 + 4xy^2$, each term contains a plain x (which is technically x raised to the first power) and y^2 (which is y raised to the second power, or y squared). All of the exponents match; the common factor is xy^2. Since the two terms are like, you can combine them to $3xy^2$.

Suppose you had the following series of terms:

$$2xy \quad + \quad xy^2 \quad - \quad 4x^2y \quad + \quad x^2y^2$$

Two $x\,y$ plus $x\,y$ squared minus four x squared y plus x squared y squared

None of the terms are like terms. They all have different combinations of variables and exponents. For now, you're stuck. (Later in this chapter, you'll see that there's *something else* you can do with that expression, but you can't combine like terms.)

The two terms in the following expression *are* like:

$$xy^2 \quad + \quad 3y^2x$$

$x\,y$ squared plus three y squared x

The order of the variables does not matter, since you can multiply in any order. All that matters is that the variables and exponents all match; the common factor is still xy^2. You can flip around $3y^2x$ to $3xy^2$ to get:

$$xy^2 \quad + \quad 3xy^2 \quad = \quad 4xy^2 \quad = \quad 4y^2x$$

$x\,y$ squared plus three $x\,y$ squared equals four $x\,y$ squared, and that four y squared x
                                                                    equals

In general, be ready to flip around products as you deal with numbers times variables. The order of multiplication does not matter. For example:

$$x(-3) \quad = \quad -x(3) \quad = \quad -3x$$

x times negative three equals negative x times three equals negative three x

The last form, $-3x$, is the standard way to write this kind of expression: The coefficient is generally written before the variable(s).

For the purposes of like terms, treat the square root of a number like a variable; the square root is the common factor:

$$\sqrt{2} + 3\sqrt{2} = 1\sqrt{2} + 3\sqrt{2} = (1 + 3)\sqrt{2} = 4\sqrt{2} \qquad \text{Common factor: } \sqrt{2}$$

Likewise, the common factor could include π:

$$2\pi r + 9\pi r = (2 + 9)\pi r = 11\pi r \qquad \text{Common factor: } \pi r$$

When terms are *not* like, tread carefully. Don't automatically combine everything; see what you can combine and what you cannot.

As you practice simplifying expressions, keep in mind that your main goal is to reduce the overall number of terms by combining like terms.

PEMDAS becomes more complicated when an expression contains terms that are not like and so cannot be combined. Be especially careful when you see terms buried within part of an expression, as in the following cases that you'll come back to later:

Terms inside parentheses:

$$-3(x - 2)$$

x and 2 are not like terms.

Terms in the numerator or denominator of a fraction:

$$\frac{1}{1 - x} = 2$$

x and 1 are not like terms.

Terms involving exponents:

$$\frac{x^{-3} + \left(x^2\right)^4}{x^5}$$

x^{-3} and $(x^2)^4$ are not like terms.

Terms under a root sign:

$$\sqrt{x^2 + y^2}$$

x^2 and y^2 are not like terms.

Terms in parentheses, with the parentheses raised to an exponent:

$$(x + y)^2$$

x and y are not like terms.

Check Your Skills

Combine as many like terms as possible in each of the following expressions.

24. $-3 + 4\sqrt{2} + 6$
25. $4\pi r^2 - 3\pi r + 2\pi r$
26. $8ba + ab^2 - 5ab + ab^2 - 2ba^2$

Answers can be found on page 39.

Distribution

Things become more complicated when multiple terms are found within a set of parentheses.

Start by distributing the example from the previous section: $-3(x - 2)$. Multiply -3 by $(x - 2)$. To keep track of minus signs as you distribute, you can think of $(x - 2)$ as $[x + (-2)]$. The following example shows the multiplication sign ($\times$) to make it clear that you're multiplying:

$$-3 \quad \times \quad (x - 2) \quad = \quad -3 \times x \quad + \quad -3 \times -2 \quad = \quad -3x + 6$$

| Negative three | times | the quantity x minus two | equals | negative three times x | plus | negative three times negative two, | and that equals | negative three x plus six |

Reminder: The negative sign (on -3) distributes across both terms in the parentheses.

When you do all this on your paper, don't use the multiplication symbol $\times$ to show multiplication, because you could confuse it with the variable x. Use parentheses, such as $(-3)(-2)$. You might also put parentheses around the second product to help keep track of sign:

$$-3 \quad \times \quad (x - 2) \quad = \quad -3x \quad + \quad (-3)(-2) \quad = \quad -3x + 6$$

| Negative three | times | the quantity x minus two | equals | negative three times x | plus | negative three times negative two, | and that equals | negative three x plus six |

1

How can you simplify this expression?

$$4y^2 - y(5 - 2y)$$

First, distribute negative y, $(-y)$, to both terms in the parentheses:

$$4y^2 - y(5 - 2y) = 4y^2 - 5y + 2y^2$$

Notice that $-y$ times $-2y$ becomes $+2y^2$.

Next, combine $4y^2$ and $2y^2$ because they are like terms:

$$4y^2 - y(5 - 2y) = 4y^2 - 5y + 2y^2 = 6y^2 - 5y$$

Sometimes the term being distributed involves a root or π. Consider this tougher example:

$$\sqrt{2}\left(1 - x\sqrt{2}\right)$$

The principle is the same. Distribute the first $\sqrt{2}$ to both terms in the parentheses:

$\sqrt{2}$	$\times$	$\left(1 - x\sqrt{2}\right)$	$=$	$\left(\sqrt{2}\right)(1)$	$+$	$\left(\sqrt{2}\right)\left(-x\sqrt{2}\right)$	$=$	$\sqrt{2} - 2x$
Root two	times	the quantity one minus x root two	equals	root two times one	plus	root two times negative x root two,	and that equals	root two minus two x

It turns out that $\sqrt{2}$ times $\sqrt{2}$ is equal to 2:

$$\sqrt{2} \times \sqrt{2} = \sqrt{2 \times 2} = \sqrt{4} = 2$$

You'll learn more about multiplying roots in Chapter 3. If you're curious now, feel free to jump ahead to that chapter; look for the section titled *Multiply Square Roots: Put Everything Under the Root*. Curiosity is a really powerful learning tool; when something captures your attention, follow that spark. You'll have a better chance of remembering that material for future use.

Here's an example with π:

$$\pi(1 + r)$$

Distribute the π:

π	$\times$	$(1 + r)$	$=$	$(\pi)(1)$	$+$	$(\pi)(r)$	$=$	$\pi + \pi r$
pi	times	the quantity one plus r	equals	pi times one	plus	pi times r,	and that equals	pi plus pi r

Check Your Skills

27. $x(3 + x)$
28. $4 + \sqrt{2}\left(1 - \sqrt{2}\right)$

Answers can be found on pages 39–40.

Pulling Out a Common Factor

Earlier, you saw the long expression below:

$$3x^2 + 7x + 2x^2 - x$$

This expression has four terms. By combining two pairs of like terms, you can simplify this expression to $5x^2 + 6x$, which has only two terms.

This expression can't go below two terms. The two remaining terms ($5x^2$ and $6x$) aren't *like*, because the variable parts aren't identical. However, these two terms do still have a common factor—namely, x. Each term is x times something, and you can use this fact to rewrite $5x^2 + 6x$:

x is a factor of $6x$, because $6x = 6$ times x.

x is also a factor of $5x^2$, because $x^2 = x$ times x, so $5x^2 = 5x$ times x.

Since x is a factor of both $5x^2$ and $6x$, you can factor it out and group what's left as a sum within parentheses:

$$5x^2 \qquad + \qquad 6x \qquad = \qquad x(5x + 6)$$

Five x squared $\quad$ plus $\quad$ six x $\quad$ equals $\quad$ x times the quantity
five x plus six

If you distribute the x back through, you'll end up back where you started:

$$x(5x + 6) \qquad = \qquad 5x^2 \qquad + \qquad 6x$$

x times the quantity $\quad$ equals $\quad$ five x squared $\quad$ plus $\quad$ six x
five x plus six

In addition, $x(5x + 6)$ can be written as $(5x + 6)x$. Either way, it may or may not be truly "simpler" than $5x^2 + 6x$. However, pulling out a common factor can be the key move when you solve a GMAT problem.

Sometimes the common factor is hidden among more complicated variable expressions. In the example below, the common factor is xy:

$$x^2y \qquad - \qquad xy^2 \qquad = \qquad xy(x - y)$$

x squared y $\quad$ minus $\quad$ $x\,y$ squared $\quad$ equals $\quad$ $x\,y$ times the
quantity x minus y

Sometimes the common factor involves a root or π:

$$\sqrt{2} \qquad + \qquad \pi\sqrt{2} \qquad = \qquad \sqrt{2}(1 + \pi)$$

Root two $\quad$ plus $\quad$ pi times root $\quad$ equals $\quad$ root two times
two $\qquad\qquad\qquad$ the quantity one plus pi

Here, the common factor is $\sqrt{2}$. Notice that the first term ($\sqrt{2}$) is the same as the common factor; the term $\sqrt{2}$ is really the equivalent of $1\sqrt{2}$. In this case, once you pull out the common term, you'll end up with a 1 in the parentheses. Here's another example:

$$\pi r^2 \qquad - \qquad \pi \qquad = \qquad \pi(r^2 - 1)$$

pi r squared $\quad$ minus $\quad$ pi $\quad$ equals $\quad$ pi times the quantity
r squared minus one

In the example above, the common factor is π. Again, when you pull π out of the second term (which is 1π), you'll end up with a 1 in the parentheses. You can check that this works by distributing π back through.

You can also factor out an integer:

$$2 \quad + \quad 4x \quad = \quad 2(1 + 2x)$$

Two   plus   four x equals   two times the quantity one plus two x

You can even pull out a -1:

$$3 \quad - \quad x \quad = \quad -(x - 3)$$

Three   minus x equals   the negative of quantity x minus three

In the example above, notice that the expression reverses itself: It goes from $3 - x$ to $x - 3$. You can take any similar expression and reverse it by pulling out a -1.

Remember this monster from earlier in this chapter?

$$2xy \quad + \quad xy^2 \quad - \quad 4x^2y \quad + \quad x^2y^2$$

Two $x\,y$ plus $x\,y$ squared   minus   four x squared y plus x squared y squared

You can't combine like terms but you can pull out a common factor. What is it?

The common factor is xy:

$$2xy + xy^2 - 4x^2y + x^2y^2 = xy(2 + y - 4x + xy)$$

Check Your Skills

29. Factor a negative x, $(-x)$, out of the expression $-2x^3 + 5x^2 + 3x$.

30. Factor the following expression: $4x^2 + 3xy - yx + 6x$

Answers can be found on page 40.

Long Addition and Subtraction

Sometimes you'll need to add or subtract larger numbers. If you were taking a real math test, you'd have to do this the "long" way, since you don't have a calculator. On a multiple-choice exam, however, you can usually get away with using various shortcuts.

Try this problem:

$$\begin{array}{r} 283 \\ + \; 654 \end{array}$$

Take a look at these sets of possible answer choices. How would they change your approach to the problem?

Set 1	Set 2	Set 3
(A) 717	(A) 935	(A) 917
(B) 801	(B) 936	(B) 927
(C) 937	(C) 937	(C) 937
(D) 1,042	(D) 938	(D) 947
(E) 1,137	(E) 939	(E) 957

At a glance, the first set is spread pretty far apart, while the second and third sets are quite close together. If you are given the first set of answers, you can estimate. Take a look at the problem; what's the best way to estimate long addition?

When the answers are as far apart as in set 1, estimate the first digit of the answer. In this case, the first digit is at least $2 + 6 = 8$. In fact, it will cross over into 9, because the next digit over for each is 5 or greater. So the answer will be 9-something. Only answer choice (C) fits.

What about set 2? The answers are so close together that it may seem like you will have to solve the long way—but you don't. This time, look at the units digits. Each one is different. How can you find the units digit of the given sum?

To find the units digit of the sum, you need to process only the units digits of the original numbers: $3 + 4 = 7$. Again, only answer choice (C) fits.

And what about set 3? Each answer starts with 9 and ends with 7, so this time it looks like you really do have to do long addition, right?

Sort of—but not entirely. You already know the answer must start with 9 (since all of the answers start with 9!), so only do as much work as you have to do.

Here's how to set that up:

$$\begin{array}{r} 283 \\ + 654 \\ \hline \end{array}$$
Begin with the right-most column of numbers and work your way to the left. The right-most column contains the units digits 3 and 4.

$$\begin{array}{r} 283 \\ + 654 \\ \hline 7 \end{array}$$
Sum the units digits: $3 + 4 = 7$. (If you hadn't already, glance at your answers. For set 3, you can't cross off any answers.)

$$\begin{array}{r} {}^{1}283 \\ + 654 \\ \hline 37 \end{array}$$
Move to the next column, the tens digits 8 and 5, where $8 + 5 = 13$. Place the 3 below and *carry* the 1 (from 13) above the top number in the next column, the hundreds digits. Glance at your answers again. For set 3, only choice (C) can work.

$$\begin{array}{r} {}^{1}283 \\ + 654 \\ \hline 937 \end{array}$$
Here's the rest of the math, just to show you how carrying works. Move to the next column, the hundreds digit. Now, there are three numbers: the carried 1, as well as the 2 and 6. Add all of these together to get 9.

When conducting long addition, add the columns of numbers: units digit + units digit, tens digit + tens digit, and so on. If one of those sums is a number greater than 9, you'll need to carry over part of the number to the next column. For example, if the sum is 15, then place the units digit of the sum (the 5) below, but carry the extra part (the tens digit, 1) over to the next column and add it there instead.

On a multiple-choice test, keep glancing back at the answers. Do only as much work as you absolutely need to do.

Subtraction works similarly, although there is one special circumstance to note.

Try this problem:

$$\begin{array}{r} 653 \\ - 472 \\ \hline \end{array}$$

What can you estimate about this answer?

At most, it will start with a 2. In fact, that first digit will drop below 2 since the 7 is larger than the 5. And the units digit will be 1.

If you need to be more precise than that, you can also count up the distance between the two. From 472 to 672 would be 200. The actual greater number is 653, so the distance is a bit less than 200. How much less? About 20 less. Since you also know the units digit must be 1, you've got the answer: 181.

You can also do long subtraction—but, again, take the math only as far as you have to based upon the given answer choices:

$$\begin{array}{r} 653 \\ -\ 472 \\ \hline \end{array}$$

Begin with the right-most column of numbers and work your way to the left. The right-most column contains the units digits 3 and 2.

$$\begin{array}{r} 653 \\ -\ 472 \\ \hline 1 \end{array}$$

Subtract: $3 - 2 = 1$. If you had answers, you would now cross off any that have a different units digit.

$$\begin{array}{r} {}^5\!\!\not{6}{}^1\!53 \\ -\ 472 \\ \hline 81 \end{array}$$

$5 - 7 = -2$, but you aren't allowed to put -2 down below. Instead, *borrow* a 10 from the next column. The tens column becomes $15 - 7 = 8$. The 6 in the hundreds column turns into a 5. (Check the answers again.)

$$\begin{array}{r} {}^5\!\!\not{6}{}^1\!53 \\ -\ 472 \\ \hline 181 \end{array}$$

Finally, subtract the hundreds column: $5 - 4 = 1$. (On a multiple-choice test, you usually don't have to go all the way to this point.)

As long as the top number is greater than the bottom one, you can subtract normally. If the top number is smaller, though, then borrow a 10 from the next number to the left. Once you've done that, proceed normally.

Long Multiplication

This section reviews the basics of long multiplication. As with long addition and long subtraction, you can usually get away with doing only some of this work.

For multiplication in particular, set up the whole problem before you try to solve. Later steps might require you to divide. Since you can multiply and divide in any order, divide first to make the numbers smaller. (You may have learned this mantra in school: *Simplify before you multiply.*)

Try this:

What is 8 multiplied by 57 ?

As with addition, the units digit of the product is entirely determined by the units digits of the starting numbers. In this case, the units digits of the two starting numbers are 7 and 8:

$7 \times 8 = 56$

The units digit of the resulting product, 6, will be the units digit of the entire product 8×57. Use that knowledge to cross off as many answer choices as you can. (In many cases, that will be all four incorrect answers!)

You can also estimate the answer in this way:

$8 \times 57 = ?$

$8 \times 50 = 400$ Do $8 \times 5 = 40$, then add a 0.

$8 \times 60 = 480$ Do $8 \times 6 = 48$, then add a 0.

The correct answer must be between 400 and 480, and it will be closer to 480, since 57 is closer to 60 than to 50.

If you do need to do some level of long multiplication, always put the smaller number in the bottom row:

$$\begin{array}{r} 57 \\ \times\ 8 \end{array}$$

Multiply the two numbers in the right-most column: $7 \times 8 = 56$. Put the 6 underneath, then carry the 5.

$$\begin{array}{r} ^5 57 \\ \times 8 \\ \hline 6 \end{array}$$

Multiply the next number over in the top row by the number in the bottom row: $5 \times 8 = 40$. Then add the 5 you carried: $40 + 5 = 45$.

$$\begin{array}{r} 57 \\ \times\ 8 \\ \hline 456 \end{array}$$

Because you're at the end, put the whole 45 underneath.

You may also need to multiply two two-digit numbers, such as 12×85. Whenever possible, break this up into two easier parts. Here, break 12 into 10 and 2:

$$\begin{aligned} 12 \times 85 &= (10 + 2) \times 85 \\ &= 10 \times 85 + 2 \times 85 \\ &= 850 + 170 \end{aligned}$$

Finally, add the numbers from the two parts. Don't forget to keep an eye on the answers and estimate as appropriate. In this case, $850 + 170$ is about $850 + 200 = 1,050$. That sum includes an extra 30, so, if needed, subtract 30 to get to the exact answer: $1,020$.

Some combination of breaking the math into parts and estimating is usually enough to get you to the answer. In the interest of thoroughness, the example below shows how to do long multiplication with two two-digit numbers, but if this is the only way that you can think to solve, it may be a better decision to guess and move on. Save your limited time for use elsewhere.

$$\begin{array}{r} ^1 85 \\ \times 12 \\ \hline 0 \end{array}$$

Start with the 2 in the bottom row: $5 \times 2 = 10$. Put the 0 underneath, then carry the 1.

$$\begin{array}{r} \cancel{1} 85 \\ \times 12 \\ \hline 170 \end{array}$$

$8 \times 2 = 16$. Then add the 1 you carried to get 17 and cross off the 1. Because you are done multiplying by 2, place the 17 underneath.

$$\begin{array}{r} \cancel{1} 85 \\ \times 12 \\ \hline 170 \\ 0 \end{array}$$

Now deal with the 1 in the second row. That 1 actually represents 10, since it's the tens digit. Place a 0 underneath the right-most column. (You'll always place a 0 here, no matter what the starting numbers are.)

$$\begin{array}{r} \cancel{1} 85 \\ \times 12 \\ \hline 170 \\ 50 \end{array}$$

Next, do $5 \times 1 = 5$. Place the 5 in the second row to the left of the 0.

$\cancel{1}85$

$\times\ 12$

$\overline{170}$

850

Next, do $8 \times 1 = 8$. Place the 8 in the second row to the left of the 5.

^{1170}

$+\ 850$

$\overline{1020}$

Finally, add the two bottom rows using the normal long addition process (that is, starting from the right). In the second column, $7 + 5 = 12$, so place the 2 underneath and carry the 1. In the third column, $1 + 8 +$ the carried $1 = 10$. Because you are done with the addition, place the 10 underneath.

Overall, don't spend a ton of time drilling long multiplication. Rather, practice estimating and breaking larger numbers down into easier parts. Just five minutes a day will make a big difference after a few weeks.

Long Division

As with all other operations, think carefully before you dive into long division. For instance, what would you do with this problem?

What is 468 divided by 26 ?

(A) 12

(B) 18

(C) 22

The answers are pretty far apart, so estimate. Use the answer choices themselves to decide how to estimate. Multiplying by 10 and by 20 is easier than using the given numbers:

$26 \times 10 = 260$

$26 \times\ ?? = 486$

$26 \times 20 = 520$

The correct answer must be between 10 and 20. Answer (C) is definitely too large. Answers (A) and (B) are both within the range. Should the answer be closer to the 10 end of the range or the 20 end of the range?

Since 486 is closer to 520 than to 260, the answer should be closer to 20 than to 10; answer (A) is too small. Therefore, the answer must be (B), 18. Done—without having to resort to long division.

On other problems, you may be able to simplify *and* estimate. Try this:

What is the result when 440 is divided by 11 and then divided by 2 ?

(A) 20

(B) 22

(C) 24

Don't be fooled by the language! When multiplying or dividing, you can do the work in any order; that's why M and D go together in the PEMDAS order of operations. You can do this work in whatever order you think is best. In this case, it's easier to divide by 2 first:

$440 \div 2 = 220$

Next, you need to divide by 11, but that's more cumbersome. Estimate a little. What is 220 divided by 10?

That's equal to 22. Glance at the answers and eliminate (B). Next, is 22 too big or too small?

You were supposed to divide by 11, but you divided by a smaller number, 10, instead. That's going to make the estimate a bit too large, so the answer also can't be (C). The correct answer must be (A), 20.

You can also do long division—but don't on this test. If the *only* way you can think to solve a problem on this type of exam is to do long division, your best call is to guess immediately and move on—that problem is not worth your limited time.

If you find math fun, here's how to do long division. If you're side-eyeing the statement "if you find math fun," you have full permission to skip the rest of this chapter and go straight to the problem set.

What is 182 divided by 13 ?

$13\overline{)182}$

First, set up 182 divided by 13 in this way: The first number, 182, goes under the division bracket. The second number, 13, goes outside to the left of the bracket. (For the purposes of division, the number 182 is called the **dividend** and the number 13 is called the **divisor**. You'll see that word *divisor* again in the next chapter, but with a different definition.)

$$\begin{array}{r} 1 \\ 13\overline{)182} \\ -13 \\ \hline 5 \end{array}$$

Starting from the left of the number, find the smallest portion of 182 that is still larger than 13. The number 1 is not larger than 13. The number 18 is larger than 13. The divisor 13 goes into 18 one time with 5 left over, so place a 1 above the unit's digit of the portion of the number you used (in this case, above the 8). Then subtract out $13 \times 1 = 13$ to get the 5 left over.

$$\begin{array}{r} 1 \\ 13\overline{)182} \\ -13 \\ \hline 52 \end{array}$$

Next, take the next digit in the number under the bracket (in this case, 2) and drop it down to the bottom, next to the 5. You now have the number 52.

$$\begin{array}{r} 14 \\ 13\overline{)182} \\ -13 \\ \hline 52 \\ -52 \\ \hline 0 \end{array}$$

Figure out how many times 13 goes into 52. In this case, it goes in 4 times with 0 left over, since $13 \times 4 = 52$. Place the 4 above the division bracket, to the right of the 1. Subtract out the 52 to get 0. The 0 tells you that you're done with the math and that $182 \div 13 = 14$. The number 14 is called the **quotient**.

In the previous example, the answer was an integer (14). However, the answer will not always be an integer.

1

Try dividing 123 by 6:

$$6\overline{)123}$$

$$\begin{array}{r} 2 \\ 6\overline{)123} \\ -12 \\ \hline 03 \end{array}$$

6 doesn't go into 1, but it does go into 12 two times. Place a 2 on top of the digit farthest to the right in 12. Because $2 \times 6 = 12$, subtract 12 from 12, then bring down the next digit in 123, which is 3.

$$\begin{array}{r} 20. \\ 6\overline{)123.0} \\ -12 \\ \hline 030 \end{array}$$

6 doesn't go into 3, so place a 0 next to the 2 above the division bracket. Add a decimal point and 0 to get 123.0, and add a decimal after the 20. Then bring down the 0 next to the 3.

$$\begin{array}{r} 20.5 \\ 6\overline{)123.0} \\ -12 \\ \hline 030 \\ -30 \\ \hline 0 \end{array}$$

6 goes into 30 five times. Place a 5 after the 20. Then subtract 30 from 30 to get 0. When you have a 0 at the bottom, and you've used all of the numbers in your original dividend (123, in this case), you know you're done: $123 \div 6 = 20.5$.

Long division is cumbersome, and this test is not really a math test. There is always a way to estimate or do something else to avoid such division. If you're facing a problem but can't think of a way around long division, consider whether that problem is worth your time at all. Your best bet may be to guess and move on.

Check Your Skills Answer Key

1. **178:** For addition, first place the larger number on the number line. Then break the smaller number into convenient pieces and add.

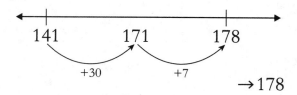

$$\rightarrow 178$$

In this case, it's easier to add 30 and then add 7 than it is to add 37 all at once.

2. **−113:** Find the positive difference, then add a negative sign (so the answer is −113, not +113).

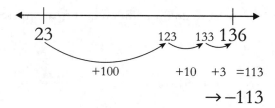

$$\rightarrow -113$$

3. **0:** The greatest negative integer is −1, and the smallest positive integer is 1.

$$-1 + 1 = 0$$

4. **5 < 16:** 5 is less than 16.

5. **−5 > −16:** Negative 5 is greater than negative 16. This one is more confusing! Visualize or sketch out a number line to make sure you don't make a careless mistake.

6. **A positive minus a negative:** If a theory question is at all confusing, test some real numbers to understand what's happening. A positive minus a negative could be $3 - (-2) = 3 + 2 = 5$. A positive minus a negative, therefore, is the same as a positive plus a positive, so $(+) - (-)$ will always lead to a positive result.

    Now try a negative minus a positive: $(-3) - 2 = -5$. You start with a negative number and then get even more negative, so the $(-) - (+)$ scenario will always have a negative result.

    Since $(+) - (-)$ will always be positive and $(-) - (+)$ will always be negative, the $(+) - (-)$ scenario will always be greater. (It's always true that any positive number is greater than any negative number.)

7. **42:** $7 \times 6 = 42$. In school, you had to memorize multiplication tables to know what something like 7×6 equals. If you've forgotten your multiplication tables, just google *times table* or *multiplication table* and find one in a format that works for you.

8. **4:** $52 \div 13 = 4$. Division is more annoying. Try this trick: Start with the smaller number (in this case, 13), and count up by multiples till you get to the larger one.

    13, 26, 39, 52

    So 13 times 4 is equal to 52. Therefore, $52 \div 13 = 4$.

9. **35:** On the real test, you would likely want to add first within the parentheses and then multiply. The instructions, though, specify to use distribution to solve.

$$5 \times (3 + 4) =$$
$$(5 \times 3) + (5 \times 4) =$$
$$15 + 20 = 35$$

10. **6(6 − 2):**

$$36 - 12 =$$
$$(6 \times 6) - (6 \times 2) =$$
$$6(6 - 2)$$

11. **−10:** $(2)(-5) = -10$

12. **48:** On the real test, you might prefer to do the addition in the parentheses first, but the instructions specify to use distribution.

$$-6 \times (-3 + (-5)) =$$
$$(-6 \times -3) + (-6 \times -5) =$$
$$18 + 30 = 48$$

13. **Division:** Sometimes an integer divided by an integer equals an integer (e.g., $6 \div 2 = 3$), and sometimes it does not (e.g., $8 \div 5 = 1.6$).

14. **0:** The number 2 divided by 7 is written $\frac{2}{7}$ in fraction form. If you multiply any number by 0, the result is 0.

15. **Multiple:** If 7 is a factor of 21, then 21 is a multiple of 7.

16. **No:** 2,284,623 ends in 3, which means that it is an odd number. It is not divisible by 2.

17. **64:** When multiplying so many of the same number together, parentheses can help to keep everything straight.

$$2^6 = (2 \times 2) \times (2 \times 2) \times (2 \times 2) = 4 \times 4 \times 4 = 64$$

Or feel free to count on your fingers: 2 times 2 is 4... times 2 is 8... times 2 is 16... times 2 is 32... times 2 is 64. Just keep track on your fingers so that you use exactly six 2's.

18. **3:** What number multiplied by itself three times equals 27? $3 \times 3 \times 3 = 27$.

19. **11:** Plug the values for each variable into the expression to find the value of the expression.

$$2x - 3y = 2(4) - 3(-1) = 8 + 3 = 11$$

20. **0:**

$$-4 + \frac{12}{3} = \qquad \text{First, divide.}$$
$$-4 + 4 = 0 \qquad \text{Then, add.}$$

21. **−37:**

$$(5 - 8) \times 10 - 7 =$$ First, do the parentheses.
$$(-3) \times 10 - 7 =$$ Next, multiply.
$$-30 - 7 =$$ Finally, subtract.
$$-30 - 7 = -37$$

22. **−74:** You can multiply and divide in whatever order you like.

$$-3 \times 12 \div 4 \times 8 + (4 - 6) =$$
$$-3 \times 12 \div 4 \times 8 + (-2) =$$ First, do the parentheses.
$$-3 \times 3 \times 8 + (-2) =$$ Then, multiply and divide in your desired order.
$$-9 \times 8 + (-2) =$$
$$-72 + (-2) = -74$$ Finally, subtract.

23. **8:**

$$\frac{2^4 \times (8 \div 2 - 1)}{(9 - 3)} =$$ Do the parentheses. For the top parentheses, divide first.

$$\frac{2^4 \times (4 - 1)}{(6)} =$$

$$\frac{16 \times (3)}{(6)} =$$ Finish the parentheses and simplify the exponent.

$$\frac{16 \times \overset{1}{\cancel{3}}}{\underset{2}{\cancel{6}}} =$$

$$\frac{\overset{8}{\cancel{16}} \times 1}{\underset{1}{\cancel{2}}} = 8$$ Divide before you multiply to keep the numbers smaller.

24. **$3 + 4\sqrt{2}$:**

$$(-3 + 6) + 4\sqrt{2} = 3 + 4\sqrt{2}$$

25. **$4\pi r^2 - \pi r$:**

$$4\pi r^2 + (-3\pi r + 2\pi r) = 4\pi r^2 - \pi r$$

26. **$3ab + 2ab^2 - 2a^2b$:**

$$(8ab - 5ab) + (ab^2 + ab^2) + (-2a^2b) = 3ab + 2ab^2 - 2a^2b$$

27. **$3x + x^2$:**

$$x(3 + x) = (3)(x) + (x)(x) = 3x + x^2$$

28. $2 + \sqrt{2}$:

$$4 + \sqrt{2}\left(1 - \sqrt{2}\right) =$$
$$4 + \left(\sqrt{2}\right)(1) + \left(\sqrt{2}\right)\left(-\sqrt{2}\right) =$$
$$4 + \sqrt{2} - 2 =$$
$$2 + \sqrt{2}$$

29. $-x(2x^2 - 5x - 3)$: Pull a -1 and an x out of each term. Switch the sign on each term in the parentheses in order to pull out the -1.

30. $2x(2x + y + 3)$:

$$4x^2 + 3xy - yx + 6x =$$

$4x^2 + 2xy + 6x =$ First, combine the two like terms.

$2x(2x + y + 3)$ Then, pull out a $2x$.

Chapter Review: Drill Sets

Drill 1

Evaluate the following expressions.

1. $39 - (25 - 17)$

2. $3(4 - 2) \div 2$

3. $15 \times 3 \div 9$

4. $(7 - 5) - (3 - 6)$

5. $14 - 3(4 - 6)$

6. $(5)(-3)(-4)(2)$

7. $5 - (4 - (3 - (2 - 1)))$

8. $-4(5) - \dfrac{12}{2 + 4}$

   (A) 22

   (B) 0

   (C) −22

9. $17(6) + 3(6)$

Drill 2

Evaluate the following expressions.

10. $\dfrac{24}{2 + 6 \div 3}$

    (A) 24

    (B) 6

    (C) 1

11. $-10 - (-3)^2$

    (A) −1

    (B) −7

    (C) −19

12. $247 + 142 - 226 =$

    (A) 615

    (B) 437

    (C) 163

13. $7(4) + 7(3) + 7(2) + 7(1)$

14. $3 \times 99 - 2 \times 99 - 1 \times 99$

Drill 3

Combine as many like terms as possible.

15. $5\sqrt{3} + 5\sqrt{2} - 2\sqrt{3}$

16. $12xy^2 - 6x^2y^2 + (2)^2x^2y^2$

17. $\sqrt{2} + x\sqrt{2} - 2\sqrt{2}$

18. $12xy - (6x + 2y)$

19. $5x - (4x + 2 - (5x - 3))$

20. $\pi^2r^2 - (\pi r + 2\pi r^2) + \pi r^2 + \pi^2r^2 + 2\pi r$

21. $2x^2 - (2x)^2 - 2^2 - x^2$

22. $4x^2 + 2x - \left(2\sqrt{x}\right)^2$

Drill 4

Distribute the following expressions. Simplify as necessary.

23. $-(a - b)$

24. $(m + 2n)4m$

25. $52r(2t - 10s)$

26. $(-37x + 63)10^2$

27. $6kl(k - 2l)$

28. $-\sqrt{2}(18 - 8x)$

29. $d(d^2 - 2d + 1)$

30. $xy^2z(x^2z + yz^2 - xy^2)$

Drill Sets Solutions

Drill 1

1. **31:** Start by approximating what the answer will be. Not only will this give you a way to catch any errors you could make in your computations, but it may also enable you to eliminate several answer choices on a multiple-choice question.

    $$39 - (25 - 17) \approx \qquad \text{Round to easier numbers.}$$
    $$40 - (25 - 20) =$$
    $$40 - 5 \quad = 35$$

    The answer is approximately 35. Use this estimation method whenever possible and complete the following exact calculations only if the problem requires it.

    $$39 - (25 - 17) = \qquad \text{First, parentheses.}$$
    $$39 - \quad (8) \quad = 31 \qquad \text{Then, subtract.}$$

    If you prefer, you could first distribute the minus sign: $(39 - 25 + 17) = 14 + 17 = 31$.

2. **3:** Follow PEMDAS.

    $$3 \times (4 - 2) \div 2 = \qquad \text{First, parentheses.}$$
    $$3 \times \quad (2) \quad \div 2 = \qquad \text{Then, divide.}$$
    $$3 \times 1 = 3 \qquad \text{Finally, multiply.}$$

3. **5:** Approximate an answer first. 15 is multiplied by one number and divided by another, but the divisor is larger than the multiplier. Because of this, the final number should be less than 15. All of the numbers are positive, so the answer must be between 0 and 15.

    $$15 \times 3 \div 9 =$$
    $$45 \div 9 = 5$$

4. **5:**

    $$(7 - 5) - (3 - 6) =$$
    $$(2) - (-3) =$$
    $$2 + 3 = 5$$

5. **20:** Estimate first. The terms in the parentheses simplify to a negative number, so you're subtracting a negative number from 14. That translates to adding a positive number to 14, resulting in a value larger than 14.

    $$14 - 3(4 - 6) = \qquad \text{First, apply the parentheses.}$$
    $$14 - 3(-2) = \qquad \text{Then, multiply.}$$
    $$14 + 6 = 20 \qquad \text{Finally, add.}$$

6. **120:** First, estimate. The answer should be positive, since an even number of negative terms multiplied or divided together will produce a positive value.

    Simplify the computations by combining 2's and 5's.

    $$(5)(-3)(-4)(2) = (10)(12) = 120$$

7. **3:** Start with the innermost parentheses. To avoid careless mistakes, write everything down and don't try to do multiple steps at once.

$$5 - \left(4 - \left(3 - \left(2 - 1\right)\right)\right) =$$
$$5 - \left(4 - \left(3 - \ \ (1)\right)\right) \ \ =$$
$$5 - \left(4 - (2)\right) \qquad =$$
$$5 - \ \ (2) \qquad\qquad = 3$$

8. **(C) −22:** The answer choices are significantly far apart from each other, so estimate. The first term is a negative multiplied by a positive, which will be negative. The second term is positive, but is subtracted from the first. A negative minus a positive is always negative, so the answer must be −22.

Here's how to compute the answer directly.

$$-4(5) - \frac{12}{2+4} =$$
$$-20 - \frac{12}{6} =$$
$$-20 - 2 = -22$$

In order to avoid a careless mistake, consider writing the last line as $-20 + (-2) = -22$.

9. **120:** You could follow PEMDAS order.

$$17(6) + 3(6) =$$
$$102 + 18 = 120$$

However, if you find that math cumbersome (most people would), then don't just start doing it. Take a moment: Is there an easier way to do this math? Yes! Factor a 6 out of both terms, and then follow PEMDAS order.

$$17(6) + 3(6) =$$
$$6(17 + 3) =$$
$$6(20) = 120$$

Drill 2

10. **(B) 6:** The answers are far apart, so estimate. The denominator can be estimated to be a positive number greater than 2, but significantly less than 24. Answer choice (C) would be correct if the denominator were 24, and (A) would be correct if the denominator were 1. Since the denominator is somewhere between those two numbers, the answer must be 6.

To compute the answer exactly, start with the division within the fraction.

$$\frac{24}{2 + 6 \div 3} =$$
$$\frac{24}{2 + 2} =$$
$$\frac{24}{4} = 6$$

1

11. **(C) −19:** The problem begins with the value −10. Two of the answers are greater and one is less than −10, so estimate to determine whether the final answer will be higher or lower. Squaring a negative number results in a positive number, leaving you with −10 minus a positive. The result would be a *more* negative number, or a number farther away from zero. The only answer choice that fits is (C).

Here's the exact math.

$$-10 - (-3)^2 = \qquad \text{First, exponents.}$$
$$-10 - (-3)(-3) =$$
$$-10 - (9) = -19 \qquad \text{Then, subtract.}$$

You could also write the last line as $-10 + (-9) = -19$.

12. **(C) 163:** The answers differ by hundreds, so estimate. The numbers 247 and 226 are about 250 and 225, and $250 - 225 = 25$. Then add 142. The only answer choice that is reasonable is (C).

If you have to solve precisely, consider subtracting 226 from 247 first since that computation is relatively straightforward and results in more manageable numbers.

$$247 + 142 - 226 =$$
$$(247 - 226) + 142 =$$
$$21 + 142 = 163$$

13. **70:** Factor a 7 out of each term.

$$7(4) + 7(3) + 7(2) + 7(1) =$$
$$7(4 + 3 + 2 + 1) =$$
$$7(10) = 70$$

You can also multiply out each term and then add, as shown below, but that will generally take longer when the numbers are at all large or cumbersome.

$$7(4) + 7(3) + 7(2) + 7(1) =$$
$$28 + 21 + 14 + 7 = 70$$

14. **0:** Estimate first. Multiplying by 99 is a pain, but it's nearly the same as multiplying by 100, which is much easier.

$$3 \times 99 - 2 \times 99 - 1 \times 99 \approx$$
$$3 \times 100 - 2 \times 100 - 1 \times 100 =$$
$$300 - 200 - 100 =$$
$$100 - 100 = 0$$

If you do need to compute an exact answer, plan before diving in. Multiplying by 99 three times is too cumbersome. Instead, factor 99 out of each term.

$$3 \times 99 - 2 \times 99 - 1 \times 99 =$$
$$99(3 - 2 - 1) =$$
$$99(0) = 0$$

Drill 3

15. $3\sqrt{3} + 5\sqrt{2}$: Group and combine like terms.

$$5\sqrt{3} + 5\sqrt{2} - 2\sqrt{3}$$
$$\left(5\sqrt{3} - 2\sqrt{3}\right) + 5\sqrt{2}$$
$$3\sqrt{3} + 5\sqrt{2}$$

16. $12xy^2 - 2x^2y^2$: Simplify before grouping and combining like terms.

$$12xy^2 - 6x^2y^2 + (2)^2 x^2 y^2 =$$
$$12xy^2 - 6x^2y^2 + 4x^2 y^2 =$$
$$12xy^2 + (-6 + 4)x^2 y^2 =$$
$$12xy^2 + (-2)x^2 y^2 =$$
$$12xy^2 - 2x^2 y^2$$

17. $(x - 1)\sqrt{2}$:

$$\sqrt{2} + x\sqrt{2} - 2\sqrt{2} =$$
$$1\sqrt{2} + x\sqrt{2} - 2\sqrt{2} =$$
$$(1 + x - 2)\sqrt{2} =$$
$$(-1 + x)\sqrt{2} = (x - 1)\sqrt{2}$$

18. $12xy - 6x - 2y$: The terms in the parentheses are not like terms, so distribute the negative sign before grouping and combining like terms.

$$12xy - (6x + 2y) =$$
$$12xy - 6x - 2y$$

Note that you cannot actually combine any of the terms in this problem; none are like terms!

19. $6x - 5$: Work from the innermost parentheses out.

$$5x - \left(4x + 2 - (5x - 3)\right) =$$
$$5x - (4x + 2 - 5x + 3) =$$
$$5x - (4x - 5x + 2 + 3) =$$
$$5x - (-x + 5) =$$
$$5x + x - 5 = 6x - 5$$

20. $2\pi^2 r^2 + \pi r - \pi r^2$:

$$\pi^2 r^2 - \left(\pi r + 2\pi r^2\right) + \pi r^2 + \pi^2 r^2 + 2\pi r =$$
$$\pi^2 r^2 - \pi r - 2\pi r^2 + \pi r^2 + \pi^2 r^2 + 2\pi r =$$
$$\left(1\pi^2 r^2 + 1\pi^2 r^2\right) + (-1\pi r + 2\pi r) + \left(-2\pi r^2 + 1\pi r^2\right) =$$
$$2\pi^2 r^2 + 1\pi r - \pi r^2$$

1

21. $-3x^2 - 4$:

$$2x^2 - (2x)^2 - 2^2 - x^2 =$$
$$2x^2 - 2^2x^2 - 4 - 1x^2 =$$
$$2x^2 - 4x^2 - 1x^2 - 4 =$$
$$(2 - 4 - 1)x^2 - 4 =$$
$$(-3)x^2 - 4 =$$
$$-3x^2 - 4$$

22. $4x^2 - 2x$:

$$4x^2 + 2x - \left(2\sqrt{x}\right)^2 =$$
$$4x^2 + 2x - 2^2\left(\sqrt{x}\right)^2 =$$
$$4x^2 + 2x - 4x =$$
$$4x^2 + (2 - 4)x =$$
$$4x^2 + (-2)x =$$
$$4x^2 - 2x$$

Drill 4

23. **$-a + b$ OR $b - a$:** The minus sign in front of the left parenthesis can be interpreted as -1 times the expression $(a - b)$. Distribute the -1 to both terms in the parentheses: $-(a - b) = (-1) \times (a - b) = -a + b$.

24. **$4m^2 + 8mn$:** Ordinarily, you see the Distributive Property in this form:

$$a(b + c) = ab + ac$$

If you place a to the right of the parentheses, you can still distribute in the same way:

$$(b + c)a = ba + ca$$

This works because the order in which numbers are multiplied does not matter. The GMAT sometimes disguises a possible distribution by presenting it in this alternative form.

$$(m + 2n)4m =$$
$$4m^2 + 8mn$$

25. **$104rt - 520rs$:** First, distribute normally. Then, combine like terms.

$$52r(2t - 10s) =$$
$$(52r)(2t) - (52r)(10s) =$$
$$104rt - 520rs$$

26. **$-3{,}700x + 6{,}300$:**

$$(-37x + 63)10^2 =$$
$$(-37x)(100) + (63)(100) =$$
$$-3{,}700x + 6{,}300$$

27. **$6k^2l - 12kl^2$:** First, distribute normally. Then, combine like terms.

$$6kl\,(k - 2l) =$$
$$6kl\,(k) - 6kl\,(2l) =$$
$$6k^2l - 12kl^2$$

28. **$-18\sqrt{2} + 8x\sqrt{2}$:** Distribute carefully to keep track of the negative signs.

$$-\sqrt{2}(18 - 8x) =$$
$$\left(-\sqrt{2}\right)(18) + \left(-\sqrt{2}\right)(-8x) =$$
$$-18\sqrt{2} + 8x\sqrt{2}$$

29. **$d^3 - 2d^2 + d$:** Even though there are three terms inside the parentheses, distribution works exactly the same. Multiply d by every term in the parentheses.

$$d\left(d^2 - 2d + 1\right) =$$
$$\left(d \times d^2\right) - (d \times 2d) + (d \times 1) =$$
$$d^3 - 2d^2 + d$$

30. **$x^3y^2z^2 + xy^3z^3 - x^2y^4z$:** The term xy^2z on the outside of the parentheses must be multiplied by each of the three terms inside the parentheses. Then simplify the expression as much as possible.

$$xy^2z\left(x^2z + yz^2 - xy^2\right) =$$
$$\left(xy^2z\right)\left(x^2z\right) + \left(xy^2z\right)\left(yz^2\right) - \left(xy^2z\right)\left(xy^2\right) =$$
$$x^3y^2z^2 + xy^3z^3 - x^2y^4z$$

Divisibility

In This Chapter

- Divisibility Rules

- Factors Are Divisors

- Prime Number: Only Divisible by 1 and Itself

- Prime Factor Trees

- Every Number Is Divisible by the Factors of Its Factors

- Factors: Built Out of Primes

- Factor Tree of a Variable: Contains Unknowns

- Factors of x with No Common Primes: Combine

- Factors of x with Primes in Common: Combine to Least Common Multiple

In this chapter, you will learn how to recognize problems testing divisibility, primes, and factors, as well as how to set up and solve such problems efficiently.

CHAPTER 2 Divisibility

Divisibility has to do with **integers**. Recall that integers are the counting numbers (1, 2, 3, etc.), their opposites (−1, −2, −3, etc.), and 0. Integers have no decimals or fractions attached.

Most integer arithmetic is boring:

$$\text{Integer} + \text{Integer} = \text{Always an integer} \qquad 4 + 11 = 15$$

$$\text{Integer} - \text{Integer} = \text{Always an integer} \qquad -5 - 32 = -37$$

$$\text{Integer} \times \text{Integer} = \text{Always an integer} \qquad 14 \times 3 = 42$$

However, when you divide an integer by another integer, sometimes you get an integer ($18 \div 3 = 6$), and sometimes you don't ($12 \div 8 = 1.5$).

If you get an integer out of the division, then the first number is **divisible by** the second. For example, 18 is divisible by 3 because $18 \div 3$ is equal to an integer. On the other hand, 12 is *not* divisible by 8 because $12 \div 8$ is not equal to an integer.

Divisibility Rules

It's very useful to memorize divisibility rules for small integers—you'll save yourself a lot of time on the exam. An integer is divisible by:

2 if the integer is even.

> Even numbers, such as the number 12, are integers that end in 0, 2, 4, 6, or 8. The number 12 is divisible by 2 because $12 \div 2 = 6$, an integer. Any even number is always divisible by 2.

3 if the sum of the integer's digits is divisible by 3.

> Take the number 147. Its digits are 1, 4, and 7. Add those digits to get the sum: $1 + 4 + 7 = 12$. The sum, 12, is divisible by 3, so 147 is divisible by 3.

5 if the integer ends in 0 or 5.

> The numbers 75 and 80 are divisible by 5, but 77 and 84 are not. Any integer that ends in 0 or 5 is divisible by 5. Any integer that ends in something other than 0 or 5 is *not* divisible by 5.

9 if the sum of the integer's digits is divisible by 9.

> This rule is very similar to the divisibility rule for 3. Take the number 288. Add the digits: $2 + 8 + 8 = 18$. The sum, 18, is divisible by 9, so 288 is divisible by 9.

10 if the integer ends in 0.

> The number 8,730 is divisible by 10, but 8,753 is not. Any integer that ends in 0 is divisible by 10. Any integer that ends in something other than 0 is *not* divisible by 10.

Check Your Skills

1. Is 123,456,789 divisible by 2 ?

2. Is 732 divisible by 3 ?

3. Is 989 divisible by 9 ?

Answers can be found on page 70.

Factors Are Divisors

The integer 6 is divisible by what positive integers?

Test the positive integers less than or equal to 6: 1, 2, 3, 4, 5, and 6:

$6 \div 1 = 6$ Any number divided by 1 equals itself, so an integer divided by 1 always equals an integer.

$6 \div 2 = 3$

$6 \div 3 = 2$ 6 is divisible by 2, and 6 is divisible by 3. Note that these form a pair: $3 \times 2 = 6$.

$6 \div 4 = 1.5$

$6 \div 5 = 1.2$ These are not integers, so 6 is *not* divisible by 4 or by 5.

$6 \div 6 = 1$ Any number divided by itself equals 1, so an integer is always divisible by itself.

Therefore, 6 is divisible by 1, 2, 3, and 6. These numbers 1, 2, 3, and 6 are called **factors** of 6. Here are various ways you might see this relationship expressed on the GMAT:

2 is a factor of 6	2 is a divisor* of 6
6 is a multiple of 2	6 is divisible by 2
2 goes into 6 evenly	

*Note: In the prior chapter, you saw the word *divisor* used to mean *the number that you divide another number by*. In that context, the numbers don't need to divide evenly—but that definition is a secondary definition used only in the context of performing long division.

The primary usage is as follows: If a problem says that *one number is a divisor of another*, the problem is referring to the concept of divisibility as presented in this chapter. If you are told that *2 is a divisor of 6*, then 2 divides evenly into 6; therefore, 2 is a factor of 6. Similarly, if you are told 2 is a divisor of *y*, then 2 is a factor of *y*.

To find all the factors of a small number, use **factor pairs**. For example, 2 and 3 are factor pairs of 6 because $2 \times 3 = 6$. Similarly, 1 and 6 are factor pairs of 6 because $1 \times 6 = 6$.

For example:

What are the factor pairs of 60 ?

Here's an organized way to find the factor pairs of 60:

1. Label two columns Small and Large (or S and L).

2. The first set of factor pairs will always be 1 and the number itself. Put 1 in the small column and 60 in the large column.

2

3. After 1, try the next smallest integer: 2. Since 2 is a factor of 60, write 2 underneath the 1 in your table. Divide 60 by 2 to find 2's "sibling" in the pair: $60 \div 2 = 30$. Write 30 in the large column.

4. Repeat this process until the numbers in the small and the large columns run into each other. In this case, 6 and 10 are a factor pair, but 7, 8, and 9 are not factors of 60. The next number after 9 is 10, which already appears in the large column, so you can stop.

If you...	Then you...	Like this:	
		Small	Large
Want all the factors of 60	Make a table of factor pairs, starting with 1 and 60	1	60
		2	30
		3	20
		4	15
		5	12
		6	10

You do not have to draw a table exactly like this one. Just keep your work organized. Start with the number 1 and count up.

Check Your Skills

4. What are all of the factors of 90 ?

5. What are all of the factors of 72 ?

6. What are all of the factors of 105 ?

Answers can be found on page 70.

Prime Number: Only Divisible by 1 and Itself

The integer 7 is divisible by what positive integers?

Test out the positive integers less than or equal to 7: 1, 2, 3, 4, 5, 6, and 7:

$7 \div 1 = 7$ Every integer is divisible by 1.

$7 \div 2 =$ not an integer
$7 \div 3 =$ not an integer
$7 \div 4 =$ not an integer   7 is not divisible by *any* integer besides 1 and itself.
$7 \div 5 =$ not an integer
$7 \div 6 =$ not an integer

$7 \div 7 = 1$ Every integer is divisible by itself.

So 7 has exactly two factors—1 and itself. Integers that have exactly two factors are called **prime numbers**. Primes are commonly tested on this exam.

There are a few key details to note. First, the concept of prime applies only to positive integers. Second, note that 1 is *not* prime because it has exactly one factor (itself!), not two factors.

Finally, 2 is the only even prime number. Every even number greater than 2 has at least one more factor besides 1 and itself, namely the number 2.

Every positive integer can be placed into one of two categories—prime or not prime. Memorize the smaller primes: 2, 3, 5, 7, 11, 13, 17, and 19.

Check Your Skills

7. List all the prime numbers between 20 and 50.

Answer can be found on page 70.

Prime Factor Trees

Take another look at the factor pairs of 60:

$60 = 1 \times 60$ Always the first factor pair: 1 and itself.

and 2×30
and 3×20
and 4×15 5 other factor pairs. Let's look at these in a little more detail.
and 5×12
and 6×10

Consider the pair 4×15. One way to think about this pair is that 60 breaks down into 4 and 15. Use a **factor tree** to show this relationship:

Neither 4 nor 15 is prime, so they can both be broken down into other factor pairs: 4 breaks down into 2×2, and 15 breaks down into 3×5. Here's what that looks like:

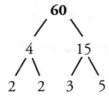

Can you break it down any further? Sort of, but only if you use the number 1—for example, $2 = 2 \times 1$. The numbers you have reached (2, 2, 3, and 5) are all primes.

When you find a prime factor, that branch on the factor tree has reached the end—don't keep going. You'll never see a 1 in a factor tree. Circle prime numbers as you go, as if they were fruit on the tree. The factor tree for 60 looks like this:

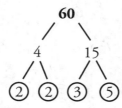

After you break down 60 into 4 and 15, and then break 4 and 15 down further, you end up with $60 = 2 \times 2 \times 3 \times 5$.

What if you start with a different factor pair of 60? Create a factor tree for 60 in which the first breakdown is 6×10:

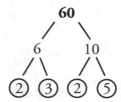

According to this factor tree, $60 = 2 \times 3 \times 2 \times 5$. These are the same primes as before (though in a different order). Any way you break down 60, you always end up with the same prime factors: two 2's, one 3, and one 5. The **prime factorization** of 60 is always $2 \times 2 \times 3 \times 5$.

Note: The factor tree represents only the *prime* factors of 60. It doesn't show *all* factors of 60; if you need all of the factors, make a table of factor pairs.

Prime factors are like the DNA or the fingerprint of a number. Every number has a unique prime factorization. Sixty is the only number that can be written as $2 \times 2 \times 3 \times 5$.

On divisibility problems, always look to break down numbers to their prime factors. A factor tree is the best way to find a prime factorization.

If you need to find the prime factorization of 630, one way to start is by finding the smallest prime factor of 630. Check 2 first: 630 is even, so it is divisible by 2. On your factor tree, break down 630 into 2 and 315:

The left branch is done: 2 is a prime. You still need to factor 315, though. It's not even, so it's not divisible by 2. Is it divisible by 3? The digits add up to $3 + 1 + 5 = 9$, which is divisible by 3, so 315 is divisible by 3. Add this next level of information, $315 = 3 \times 105$, to your factor tree:

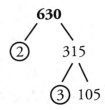

The number 105 might still be divisible by another integer. Check it out: $1 + 0 + 5 = 6$, so 105 is divisible by 3. Since $105 \div 3 = 35$, the tree now looks like this:

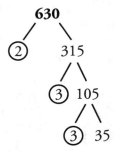

The number 35 is not divisible by 3 ($3 + 5 = 8$, which is not a multiple of 3), so the next number to try is 5. The number 35 ends in a 5, so it is divisible by 5. Since $35 \div 5 = 7$, the tree now looks like this:

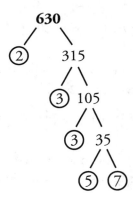

Every number on the tree has now been broken down as far as it can go. The prime factorization of 630 is $2 \times 3 \times 3 \times 5 \times 7$.

Reflect on what you just did. Why was it cumbersome to break down 630 in that particular way?

You have no calculator, so repeatedly having to figure out how to break down large numbers (315, 105) is not a fantastic way to solve. How else could you break it down?

If the number ends in 0, as 630 does, break it down by 10 or 100 or whatever is appropriate for that number. For example, 630 can be broken down into 63 and 10. From there, the numbers are already much smaller, so it's easier to continue to break them down further. The number 10 breaks down into 2 and 5 (which are both primes), and the number 63 breaks into 7 and 9. The number 9 can be broken down one more time, into 3 and 3. The full prime factorization of 63 is 2, 3, 3, 5, and 7.

However you break down a number, you will always get to the same set of prime factors in the end, so pause to think for a moment about the best way to break down the particular number in front of you right now.

2

If you...	Then you...	Like this:
Want the prime factorization of 96	Break 96 down to primes using a tree	96 2 48 12 4 3 4 2 2 2 2

Check Your Skills

8. Find the prime factorization of 90.

9. Find the prime factorization of 72.

10. Find the prime factorization of 105.

Answers can be found on pages 70–71.

Every Number Is Divisible by the Factors of Its Factors

If *a* is divisible by *b*, and *b* is divisible by *c*, then *a* is divisible by *c* as well. For instance, 12 is divisible by 6, and 6 is divisible by 3. Therefore, 12 is divisible by 3 as well as by 6.

This **factor foundation rule** also works in reverse to a certain extent. If *d* is divisible by two different *primes*, *e* and *f*, then *d* is also divisible by *e* × *f*. In other words, if 20 is divisible by 2 and 5, then 20 is also divisible by 2 × 5 (or 10).

Divisibility travels up and down the factor tree. Consider the factor tree of 150:

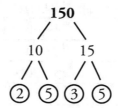

2

Prime factors are *building blocks*. When you first create the tree, you're moving down—from 150 down to 10 and 15, and then down again to 2, 5, 3, and 5. But you can also build upward, starting with the four building blocks; in this case, you have one 2, one 3, and two 5's at your disposal to build other factors of 150.

For example, $2 \times 3 = 6$, and $5 \times 5 = 25$, so 6 and 25 are also factors of 150. In fact, you could start from the four building blocks 2, 3, 5, and 5 and put them in a different order to build "up" the tree in a different way:

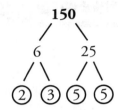

Note: Even though 5 and 5 are the same number, 5 appears twice on 150's tree. So you are allowed to multiply those two 5's together to produce another factor of 150, namely 25.

The tree above isn't even the only other possibility. Here are more:

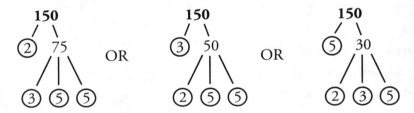

Beginning with the four prime factors of 150 (2, 3, 5, and 5), you build different factors by multiplying any two, any three, or even all four of those primes together in different combinations. In fact, all of the factors of a number (except for 1) can be built with different combinations of that number's prime factors.

Factors: Built Out of Primes

Take one more look at the number 60 and its factors (2, 2, 3, and 5). You can use those prime factors as building blocks to find both numbers in each factor pair (except for 1, which is not prime).

In any one factor pair, all of the prime factors of 60 must be used (and no extra factors can be used):

Building Blocks		Small	Large		Building Blocks
1	=	1	60	=	$2 \times 2 \times 3 \times 5$
2	=	2	30	=	$2 \times 3 \times 5$
3	=	3	20	=	$2 \times 2 \times 5$
2×2	=	4	15	=	3×5
5	=	5	12	=	$2 \times 2 \times 3$
2×3	=	6	10	=	2×5

In the pairing (1, 60), all four prime factors are used to make 60. In the pairing (2, 30), one of the 2's is used for the 2, and the other three primes (2, 3, and 5) are used to make the 30. This continues for all of the factor pairs of 60.

As a result, every factor of a number (except 1) can be expressed as the product of some or all of that number's prime factors. This relationship between factors and prime factors is true of every number, so if you have to find the factor pairs of a large number, it may be most efficient to find the prime factors and use them as building blocks to create the factor pairs.

To recap:

1. If a is divisible by b, and b is divisible by c, then a is divisible by c as well. For instance, 100 is divisible by 20, and 20 is divisible by 4, so 100 is also divisible by 4.

2. If d has e and f as prime factors, then d is divisible by e, by f, and by $e \times f$. For instance, 90 is divisible by 5 and by 3, so 90 is also divisible by $5 \times 3 = 15$. You can let e and f be the same prime, as long as there are at least two copies of that prime in d's factor tree.

3. Every factor of a number (except 1) is the product of a different combination of that number's prime factors. For example, $30 = 2 \times 3 \times 5$. The factors of 30 are 1, 2, 3, 5, 6 (2×3), 10 (2×5), 15 (3×5), and 30 $(2 \times 3 \times 5)$.

4. To find all of the *prime* factors of a number, use a factor tree.

5. To find *all* of the factors of a number (prime and not prime) in a methodical way, write out the factor pairs in order, counting up from 1. For example, 30 has the factor pairs (1, 30), (2, 15), (3, 10), and (5, 6). If the number is a large one, consider finding the prime factors first, then use those prime factors to make the factor pairs.

Check Your Skills

11. The prime factorization of a number is 3×5. What is the number and what are all of its factors?

12. The prime factorization of a number is $2 \times 5 \times 7$. What is the number and what are all of its factors?

13. The prime factorization of a number is $2 \times 3 \times 13$. What is the number and what are all of its factors?

Answers can be found on page 71.

Factor Tree of a Variable: Contains Unknowns

Imagine that you are told that some unknown positive number x is divisible by 6. You can represent this fact on paper in several different ways. For instance, you could write "$x =$ multiple of 6" or "$x = 6 \times$ integer." You could also write the information as the result of division: $\frac{x}{6} =$ integer.

You could also represent the information with a factor tree. Since the top of the tree is a variable, add in a branch to represent what you *don't* know about the variable. Label this branch with a question mark (?), three dots (…), or something to remind yourself that you have *incomplete* information about *x*:

What *else* do you know about *x*? What can you definitely say about *x* right now?

The variable *x* must also be divisible by the prime factors of 6, namely 2 and 3:

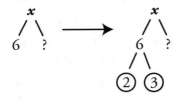

The purpose of the tree is to break integers down into primes, which are the building blocks of greater integers. When you're given some information, continue the factor tree until everything is broken down as far as it will go.

Try this problem:

> *x* is divisible by 6. Decide whether each statement below *must be true, could be true,* or *cannot be true.*
>
> I.   *x* is divisible by 3.
> II. *x* is even.
> III. *x* is divisible by 12.

First, the question stem establishes that *x* is divisible by 6, so *x* is also divisible by 2 and 3.

Begin with statement I: *x* is divisible by 3. Statement I *must be true.*

Statement II says *x* is even. Must that be true? Return to your factor tree:

Since *x* is divisible by 2, *x* is even. Statement II *must be true.*

Statement III says *x* is divisible by 12. Compare the factor tree of *x* with the factor tree of 12:

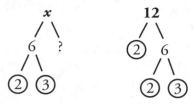

What would you have to know about *x* to guarantee that it is divisible by 12?

The number 12 is $2 \times 2 \times 3$, so 12's building blocks are two 2's and a 3. To guarantee that *x* is divisible by 12, you need to know for sure that *x* has two 2's and one 3 among its prime factors. That is, *x* would have to be divisible by everything that 12 is divisible by.

Look at the factor tree for *x*. There is a 3 but only *one* 2. So you can't claim that x *must be* divisible by 12. But *could x* be divisible by 12?

Consider the question mark on *x*'s factor tree. That question mark is there to remind you that you *don't* know everything about *x*. After all, *x could* have other prime factors. If one of those unknown factors were another 2, your tree would look like this:

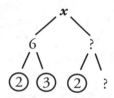

If an unknown factor were a 2, then *x* would indeed be divisible by two 2's and a 3. So *x could* be divisible by 12.

To prove to yourself that this thinking is valid, list out a few numbers that are divisible by 6 to check whether they are also divisible by 12:

x is a number on this list.
$$\left.\begin{cases} 6 & 6 \div 12 = 0.5 \\ 12 & 12 \div 12 = 1 \\ 18 & 18 \div 12 = 1.5 \\ 24 & 24 \div 12 = 2 \\ \dots & \dots \end{cases}\right\}$$
Some, but not all, of these numbers are also divisible by 12.

Since some of the possible values of *x* are divisible by 12, and some aren't, *x could be* divisible by 12. Statement III *could be true* but doesn't have to be true.

Whenever you are told that an unknown value is divisible by a specific number, it's also possible that the unknown is divisible by other numbers. So it *could be* divisible by anything.

If you...	Then you...	Like this:
Use a factor tree with a variable on top	Put in a question mark (or something similar) to remind yourself what you don't know	*x* / \ 6 ?

Check Your Skills

For each problem, the following is true: x is divisible by 24. Determine whether each statement below *must be true, could be true,* or *cannot be true.*

14. x is divisible by 6.

15. x is divisible by 9.

16. x is divisible by 8.

Answers can be found on page 72.

Factors of x with No Common Primes: Combine

Try this problem:

x is divisible by 3 and by 10. Decide whether each statement below *must be true, could be true,* or *cannot be true.*

I. x is divisible by 2.

II. x is divisible by 15.

III. x is divisible by 45.

First, create two factor trees to represent the given information:

Why not write them together at once? "x is divisible by 3" is a different fact than "x is divisible by 10." *Initially, always write separately the given facts about a variable.* That way, you can think carefully about how to combine those facts.

Continue to break down the factors until you have only primes:

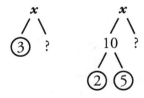

Now, examine statement I: x is divisible by 2. Since x is divisible by 10, and 10 is divisible by 2, x is definitely divisible by 2. Statement I *must be true.*

Statement II (*x* is divisible by 15) is more difficult. Study the trees. Neither one provides complete information about *x*, but you know for certain that *x* is divisible by 3 and that *x* is divisible by 2 and by 5. These primes are all different. *When the primes from two trees do not overlap, you can combine all the primes into one tree:*

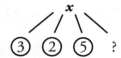

(Note: You cannot do this if the different starting trees have any factors in common. In the next section, you'll learn what to do when this happens.)

Return to the statement: *x* is divisible by 15. If *x* has all the prime factors that 15 has, then *x* is divisible by 15. The prime factors of 15 are 3 and 5. Since *x* has both a 3 and a 5, it is definitely divisible by 15. Statement II *must be true.*

Since *x* is divisible by 2, 3, and 5, you can combine these primes to form other definite factors of *x*.

For instance, $2 \times 3 = 6$ is a factor of *x*, as is $2 \times 5 = 10$. Here are some other possible factor trees for *x*, given that 2, 3, and 5 are all factors of *x*:

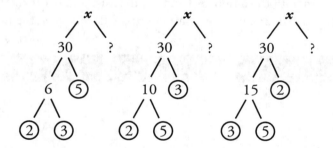

Notice what all three trees have in common. No matter how you combine the prime factors, each tree ultimately leads to 30, which is $2 \times 3 \times 5$.

Statement III says that *x* is divisible by 45. What do you need to know in order to claim that *x* is divisible by 45? Build a factor tree of 45:

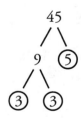

The number 45 is divisible by 3, 3, and 5. For *x* to be divisible by 45, you need to know that *x* has all the same prime factors. Does it?

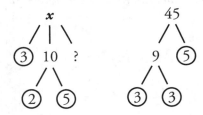

The unknown *x* has a 5, but only *one* 3 for sure. Since you don't know whether *x* has the second 3 that the 45 needs, you can't say for certain whether *x* is divisible by 45. It *could* be the case that *x* is divisible by 45, if the question mark contains another 3, but this isn't guaranteed. If the question mark contains a 3, then *x* is divisible by 45. If not, then *x* is not divisible by 45. Statement III *could be true* but does not have to be.

When the two trees do not share any common prime factors, you can keep all of their prime factors in the combined tree. If you multiply those prime factors together, you'll find the **least common multiple**, or **LCM**. In the 3 and 10 example, the LCM is $3 \times 10 = 30$.

The least common multiple of two numbers, say *A* and *B*, is defined as the least number that is a multiple of both *A* and *B*. When you start with the numbers 3 and 10, the least number that is a multiple of both is 30. So when you know that *A* and *B* don't share any common prime factors, then the LCM of *A* and *B* is equal to $A \times B$.

If you...	Then you...	Like this:
Know two factors of *x* that have no primes in common	Combine the two trees into one	(diagram)

Check Your Skills

For each problem, the following is true: *x* is divisible by 28 and by 15. Determine whether each statement below *must be true*, *could be true*, or *cannot be true*.

17. *x* is divisible by 14.

18. *x* is divisible by 20.

19. *x* is divisible by 24.

Answers can be found on pages 72–73.

Factors of *x* with Primes in Common: Combine to Least Common Multiple

In the last section, you learned what to do when the different factor trees do not contain any factors in common.

Now, consider a different situation:

If *x* is divisible by 6 and by 9, is *x* divisible by 54 ?

At first glance, it seems that x must be divisible by 54, since $6 \times 9 = 54$. But things are more complicated.

Here is the problem in tree form:

Given: Question: Is this true?

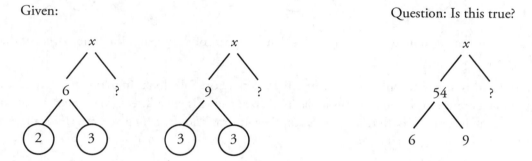

In this problem, 6 and 9 share a prime factor, namely a 3. That shared factor makes all the difference in how you combine the information into one overall tree. This time, you *cannot* include everything from both trees.

Why? List out the multiples of each number, 6 and 9, to find the least common multiple, which is the least number that appears on both lists:

Multiples of 6	Multiples of 9
6	
	9
12	
	18
18	
	27
24	

Note: Listing the two sets of multiples to find the least number on both lists works well for small numbers, but it can be messy when the numbers are greater—which is often the case on the test.

The LCM of 6 and 9 is 18. This is *not* the same number as 6×9, which equals 54. It is the case that 54 is *a* common multiple of 6 and 9, but it is not the *least* common multiple of 6 and 9. When the two starting numbers share a common factor, you can count the common factor only *once*, not twice.

What is the minimum number of 3's needed to create the LCM? The 6 needs just one 3, while the 9 needs two 3's, so you need at minimum two 3's—that allows you to reproduce a 6 or a 9. Here's how that looks visually:

Given: Conclusion:

The conclusion contains only two of the three 3's that were part of the given information. One of the 3's has to be dropped.

Why? Imagine this scenario. On a table in front of you sits a box. Sally looks in that box and tells you, "I see an apple and an orange in the box." Linus looks in the box and tells you, "I see two oranges in the box." (Assume Sally and Linus are both telling the truth; they just may not be telling you *everything* they see in the box.)

If Sally sees an apple and an orange, and Linus sees two oranges, does that mean that the box must contain three oranges?

It might, but it doesn't have to. Linus did see two oranges. But Sally's orange could have been one of those same two that Linus saw. Therefore, at the *least*, the box contains two oranges and one apple. You have to strip out one orange, because it represents overlap between the two pieces of information. (Note: The box *could* contain more oranges or apples; it could even contain bananas! But you don't know for sure that it does.)

In the same way, you need to strip out one 3 from the factor trees of 6 and 9, because that 3 could be an overlap between the two trees—it could be the "same" 3. At the *least*, you have a 2 and two 3's, so the *least* common multiple of 6 and 9 is $2 \times 3 \times 3 = 18$. As a result, x must be divisible by 18, but it does not have to be divisible by 54.

Here's why the LCM is important: *If x is divisible by A and by B, then x is divisible by the LCM of A and B.*

For instance, if you are told that x is divisible by 3 and by 10, then you can conclude that x is definitely divisible by the LCM of 3 and 10, which equals 30 (since 3 and 10 don't overlap at all).

The same principle holds true for 6 and 9, but you first have to *strip out any common factors*. If *x* is divisible by 6 and by 9, then *x* is definitely divisible by 18, the LCM of 6 and 9:

Given: Conclusion:

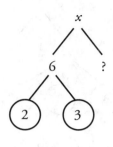 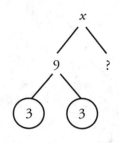 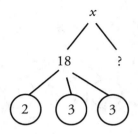

Fact 1:	Fact 2:	Conclusion:
"*x* definitely contains a 2 and a 3."	"*x* definitely contains two 3's."	"*x* definitely contains a 2 and two 3's."

When two numbers *don't* share any prime factors, their LCM is always equal to their product. For example:

> 3 and 10 don't share any prime factors, so their LCM = 3 × 10 = 30.

However, when two numbers *do* share prime factors, their LCM will always be *less* than their product, because you have to strip out the overlap:

> 6 and 9 share prime factors, so their LCM is not 6 × 9 = 54. In fact, their LCM (18) is less than 54.

Here's a shorthand way of thinking about what you just learned: Break the numbers into their primes and then take only the *greater number of instances of* any one particular prime. For example:

> 6 = 2 × 3 and 9 = 3 × 3
>
> How many 2's should you take? The number 6 has one 2 and 9 has no 2's, so take one 2.
>
> How many 3's should you take? The number 6 has one 3 and 9 has two 3's, so ignore the 6 and take the two 3's from the 9.
>
> The LCM is 2 × 3 × 3 = 18.

Here's a harder example:

> If *x* is divisible by 8, 12, and 45, what is the greatest number that *x* must be divisible by?

The *greatest number that* x *must be divisible by* is code for: *What is the LCM of 8, 12, and 45 ?*

2

First, draw three separate trees for the given information:

Given:

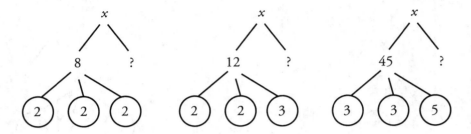

You could find this number by listing all the multiples of 8, 12, and 45 and looking for the first number on all three lists. That would be a decent amount of work, though, since there are three numbers and one of them is a lot larger than the other two. Instead, find the LCM by counting up prime factors that you *know* are in *x*, while stripping out the overlap.

Start with 2. How many 2's are guaranteed to be in *x*? There are three 2's in 8, two 2's in 12, and none in 45. To cover all the bases, there must be at least *three* 2's in *x*.

Take 3 next. Since 45 has two 3's, the most in any of the individual trees, *x* must contain at least *two* 3's. Finally, *x* must have at least *one* 5 because the 45 contains one 5. So here's the picture:

Given: Conclusion:

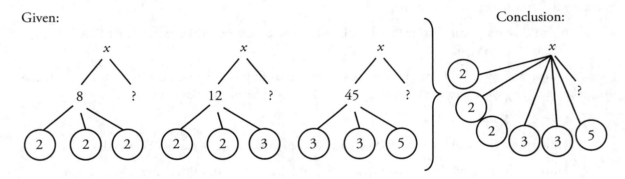

Now calculate the LCM:

$$2 \times 2 \times 2 \times 3 \times 3 \times 5 = (2 \times 2) \times (3 \times 3) \times (2 \times 5) = 4 \times 9 \times 10 = 360$$

Combine the 2 and the 5 to get a 10, then save that number for last in the calculation. Combine the remaining four numbers in whatever way seems fastest to you.

The LCM of 8, 12, and 45 is 360. It is the greatest number that *x must* be divisible by—or the *least* number that must be a multiple of 8, 12, and 45.

One final note: If the facts are about different variables (e.g., *x* and *y*), then the facts *don't* overlap when multiplying *x* and *y*. You can include all of the prime factors:

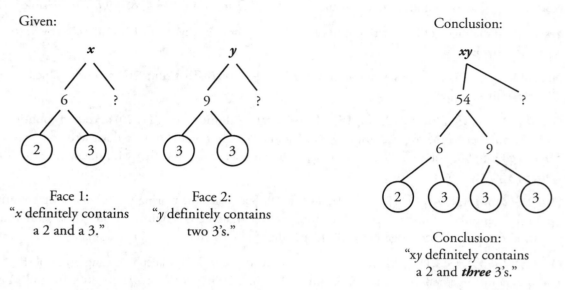

Given:

Face 1:
"*x* definitely contains a 2 and a 3."

Face 2:
"*y* definitely contains two 3's."

Conclusion:

Conclusion:
"x*y* definitely contains a 2 and ***three*** 3's."

Linus and Sally are looking at completely different boxes of fruit (*x* and *y*), so the product includes everything you see across all the trees.

If you...	Then you...	Like this:
Know two factors of *x* that have primes in common	Combine the two trees into one, eliminating the overlap = Know that *x* is divisible by the LCM of the factors	*x* is divisible by 6 *x* is divisible by 9 becomes *x* is divisible by 18, the LCM of 6 and 9
Know factors of *x* and of *y* and are asked to find the factors of *xy*	Combine all of the factors for each variable; do not eliminate overlap	*x* is divisible by 6 *y* is divisible by 9 becomes *xy* is divisible by 54

Check Your Skills

For each problem, the following is true: *x* is divisible by 6 and by 14. Determine whether each statement below *must be true*, *could be true*, or *cannot be true*.

20. *x* is divisible by 42.

21. *x* is divisible by 84.

Answers can be found on page 73.

Check Your Skills Answer Key

1. **No:** 123,456,789 is an odd number because it ends in 9. Therefore, 123,456,789 is *not* divisible by 2.

2. **Yes:** Add up the digits of the starting number: $7 + 3 + 2 = 12$. That sum, 12, is divisible by 3, so 732 is divisible by 3.

3. **No:** Add up the digits of the starting number: $9 + 8 + 9 = 26$. That sum, 26, is not divisible by 9, so 989 is *not* divisible by 9.

4. **(1, 90), (2, 45), (3, 30), (5, 18), (6, 15), (9, 10):** Start with the pairing (1, 90) and keep counting up until the numbers come together. Both 2 and 3 go into 90, but 4 does not, so don't write 4 down. Both 5 and 6 go into 90, but 7 and 8 do not. Finally, the numbers come together with the pairing (9, 10), so you know you're done.

5. **(1, 72), (2, 36), (3, 24), (4, 18), (6, 12), and (8, 9):** Start with the pairing (1, 72) and keep counting up until the numbers come together. The numbers 2, 3, 4, and 6 all go into 72, but 5 and 7 do not. Finally, the numbers come together with the pairing (8, 9), so you know you're done.

6. **(1, 105), (3, 35), (5, 21), (7, 15):** Start with the pairing (1, 105) and keep counting up until the numbers come together. The values 3, 5, and 7 all go into 105. After 7, the values 8, 9, 10, and 11 don't go into 105; at that point, the numbers come together since you're also counting down from 15, so you can stop.

7. **23, 29, 31, 37, 41, 43, 47:** Test all the numbers between 20 and 50 to find the prime numbers. This isn't as cumbersome as it sounds. First, ignore all of the even numbers (ending in 0, 2, 4, 6, and 8) as well as the numbers that end in 5. For the rest, if you can find one number between 1 and itself that divides into the number, cross it off the list. For example, 21 is divisible by 3, so it is not prime, but 23 is not divisible by anything other than 1 and itself, so 23 is prime.

8. $2 \times 3 \times 3 \times 5$: Here's one path.

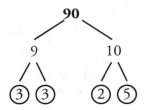

9. $2 \times 2 \times 2 \times 3 \times 3$: Here's one path.

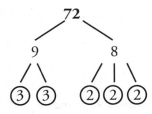

10. **3 × 5 × 7:** Here's one path.

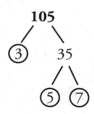

11. **15 and {1, 3, 5, 15}:** Multiply the factors to get the number: 3 × 5 = 15. Write out the factor pairs to find all of the factors: (1, 15), (3, 5).

12. **70 and {1, 2, 5, 7, 10, 14, 35, 70}:** Multiply the factors to get the number: 2 × 5 × 7 = 70. Write out the factor pairs to find all of the factors. Since 70 is a large number, consider using the prime factors as building blocks to find the greater numbers in the factor pairs. Each pair must use all of the primes 2, 5, and 7 (and no additional primes).

Building Blocks	Small	Large	Building Blocks
1	1	70	2 × 5 × 7
2	2	35	5 × 7
5	5	14	2 × 7
7	7	10	2 × 5

13. **78 and {1, 2, 3, 6, 13, 26, 39, 78}:** Multiply the factors to get the number: 2 × 3 × 13 = 78. Write out the factor pairs to find all of the factors. Since 78 is a large number, consider using the prime factors as building blocks to find the greater numbers in the factor pairs. Each pair must use all of the primes 2, 3, and 13 (and no additional primes).

Building Blocks	Small	Large	Building Blocks
1	1	78	2 × 3 × 13
2	2	39	3 × 13
3	3	26	2 × 13
2 × 3	6	13	13

For problems 14–16, x is divisible by 24.

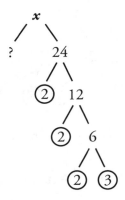

14. **Must be true:** As shown in x's factor tree, 6 is one of the factors of x.

15. **Could be true:** The number 9 has the prime factors 3 and 3. For x to be divisible by 9, it must contain these same prime factors. It does contain one 3; the question mark signals that it may or may not contain a second 3. For this reason, x could be divisible by 9 but does not have to be.

16. **Must be true:** The number 8 has the prime factors 2, 2, and 2. For x to be divisible by 8, it must contain these same prime factors. The factor tree for x does contain three 2's, so x must be divisible by 8.

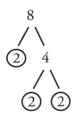

For problems 17–19, x is divisible by 28 and by 15.

First, expand the factor trees for 28 and 15. Next, check whether they have primes in common. In this case, they don't, so combine them all into one big tree.

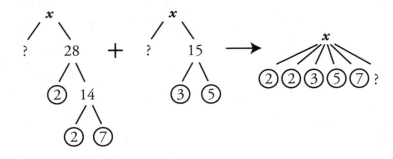

17. **Must be true:** According to the combined tree, 2 and 7 are factors of x, so $2 \times 7 = 14$ must also be a factor of x. Alternatively, examine x's factor tree for 28. The value 14 is one of the factors listed in the tree.

18. **Must be true:** For *x* to be divisible by 20, it must contain the same prime factors as 20. The factor tree for 20 contains prime factors 2, 2, and 5. Since *x* also contains two 2's and a 5, *x* must be divisible by 20.

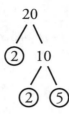

19. **Could be true:** For *x* to be divisible by 24, it must contain the same prime factors as 24. The factor tree of 24 contains prime factors 2, 2, 2, and 3. The tree for *x* contains a 2, a 2, and a 3, but it does not contain a third 2 for sure. The question mark indicates that *x* may have another 2 but does not have to have one. For this reason, *x* could be divisible by 24 but does not have to be.

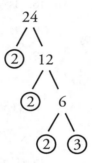

For problems 20–21, *x* is divisible by 6 and by 14.

$6 = 2 \times 3$ and $14 = 2 \times 7$. The two numbers have an overlapping factor, 2, so strip this out before you combine the factors. The LCM of 6 and 14 is $2 \times 3 \times 7 = 42$.

20. **Must be true:** Since the LCM of 6 and 14 is 42, *x* must be divisible by 42.

21. **Could be true:** The LCM is 42. Therefore, *x* could be divisible by 84 (if *x* has another factor of 2), but *x* does not have to be divisible by 84.

Chapter Review: Drill Sets

Drill 1

1. Is 4,005 divisible by 5 ?

2. If 33 is a factor of 594, is 11 a factor of 594 ?

3. Is 6,750 divisible by 18 ?

Drill 2

4. Does 23 have any factors other than 1 and itself?

5. $x = 100$

    The prime factors of x are:

    The factors of x are:

6. If 2,499 is divisible by 147, is 2,499 divisible by 49 ?

7. What are all of the positive multiples of 18 that are less than 60 ?

Drill 3

8. Is 285,284,901 divisible by 10 ?

9. Is 539,105 prime?

10. If $x = 36$, what are all of the factors of x ?

    (A) 2, 2, 3, 3

    (B) 2, 3, 4, 6, 9, 12, 18

    (C) 1, 2, 3, 4, 6, 9, 12, 18, 36

11. Find at least four even divisors of 84.

12. What are the prime factors of 30×49 ?

Drill 4

13. Is 9,108 divisible by 9 and/or by 2 ?

14. $x = 39$

    The prime factors of x are:

    The factors of x are:

15. How many more prime factors does the product of 42×120 have than the product of 21×24 ?

Drill 5

16. Is 43,360 divisible by 5 and/or by 3 ?

17. Is 513,501 prime?

18. $x = 37$

    The prime factors of x are:

    The factors of x are:

19. What are the two greatest odd factors of 90 ?

Drill 6

20. Determine which of the following numbers are prime numbers. A prime number has exactly two factors: 1 and itself.

2	3	5	6
7	9	10	15
17	21	27	29
31	33	258	303
655	786	1,023	1,325

Drill 7

21. If x is divisible by 34, what other numbers is x divisible by?

    (A)   1, 2, 17
    (B)   1, 2, 17, 68
    (C)   17, 68, 102

22. The prime factorization of a number is $3 \times 3 \times 7$. What is the number, and what are all of its factors?

23. If 6 and 14 are factors of y, must y be divisible by 21 ?

24. If $7y$ is divisible by 210, must y be divisible by 12 ?

25. If integer a is *not* divisible by 30, but ab is, what is the least possible value of integer b ?

    (A)   20
    (B)   15
    (C)   2

Drill 8

26. If 40 is a factor of x, what other numbers must be factors of x ?

27. If q is divisible by 2, 6, 9, 12, 15, and 30, is q divisible by 8 ?

28. If p is a prime number, and q is a non-prime integer, what are the minimum and maximum numbers of factors p and q can have in common?

Drill 9

29. The prime factorization of a number is $2 \times 3 \times 11$. What is the number, and what are all of its factors?

30. 14 and 3 divide evenly into n. Is 12 a factor of n ?

31. The sum of the positive integers x and y is 17. If x has only two factors and y is divisible by 5, which of the following is a possible value of x ?

    (A)   3
    (B)   7
    (C)   12

32. If n is the product of 2, 3, and a two-digit prime number, how many of its factors are greater than 6 ?

Drill 10

33. If n is divisible by both 35 and 44, is 14 a divisor of n ?

34. 4, 21, and 55 are factors of n. Is n divisible by 154 ?

35. If n is divisible by both 196 and 15, is 270 a factor of n ?

Drill Sets Solutions

Drill 1

1. **Yes:** 4,005 ends in 5, so it is divisible by 5.

2. **Yes:** You could divide 594 by 11 to determine divisibility, but it is faster to use the factor foundation rule. If 594 is divisible by 33, 594 is also divisible by all of the factors of 33. The number 11 is a factor of 33 ($33 = 11 \times 3$). Therefore, 594 is also divisible by 11.

3. **Yes:** In order to be divisible by 18, a number must be divisible by both 2 and 9, a factor pair that makes up 18. Because 6,750 ends in a 0, it is even, so it is divisible by 2. The digits of 6,750 sum to 18 ($6 + 7 + 5 = 18$), and 18 is divisible by 9, so 6,750 is also divisible by 9. Because 6,750 is divisible by both 2 and 9, it is also divisible by 18.

Drill 2

4. **No:** 23 is a prime number. It has no factors other than 1 and itself.

5. **Prime factors 2, 2, 5, 5; factors 1, 2, 4, 5, 10, 20, 25, 50, 100:** The prime factors of x are:

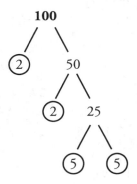

The factors of x are: (1, 100), (2, 50), (4, 25), (5, 20), and (10, 10).

6. **Yes:** The factor foundation rule is helpful in this question. The problem states that 2,499 is divisible by 147. The factor foundation rule states that if 2,499 is divisible by 147, then 2,499 is also divisible by all of the factors of 147. Check 147: It is divisible by 49 ($147 \div 49 = 3$). Since 2,499 is divisible by 147, it is also divisible by 49.

7. **18, 36, and 54:** In order to generate positive multiples of 18 that are less than 50, multiply 18 by small positive integers.

$$18 \times 1 = 18$$
$$18 \times 2 = 36$$
$$18 \times 3 = 54$$

All other positive multiples of 18 are greater than 60.

Drill 3

8. **No:** 285,284,901 ends in a 1, not a 0. It is not divisible by 10.

9. **No:** 539,105 ends in a 5, so 5 is a factor of 539,105. A prime number has only two factors, itself and 1, so 539,105 is not prime.

10. **(C) 1, 2, 3, 4, 6, 9, 12, 18, 36:** The question asks for *all* of the factors of *x*, not just the prime factors. Find the factor pairs of 36. Note: 1 and the number itself are always your first factor pair; don't forget them! Because (C) is the only answer that includes 1 and 36, it must be the correct answer. Trap answer (A) represents only the prime factors, not all factors. The factor pairs are (1, 36), (2, 18), (3, 12), (4, 9), and (6, 6).

11. **Any four of 2, 4, 6, 12, 14, 28, 42, and 84:** List the factor pairs to see which factors are even: (1, **84**), (**2**, **42**), (3, **28**), (**4**, 21), (**6**, **14**), and (7, **12**).

12. **2, 3, 5, and 7:** While you could multiply the numbers together to find the prime factors (annoying because the numbers are large!), there is a faster way. The prime factors of the product of 30 and 49 will consist of the prime factors of 30 and the prime factors of 49. The prime factors of 30 are 2, 3, and 5. The prime factors of 49 are 7 and 7. Therefore, the prime factors of 30 × 49 are 2, 3, 5, 7, and 7.

Drill 4

13. **9,108 is divisible by 9 AND by 2:** The sum of the digits of 9,108 is divisible by 9 ($9 + 1 + 0 + 8 = 18$), so 9,108 is a multiple of 9. Because 9,108 ends in 8, it is even, which means it is divisible by 2.

14. **Prime factors 3, 13; factors 1, 3, 13, 39:** The prime factors of *x* are:

The factor pairs of *x* are (1, 39) and (3, 13).

15. **Two:** You could multiply these products out or identify all of the prime factors of each number, but there is a more efficient way. Because the question asks you to make a comparison, focus only on the *differences* between the two products.

    42×120

    21×24

Compare 42 and 21: $42 = 21 \times 2$. That is, 42 contains 21 (and all of its factors), but 42 also has one additional factor of 2.

Compare 120 and 24: $120 = 24 \times 5$. That is, 120 contains 24 (and all of its factors), but 120 also has one additional factor of 5.

Therefore, the only *additional* prime factors in 42×120 are the 2 in 42 and the 5 in 120. The product of 42×120 has two more prime factors than the product of 21×24.

Drill 5

16. **43,360 is divisible by 5 but is NOT divisible by 3:** 43,360 ends in 0, so it is divisible by 5. The digits of 43,360 do not sum to a multiple of 3 ($4 + 3 + 3 + 6 + 0 = 16$), so it is not divisible by 3.

17. **No:** The sum of the digits is $5 + 1 + 3 + 5 + 0 + 1 = 15$. The number 15 is divisible by 3, so the number 513,501 is also divisible by 3. Prime numbers have only themselves and 1 as factors, so 513,501 is not prime.

18. **Prime factors 37; factors 1, 37:** The only factor pair is (1, 37). 1 is not a prime number, so the only prime factor is 37.

19. **15, 45:** Break 90 down into its factor pairs: (1, 90), (2, 45), (3, 30), (5, 18), (6, 15), and (9, 10). The two greatest odd factors of 90 are 45 and 15.

Drill 6

20. **2, 3, 5, 7, 17, 29, 31:** Eliminate all the numbers that are not prime. Those that remain are prime.

    All of the even numbers other than 2 (6, 10, 258, 786) are not prime, since they are divisible by 2.

    All of the numbers that are divisible by 5 other than 5 itself (10, 15, 655, 1,325) are not prime.

    Next, check whether the sum of the digits of any remaining numbers are divisible by 3. All of these numbers are divisible by 3: 9, 21 (digits sum to 3), 27 (digits sum to 9), 33 (digits sum to 6), 303 (digits sum to 6), and 1,023 (digits sum to 6).

    For the remaining numbers, you may already know that certain ones are prime. For the rest, check to see whether they are divisible by any prime numbers smaller than the number you are checking. For instance, to check 17, determine that it is not divisible by 2 (it's not even), by 3 (the digits don't sum to a multiple of 3), by 5 (it does not end in 0 or 5), or by 7 (actually check the division here: you don't get an integer).

Drill 7

21. **(A) 1, 2, 17:** If x is divisible by 34, then x is also divisible by everything 34 is divisible by. The factor pairs of 34 are (1, 34) and (2, 17).

22. **63; the factors are 1, 3, 7, 9, 21, and 63:** The number is the product of its prime factorization: $3 \times 3 \times 7 = 63$.

    The factor pairs are (1, 63), (3, 21), and (7, 9).

23. **Yes:**

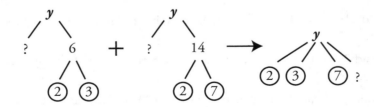

    The factors of 6 are 2 and 3. The factors of 14 are 2 and 7. The 2's do overlap, so strip out one instance of that double-counted factor. The combined factors are 2, 3, and 7. The 3 and 7 can be used to create 21 ($3 \times 7 = 21$), so y must be a multiple of 21.

24. **No:** For y to be divisible by 12, it would need to contain all of the prime factors of 12: 2, 2, and 3. Does it?

    The problem states that $7y$ is divisible by 210, so $7y$ contains the prime factors 2, 3, 5, and 7 (because $2 \times 3 \times 5 \times 7 = 210$). However, the question asks about y, not $7y$, so divide out the 7. Therefore, y must contain the remaining primes: 2, 3, and 5. Compare this to the prime factorization of 12: y does have a 2 and a 3, but it does not necessarily have *two* 2's. Therefore, y could be divisible by 12, but it doesn't have to be.

    Alternatively, you could start by dividing out the 7. If $7y$ is divisible by 210, y is divisible by 30. Therefore, y contains the prime factors 2, 3, and 5, and you can follow the remaining reasoning from above. Alternatively, since y is divisible by 30, y could be 30, which is *not* divisible by 12, or y could be 60, which *is* divisible by 12.

25. **(C) 2:** For integer a to be a multiple of 30, it would need to contain all of the prime factors of 30: 2, 3, and 5. Since a is not divisible by 30, it must be missing at least one of these prime factors. So if ab is divisible by 30, b must supply any missing prime factors. The least possible missing prime is 2. If $b = 2$ and $a = 15$ (or any odd multiple of 15), then the initial constraints will be met: ab will be divisible by 30, but a by itself will not be.

    You can also solve this problem by testing the answer choices. Determine whether there is a value of a that would make ab divisible by 30 for each answer choice. Since you're asked for the least possible value, start with the least answer choice. If b is 2, a could be 15. In that case, a is not divisible by 30, but ab is. Answer (C) is correct. Answers (A) and (B) are possible values of b, but not the least possible values.

Drill 8

26. **1, 2, 4, 5, 8, 10, and 20:** If 40 is a factor of x, then any factor of 40 is also a factor of x. List out the factors of 40: (1, 40), (2, 20), (4, 10), and (5, 8).

27. **Maybe:** To be divisible by 8, q needs three 2's in its prime factorization. Don't find the full list of (nonoverlapping) factors. Concentrate only on the 2's, since that's all you need to answer the question.

    Because there might be some overlapping factors of 2, you cannot simply count all of the numbers that contain 2. For instance, 6 is divisible by 2 and 3, so the fact that q is divisible by both 2 and 6 indicates only that there is at least one 2 (and at least one 3); there aren't necessarily two factors of 2 just because q has the factors 2 and 6.

    Instead, look for the greatest number of 2's *in one factor.* The number 12 contains two known 2's, so q must be divisible by 4, but it's unclear whether q contains three 2's. It might or it might not.

    If you ever do need to find all of the minimum factors, not just the 2's, then do this:

2	6	9	12	15	30
(2)	(2, 3)	(3, 3)	(2, 2, 3)	(3, 5)	(2, 3, 5)

    In this case, the factor 12 has the greatest number of 2's (two of them), so include two 2's in your list. The factor 9 has the greatest umber of 3's (two of them), so include two 3's in your list. Either the number 15 or the number 30 contains the greatest number of 5's (one of them), so include one 5 in your list. The value q must be divisible by $2 \times 2 \times 3 \times 3 \times 5$. (Note: You can still tell that q doesn't necessarily have to be divisible by 8, since it is only guaranteed to have two 2's, not the three 2's required to create 8.)

28. **Minimum = one; maximum = two:** Start with the more constrained variable: p. Because it is prime, it has exactly two factors—itself and 1. Therefore, the maximum number of *factors in common* cannot be more than two. Can p and q have exactly two factors in common? Certainly; q can be divisible by p. (For instance, if $p = 3$ and $q = 12$, the common factors are 1 and 3.)

What about the minimum? Can p and q have absolutely no factors in common? Try some numbers. If $p = 3$ and $q = 10$, then the two numbers don't have any prime factors in common, but notice that they are both divisible by 1. Any integer has 1 as a factor. Therefore, the minimum possible number of shared factors is one (the number 1 itself), and the maximum is two (the two factors of prime number p).

Drill 9

29. **66; the factors are 1, 2, 3, 6, 11, 22, 33, and 66:** The number is $2 \times 3 \times 11 = 66$. The factor pairs are (1, 66), (2, 33), (3, 22), and (6, 11).

30. **Maybe:**

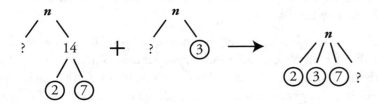

The two sets of factors don't overlap, so you can keep them all. For 12 to be a factor of n, n must contain all of the prime factors of 12 ($2 \times 2 \times 3$). In this case, n contains a 3, but only contains one 2 for sure, so 12 *could* be a factor of n but does not have to be.

31. **(B) 7:** Both x and y are positive integers and they sum to 17, so both must be less than 17. Since x has only two factors, it must be a prime number; its factors are itself and 1. Since y is divisible by 5, it must be 5, 10, or 15. List out the possible scenarios; start with y since you know there are only three possible values.

$$y + x = 17$$
$$5 + 12 = 17 \quad \text{No good: 12 isn't a prime number.}$$
$$10 + 7 = 17 \quad \text{Bingo! 10 is a multiple of 5, and 7 is a prime number.}$$

You could write out the third scenario, but wait! Check the answers whenever you have a possible solution. There is a 7 in the answers, so you're done. (It turns out that $15 + 2 = 17$ also fits the criteria given in the problem, but 2 is not among the answer choices.)

32. **Four:** Because you have been asked for a concrete answer even though the problem doesn't give a real value for n, you can infer that the answer will be the same regardless of which two-digit prime you pick. So, to make your job easier, pick the smallest one: 11.

If n is the product of 2, 3, and 11, then $n = 66$, and its factors are (1, 66), (2, 33), (3, 22), and (6, 11). There are four factors greater than 6: 11, 22, 33, and 66.

Why is the answer always four factors, even if you try a different two-digit prime number? Notice that because the other given prime factors of n (2 and 3) multiply to get exactly 6, you have to multiply by the third factor, the two-digit prime number (call it y), in order to get a number greater than 6. The factors greater than 6 will be y, $2 \times y$, $3 \times y$, and $2 \times 3 \times y$. In the example where $y = 11$, those factors are 11, 2×11, 3×11, 6×11. If you replace y with any two-digit prime, you will get the same result. (If you're not sure, test it out! That's a great way to prove a principle to yourself.)

Drill 10

33. **Yes:**

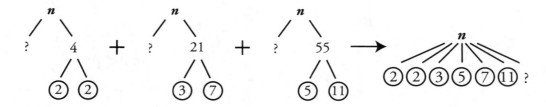

There is no overlap among the factors, so you can keep them all. In order for 14 to be a divisor of n, n has to contain all of the prime factors of 14 (2×7). Since n does contain 2 and 7, 14 is a divisor of n.

34. **Yes:**

None of the factors overlap, so you can keep all of them. For 154 to divide evenly into n, n has to contain all the same prime factors as 154 ($2 \times 7 \times 11$). n also contains 2, 7, and 11, so n is divisible by 154.

35. **Maybe:**

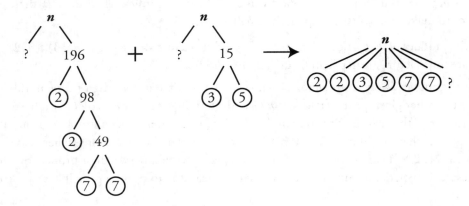

None of the factors overlap, so you can keep all of them. For 270 to be a factor of n, n must contain all the same prime factors as 270 ($2 \times 3 \times 3 \times 3 \times 5$). n contains a 2 and a 5, but only one known 3. Therefore, 270 *could* be a factor of n, but it does not have to be.

Exponents and Roots

In This Chapter

In this chapter, you will learn to apply exponent rules to solve problems. You will also learn how to manipulate square and cube roots.

CHAPTER 3 Exponents and Roots

You likely memorized the exponent and root rules in school (and you've likely since forgotten them). You could simply memorize them again, but if you understand *why* the rules work the way they do, you'll find it easier both to remember the rules and to apply them in ways that make the exam questions easier to solve.

Basics of Exponents

Exponents represent repeated multiplication. The **exponent**, or power, tells you how many **bases** to multiply together. In this example, 5 is the base and 3 is the exponent:

5^3	=	$5 \times 5 \times 5$	=	125
Five cubed	equals	three fives multiplied together, or five times five times five,	which equals	one hundred twenty-five

An **exponential expression** or term is one that contains an exponent. Exponential expressions can contain variables as well. The variable can be the base, the exponent, or even both. For example:

a^4	=	$a \times a \times a \times a$
a to the fourth	equals	four *a*'s multiplied together, or *a* times *a* times *a* times *a*

3^x	=	$3 \times 3 \times \ldots \times 3$
Three to the *x*	equals	three times three times three (there are *x* threes in the product, whatever *x* is)

Any base to the first power equals that base:

7^1	=	7
Seven to the first	equals	seven

Memorize the following powers of positive integers. Create flash cards for any that you don't already know.

Powers of 2	Powers of 3	Powers of 4	Powers of 5	Powers of 10
$2^1 = 2$	$3^1 = 3$	$4^1 = 2^2 = 4$	$5^1 = 5$	$10^1 = 10$
$2^2 = 4$	$3^2 = 9$	$4^2 = 2^4 = 16$	$5^2 = 25$	$10^2 = 100$
$2^3 = 8$	$3^3 = 27$	$4^3 = 2^6 = 64$	$5^3 = 125$	$10^3 = 1{,}000$
$2^4 = 16$	$3^4 = 81$			
$2^5 = 32$				
$2^6 = 64$				

Squares and cubes:

Squares	More Squares	Cubes
$1^2 = 1$	$10^2 = 100$	$1^3 = 1$
$2^2 = 4$	$11^2 = 121$	$2^3 = 8$
$3^2 = 9$	$12^2 = 144$	$3^3 = 27$
$4^2 = 16$	$15^2 = 225$	$4^3 = 64$
$5^2 = 25$	$20^2 = 400$	$5^3 = 125$
$6^2 = 36$	$30^2 = 900$	$10^3 = 1{,}000$
$7^2 = 49$		
$8^2 = 64$		
$9^2 = 81$		

Remember PEMDAS? Parentheses come first, then exponents. Take a look at this:

$$-(3)^2 \qquad \neq \qquad (-3)^2$$

The negative of three squared   does *not* equal   the square of negative three

The placement of the negative sign makes a significant difference:

$$-(3)^2 \qquad = \qquad -9$$

The negative of three squared   equals   negative nine

$$(-3)^2 \qquad = \qquad 9$$

The square of negative three   equals   nine

In the first case, you square before applying the negative sign, so the answer is negative. However, in the second case, you square a negative number, so the answer is positive.

Negative numbers raised to an even power are always positive. Negative numbers raised to an odd number are always negative. Here are the rules:

$$(\text{Negative})^{\text{even}} = \text{Positive} \qquad (\text{Negative})^{\text{odd}} = \text{Negative}$$

The powers of -1 alternate between 1 and -1. Even powers of -1 are always 1 (e.g., $(-1)^2 = 1$), while odd powers of -1 are always -1 (e.g., $(-1)^3 = -1$).

A positive base raised to any power is always positive, because positive times positive is positive—no matter how many times you multiply. For example, 2^{45} is positive because it represents 2 multiplied by itself 45 times.

Since an even exponent always gives a positive result, *an even exponent can hide the sign of the base.* Consider this equation:

$$x^2 = 16$$

In Chapter 6, Equations, you will cover in greater depth how to solve an equation such as this one. For now, notice that *two* different numbers for *x* would make the equation true:

$$4^2 = 16 \qquad (-4)^2 = 16$$

The value of *x* could be either 4 or -4. *Always be careful when dealing with even exponents in equations.* Look for more than one possible solution.

Check Your Skills

1. Arrange these numbers in increasing order: $(-5)^3$, 2^2, 3^1, 7^0

Answer can be found on page 108.

Multiply Terms with Same Base: Add the Exponents

Imagine that you multiply together a string of five *a*'s. Now multiply a second string of three *a*'s together. Finally, because you love multiplication, go ahead and multiply the two strings together. How many *a*'s do you end up with?

Write it all out longhand:

$$(a \times a \times a \times a \times a) \times (a \times a \times a) = a \times a \times a \times a \times a \times a \times a \times a$$

Now, use exponential notation:

a^5	$\times$	a^3	$=$	a^8
a to the *fifth*	times	*a* to the *third*	equals	*a* to the *eighth*

What happens to the exponents 5 and 3? They add up: $5 + 3 = 8$. This works as long as the bases are identical. The two terms on the left (a^5 and a^3) have the same base (*a*), so there are eight *a*'s on each side of the equation.

When you multiply exponential terms that have the same base, add the exponents.

Treat any term without an exponent as if it had an exponent of 1:

$$y(y^6) = y \times y^6 = y^1 \times y^6 = y^{(1 + 6)} = y^7$$

Adding exponents also works with numbers in the base, even weird numbers such as π. You just have to make sure that the bases are the same:

$$5^3 \times 5^6 = 5^9 \qquad \pi \times \pi^2 = \pi^3$$

The rule also works with variables in the exponent:

$$2^3 \times 2^z = 2^{(3+z)} \qquad 6(6^x) = 6^1 \times 6^x = 6^{1+x} \text{ or } 6^{x+1}$$

You can choose whether to use parentheses around the exponent, as in the first example—this can help to minimize careless mistakes. If you don't think you're likely to make a careless mistake here, then feel free to skip the parentheses, as in the second example.

Check Your Skills

Simplify the following expression.

2. $b^5 \times b^7$

Answer can be found on page 108.

Divide Terms with Same Base: Subtract the Exponents

Now, divide a string of five multiplied a's by a string of three multiplied a's. What is the result?

$$\frac{a \times a \times a \times a \times a}{a \times a \times a} = \frac{a \times \cancel{a} \times \cancel{a} \times \cancel{a} \times a}{\cancel{a} \times \cancel{a} \times \cancel{a}} = a \times a$$

In exponential notation, you have this: $\dfrac{a^5}{a^3} = a^2$

What happens to the exponents? You subtract the bottom exponent from the top exponent: $5 - 3 = 2$.

When you divide exponential terms that have the same base, subtract the exponents.

This rule works the same for numbers as for variables:

$$\frac{2^{16}}{2^{13}} = 2^{16-13} = 2^3 = 8 \qquad \frac{x^y}{x^2} = x^{y-2}$$

As before, treat any term without an exponent as if it had an exponent of 1:

$$\frac{a^9}{a} = \frac{a^9}{a^1} = a^8$$

Just make sure that the bases are the same. If the bases are not the same, then you can't combine the two terms into one in this way.

Here are the rules you've learned so far:

If you...	Then you...	Like this:
Multiply exponential terms that have the same base	Add the exponents	$a^2 \times a^3 = a^5$
Divide exponential terms that have the same base	Subtract the exponents	$\dfrac{a^5}{a^3} = a^2$

Check Your Skills

Simplify the following expressions.

3. $\dfrac{y^6}{y^2}$

4. $\dfrac{d^7}{d^8}$

Answers can be found on page 108.

Pretty Much Anything to the Power of Zero: One

Divide a string of five a's by a string of five a's. As before, each string is internally multiplied. What do you get?

Using longhand, you get 1:

$$\frac{a \times a \times a \times a \times a}{a \times a \times a \times a \times a} = \frac{\cancel{a} \times \cancel{a} \times \cancel{a} \times \cancel{a} \times \cancel{a}}{\cancel{a} \times \cancel{a} \times \cancel{a} \times \cancel{a} \times \cancel{a}} = 1$$

Using the exponent subtraction rule, you get a^0:

$$\frac{a^5}{a^5} = a^{5-5} = a^0$$

So a^0 must equal 1. That's true for practically any value of a. For example:

$$1^0 = 1 \qquad 6.2^0 = 1 \qquad (-4)^0 = 1 \qquad \left(\frac{3}{4}\right)^0 = 1 \qquad \left(\sqrt{2}\right)^0 = 1$$

The *only* base value for which this doesn't work is 0 itself. The expression 0^0 is called *undefined*. Notice that the exponent subtraction rule above required you to divide by a. Since you can't divide by 0, you can't raise 0 to the 0 power either. The GMAT will never ask you to do so.

For any nonzero value of a, $a^0 = 1$.

Now you can extend the powers of 2 to include 2^0:

Powers of 2
$$2^0 = 1$$
$$2^1 = 2$$
$$2^2 = 4$$
$$2^3 = 8$$
$$2^4 = 16$$

Notice the pattern: Each power of 2 is 2 times the previous power of 2.

Negative Power: One Over a Positive Power

What happens if you divide a string of three a's by a string of five a's?

Using longhand, you get a leftover a^2 in the denominator of the fraction:

$$\frac{a \times a \times a}{a \times a \times a \times a \times a} = \frac{\cancel{a} \times \cancel{a} \times \cancel{a}}{a \times a \times \cancel{a} \times \cancel{a} \times \cancel{a}} = \frac{1}{a \times a} = \frac{1}{a^2}$$

Using the exponent subtraction rule, you get a^{-2}:

$$\frac{a^3}{a^5} = a^{3-5} = a^{-2}$$

So those two results must be equal. *Something with a negative exponent is "1 over" that same thing with a positive exponent:*

$$a^{-2} \qquad = \qquad \frac{1}{a^2}$$

a to the negative two   equals   one over a squared

In other words, a^{-2} is equal to the **reciprocal** of a^2. The reciprocal of 5 is 1 *over* 5, or $\frac{1}{5}$. You can also think of reciprocals this way: Something times its reciprocal always equals 1. For example:

$$5 \times \frac{1}{5} = 1 \qquad\qquad a^2 \times \frac{1}{a^2} = 1 \qquad\qquad a^2 \times a^{-2} = a^{2-2} = a^0 = 1$$

Now you can extend the powers of 2 to include negative exponents:

$\underline{\text{Powers of 2}}$

$$2^{-3} = \frac{1}{2^3} = \frac{1}{8} = 0.125$$

$$2^{-2} = \frac{1}{2^2} = \frac{1}{4} = 0.25$$

$$2^{-1} = \frac{1}{2} = 0.5$$

$$2^0 = 1$$

$$2^1 = 2$$

$$2^2 = 4$$

$$2^3 = 8$$

$$2^4 = 16$$

The pattern still holds! Each power of 2 is 2 times the previous power of 2.

The rules you've seen so far work the same for negative exponents:

$$5^{-3} \times 5^{-6} = 5^{-3+(-6)} = 5^{-9}$$

$$\frac{x^3}{x^{-5}} = x^{3-(-5)} = x^8$$

Negative exponents are tricky, so it can be useful to *rewrite negative exponents using positive exponents*. A negative exponent in a term on top of a fraction becomes positive when you move the term to the bottom of the fraction:

$$\frac{5x^{-2}}{y^3} = \frac{5}{x^2 y^3}$$

Here, x^{-2} moved from the numerator to the denominator and the sign of the exponent switched from -2 to 2. Everything else stayed the same.

Likewise, a negative exponent in the bottom of a fraction becomes positive when the term moves to the top:

$$\frac{3}{z^{-4} w^2} = \frac{3z^4}{w^2}$$

Here, z^{-4} moved from the denominator to the numerator and the sign of the exponent switched from -4 to 4.

If you move the entire denominator, leave a 1 behind:

$$\frac{1}{z^{-4}} = \frac{1 \times z^4}{1} = z^4$$

The same is true for a numerator:

$$\frac{w^{-5}}{2} = \frac{1}{2w^5}$$

Don't confuse the sign of the base with the sign of the exponent. The sign of the base does not change.

A positive base raised to a negative exponent stays positive. For example:

$$3^{-3} = \frac{1}{3^3} = \frac{1}{27}$$

A negative base stays negative. Odd powers of a negative base still produce negative numbers:

$$(-4)^{-3} = \frac{1}{(-4)^3} = \frac{1}{-64} = -\frac{1}{64}$$

A negative base raised to an even power still produces a positive number:

$$\frac{1}{(-6)^{-2}} = (-6)^2 = 36$$

Here is a summary of the additional rules you've just learned:

If you...	Then you...	Like this:
Raise anything except zero to the power of zero	Get 1	$a^0 = 1$
Raise anything except zero to a negative power	Get 1 over that same thing to the corresponding positive power	$a^{-2} = \dfrac{1}{a^2}$
Move a term from top to bottom of a fraction (or vice versa)	Switch the sign of the exponent	$\dfrac{2a^{-2}}{3} = \dfrac{2}{3a^2}$

Check Your Skills

Simplify the following expressions.

5. 2^{-3}

6. $\dfrac{1}{3^{-3}}$

Answers can be found on page 108.

Apply Two Exponents: Multiply the Exponents

How do you simplify this expression?

$$(a^2)^4$$

Use the definition of exponents. First, square a. Next, multiply four separate a^2 terms together. In longhand:

$$(a^2)^4 = a^2 \times a^2 \times a^2 \times a^2 = a^{2+2+2+2} = a^8$$

What happens to the exponents 2 and 4? You multiply them: $2 \times 4 = 8$. On each side, you have eight a's multiplied together.

When you raise something that already has an exponent to another power, multiply the two exponents together.

Remember this rule? If you see two bases *multiplied* together, as in $a^2 \times a^4$, then *add* the exponents. Think of this as "dropping down" one mathematical operation—from multiplication to addition.

If, on the other hand, you see just one base raised to two *successive* powers, as in $(a^2)^4$, then *multiply* the exponents. Likewise, think of this as "dropping down" one mathematical operation—from an exponent to multiplication.

Note these rules on a flash card, as it's common to mix them up:

If you multiply the bases, *add* the exponents. $a^2 \times a^4 = a^{2+4} = a^6$

If you raise a power to a power, *multiply* the exponents. $(a^2)^4 = a^{(2)(4)} = a^8$

The "raise a power to a power" rule works the same way with negative exponents:

$$(x^{-3})^5 = x^{(-3)(5)} = x^{-15}$$
$$(4^{-2})^{-3} = 4^{(-2)(-3)} = 4^6$$

If you...	Then you...	Like this:
Raise something to two successive powers	Multiply the powers	$(a^2)^4 = a^8$

Put it all together. Simplify this expression:

$$\frac{x^{-3}(x^2)^4}{x^5}$$

First, simplify the parentheses $(x^2)^4$:

$$(x^2)^4 = x^{(2)(4)} = x^8$$

The fraction now reads:

$$\frac{x^{-3}x^8}{x^5}$$

Now, follow the rules for multiplying and dividing terms that have the same base. That is, add and subtract the exponents:

$$\frac{x^{-3}x^8}{x^5} = x^{-3+8-5} = x^0 = 1$$

If you have *different* bases that are numbers, try breaking down the bases to prime factors. You might discover that you can express everything in terms of one base. Try this problem:

$$2^2 \times 4^3 \times 16 =$$

(A) 2^6

(B) 2^{12}

(C) 2^{18}

The bases are all different, but they have something in common: Both 4 and 16 are powers of 2. Rewrite those terms with 2 as the base:

$$4 = 2^2 \text{ and } 16 = 2^4$$

Everything can now be expressed with 2 as the base:

$$\begin{aligned} 2^2 \times 4^3 \times 16 &= 2^2 \times (2^2)^3 \times 2^4 \\ &= 2^2 \times 2^6 \times 2^4 \\ &= 2^{2+6+4} \\ &= 2^{12} \end{aligned}$$

The correct answer is (B).

Check Your Skills

Simplify the following expressions.

7. $\dfrac{a^{15}}{a^0\left(a^3\right)^3}$

8. $\dfrac{3^4 9^2}{27}$

Answers can be found on pages 108–109.

Apply an Exponent to a Product: Apply the Exponent to Each Factor

Consider this expression:

$$(ab)^3$$

How can you rewrite this? Use the definition of exponents. You multiply three *ab* terms together:

$$(ab)^3 = ab \times ab \times ab$$

So you have three *a*'s and three *b*'s in the list. Since everything is multiplied, you can group the *a*'s and the *b*'s:

$$(ab)^3 = ab \times ab \times ab = (a \times a \times a)(b \times b \times b) = a^3 b^3$$

When you apply an exponent to a product, apply the exponent to each factor in the product.

This rule works with every kind of base and exponent you've seen so far. For example:

$$(3x)^4 = 3^4 x^4 = 81x^4$$

$$(wz^3)^x = w^x z^{3x}$$

$$\left(2^{-2} y^2\right)^{-3} = 2^{(-2)(-3)} y^{(2)(-3)} = 2^6 y^{-6} = 64 y^{-6} = \frac{64}{y^6}$$

Do the same thing with fractions. In particular, if you raise an entire fraction to a power, separately apply the exponent to the numerator and to the denominator:

$$\left(\frac{3}{4}\right)^{-2} = \frac{3^{-2}}{4^{-2}} = \frac{4^2}{3^2} = \frac{16}{9}$$

In this example, the top and the bottom of the fraction each have a negative exponent. In this case, you can flip the whole fraction to get rid of the negative exponents in both pieces.

The following case is different:

$$\frac{3^{-2}}{4} = \frac{1}{4 \times 3^2} = \frac{1}{36}$$

In $\dfrac{3^{-2}}{4}$, the exponent applies only to the numerator (3). Respect PEMDAS, as always.

If you...	Then you...	Like this:
Apply an exponent to a product	Apply the exponent to each factor in the product	$(ab)^3 = a^3b^3$
Apply an exponent to an entire fraction	Apply the exponent separately to the top and bottom	$\left(\dfrac{a}{b}\right)^4 = \dfrac{a^4}{b^4}$

You can write the prime factorization of big numbers without computing those numbers directly. For example:

What is the prime factorization of 18^3?

Definitely don't multiply out $18 \times 18 \times 18$. Instead, figure out the prime factorization of 18 itself, then apply the rule above:

$$18 = 2 \times 9 = 2 \times 3^2$$
$$18^3 = (2 \times 3^2)^3 = 2^3 \times 3^6 = 2^3 3^6$$

The prime factorization of 18^3 is $2^3 3^6$, or 2, 2, 2, 3, 3, 3, 3, 3, 3.

Simplify this harder problem:

$$\frac{12^2 \times 8}{18} =$$

First, break each base into its prime factors:

$$12 = 2^2 \times 3 \qquad 8 = 2^3 \qquad 18 = 2 \times 3^2$$

$$\frac{12^2 \times 8}{18} = \frac{\left(2^2 \times 3\right)^2 \times 2^3}{2 \times 3^2}$$

Next, apply the exponent to the parentheses:

$$\frac{\left(2^2 \times 3\right)^2 \times 2^3}{2 \times 3^2} = \frac{2^4 \times 3^2 \times 2^3}{2 \times 3^2}$$

Finally, combine the terms with 2 as their base. Remember that a 2 without a written exponent really has an exponent of 1. Separately, combine the terms with 3 as their base:

$$\frac{2^4 \times 3^2 \times 2^3}{2 \times 3^2} = 2^{4+3-1} \times 3^{2-2} = 2^6 \times 3^0 = 2^6 \times 1 = 2^6 = 64$$

Before you do the exponent addition and subtraction, you can simplify by crossing off common terms in the numerator and denominator of the fraction. For the previous problem, you could do this:

$$\frac{2^4 \times 3^2 \times 2^3}{2 \times 3^2} = \frac{2^4 \times \cancel{3^2} \times 2^{\cancel{3}^{2}}}{\cancel{2} \times \cancel{3^2}} = \frac{2^4 \times 2^2}{1} = 2^6 = 64$$

Occasionally, it's faster *not* to break down all the way to primes. If you spot a greater common base, feel free to use it. Try this problem:

$$\frac{36^3}{6^4} =$$

You can simplify this expression by breaking 36 and 6 down to primes—but you don't have to go that far. Pause and think about the best path. Since $36 = 6^2$, you can go much faster by breaking down just the 36 to a base of 6:

$$\frac{36^3}{6^4} = \frac{\left(6^2\right)^3}{6^4} = \frac{6^6}{6^4} = 6^2 = 36$$

Think before you act. You can often save yourself significant time and effort by pausing and choosing the best path rather than blindly following some standard "rule."

One last point: Be ready to do the reverse. For example, you can rewrite a^3b^3 as $(ab)^3$.

Consider $2^4 \times 3^4$. Here's a way to see that $2^4 \times 3^4$ equals $(2 \times 3)^4$, or 6^4:

$$2^4 \times 3^4 = (2 \times 2 \times 2 \times 2) \times (3 \times 3 \times 3 \times 3)$$
$$= (2 \times 3) \times (2 \times 3) \times (2 \times 3) \times (2 \times 3) \qquad \text{by regrouping}$$
$$= (2 \times 3)^4 = 6^4$$

More often, you'll need to change $(ab)^3$ into a^3b^3, but occasionally it's handy to go in reverse.

If you...	Then you...	Like this:
See two factors with the same exponent	Might regroup the factors as a product	$a^3b^3 = (ab)^3$

Check Your Skills

Simplify the following expressions. Write any negative exponents in positive form.

9. $\left(\dfrac{x^2 y}{z^{-3}}\right)^2$

10. $\dfrac{75^3 \times 45^3}{15^8}$

Answers can be found on page 109.

Add or Subtract Terms with the Same Base: Pull Out a Common Factor

Every case so far in this chapter has involved *only* multiplication and division. What if you are adding or subtracting exponential terms?

Consider this example:

$$13^5 + 13^3 =$$

Do not add the exponents to get 13^8. If you were given $13^5 \times 13^3$, then you would add the exponents, so the answer can't be the same when the multiplication symbol has changed to an addition symbol.

When you see addition, *look for a common factor and pull it out*. Both 13^5 and 13^3 are divisible by 13^3, since 13^5 can be written $13^3 13^2$, so 13^3 is the common factor. Here's how this works:

$$13^5 + 13^3 = 13^3 13^2 + 13^3 = 13^3(13^2 + 1)$$

Whenever you pull a term out of itself (e.g., pull 13^3 out of 13^3), you'll always be left with 1.

You could simplify further (though you would want to glance at any multiple-choice answers before taking this extra step). The expression $13^3(13^2 + 1)$ could be written as $13^3(169 + 1) = 13^3(170)$.

If you were given x's instead of 13's as bases, the factoring would work the same way:

$$x^5 + x^3 = x^3 x^2 + x^3 = x^3(x^2 + 1)$$

Now try this problem:

$3^8 - 3^7 - 3^6 =$

(A) $3^6(5)$

(B) 3^6

(C) 3^{-5}

Before you do any work, reflect. One of the three answers doesn't make logical sense. Which one and why?

The starting number, 3^8, is quite a bit larger than the other two numbers, 3^7 and 3^6. The answer should be a pretty decent-sized positive integer.

All of the choices are positive, but the third option, 3^{-5}, is a very small number; it's the equivalent of $\frac{1}{3^5}$.

It's too small, in fact, to be the correct answer. If you had to guess right now, you'd want to pick one of the other options.

Let's solve. All three terms (3^8, 3^7, and 3^6) are divisible by 3^6, so pull 3^6 out of the expression:

$$3^8 - 3^7 - 3^6 = 3^6(3^2 - 3^1 - 1) = 3^6(9 - 3 - 1) = 3^6(5)$$

The correct answer is (A).

This time, note that when you're pulling from a term whose exponent is just one number higher, you're always left with that term to the first power: Pulling 3^6 from 3^7 returns 3^1, which is the same as 3. Once you get used to that, feel free to skip the step of writing out 3^1 and just write 3 (or whatever number is in the base) directly.

Simplify this fraction:

$$\frac{3^4 + 3^5 + 3^6}{13}$$

Ignore the 13 on the bottom of the fraction for the moment. On the top, each term is divisible by 3^4:

$$\frac{3^4 + 3^5 + 3^6}{13} = \frac{3^4(1 + 3^1 + 3^2)}{13}$$

Continue to simplify the small powers of 3 in the parentheses:

$$\frac{3^4 + 3^5 + 3^6}{13} = \frac{3^4\left(1 + 3^1 + 3^2\right)}{13} = \frac{3^4(1 + 3 + 9)}{13} = \frac{3^4(13)}{13} = 3^4$$

The 13's on the top and bottom of the fraction cancel, leaving you with 3^4 as the answer.

If you...	Then you...	Like this:
Add or subtract terms with the same base	Pull out the common factor	$2^3 + 2^5$ $= 2^3(1 + 2^2)$

Check Your Skills

Simplify the following expression by factoring out a common term. (You don't need to simplify further than that.)

11. $5^5 + 5^4 - 5^3$ $5^3\left(5^2 + 5 - 1\right)$

Answer can be found on page 109.

Roots: Opposite of Exponents

Squaring a number means raising it to the second power (or multiplying it by itself). Square-rooting a number undoes that process. For example:

3^2 $=$ 9   and $\sqrt{9}$ $=$ 3

Three squared   is   nine,   and   the square root of nine   is   three

If you square-root first, then square, you get back to the original number:

$\left(\sqrt{16}\right)^2$ $=$ $\sqrt{16} \times \sqrt{16}$ $=$ 16

The square of the square root of sixteen   equals   the square root of sixteen times the square root of sixteen,   and that equals   sixteen

If you square first, then square-root, you get back to the original number if the original number is positive:

$\sqrt{5^2}$ $=$ $\sqrt{5 \times 5}$ $=$ 5

The square root of five squared   equals   the square root of five times five,   and that equals   five

If the original number is negative, you just flip the sign, so you end up with a positive:

$\sqrt{(-5)^2}$ $=$ $\sqrt{25}$ $=$ 5

The square root of the square of negative five   equals   the square root of twenty-five,   and that equals   five

If you...	Then you...	Like this:
Square a square root	Get the original number	$\left(\sqrt{10}\right)^2 = 10$
Square-root a square	Get the positive value of the original number	$\sqrt{10^2} = 10$ $\sqrt{(-10)^2} = 10$

Because 9 is the square of an integer ($9 = 3^2$), 9 is called a **perfect square**: its square root is an integer. In contrast, 2 is not the square of an integer, so its square root is an ugly decimal, as you saw in Chapter 1.

Memorize the perfect squares listed earlier in this chapter so that you can take their square roots easily. Also memorize these approximations:

$$\sqrt{2} \approx 1.4 \qquad \sqrt{3} \approx 1.7$$

Here's a neat way to remember them: February 14, or the date 2/14, is Valentine's Day, and March 17, or the date 3/17, is St. Patrick's Day.

You can approximate the square root of a non-perfect square by looking at nearby perfect squares. Try this problem:

$\sqrt{70}$ is between which two consecutive integers?

The two nearby perfect squares are 64 and 81. Since $\sqrt{64} = 8$ and $\sqrt{81} = 9$, it must be the case that $\sqrt{70}$ is between 8 and 9. It's roughly halfway in between, so you could say that $\sqrt{70} \approx 8.5$.

When you take the square root of any number greater than 1, your answer will be less than the original number:

$$\sqrt{2} < 2 \qquad \sqrt{21} < 21 \qquad \sqrt{1.3} < 1.3$$

However, the square root of a number between 0 and 1 is *greater* than the original number:

$$\sqrt{0.5} > 0.5 \qquad \sqrt{\frac{2}{3}} > \frac{2}{3}$$

$$\left(\sqrt{0.5} \approx 0.7\right) \qquad \left(\sqrt{\frac{2}{3}} \approx 0.8\right)$$

For any positive values, the square root of the number is *always closer to 1* than is the original number. If the starting number is greater than 1, the square root will be less than the starting number (and closer to 1). If the starting number is between 0 and 1, the square root will be greater than the starting number (and closer to 1).

The square root of 1 is 1, since $1^2 = 1$. Likewise, the square root of 0 is 0, since $0^2 = 0$:

$$\sqrt{1} = 1 \qquad \sqrt{0} = 0$$

You cannot take the square root of a negative number in GMAT world. This is only possible in advanced math that (thankfully) isn't tested on the GMAT.

Likewise, the square root symbol never gives a negative result. When you are given the square root symbol on the GMAT, *consider only the positive root*:

$$x = \sqrt{25} \qquad \text{Solution: } x = 5$$

In contrast, when the test gives a square, do take both the positive and negative roots. For example:

$$x^2 = 25 \qquad \text{Solutions: } x = 5 \text{ OR } x = -5$$

Why? You can plug either 5 or -5 into the original equation: $(5)^2 = 25$ and $(-5)^2 = 25$. Since no square root sign was given, you don't need to worry about ending up with a negative sign under the square root.

If you...	Then you...	Like this:
Take the square root of a number greater than 1	Get a smaller, positive number (a number that is closer to 1)	$\sqrt{25} = 5$
Take the square root of a number between 0 and 1	Get a greater, positive number (a number that is closer to 1)	$\sqrt{0.5} \approx 0.7$
Take the square root of 1 or 0	Get the number you started with	$\sqrt{1} = 1$ $\sqrt{0} = 0$

Check Your Skills

12. Simplify $\sqrt{27} \times \sqrt{27}$.

Answer can be found on page 110.

Square Root: Power of One-Half

Consider this equation:

$$\left(9^x\right)^2 = 9$$

Rewrite the expression using the tools you already have:

$$\left(9^x\right)^2 = 9$$
$$9^{2x} = 9^1$$

Since the bases are equal, the exponents must also be equal. Therefore, set the exponents equal to each other: $2x = 1$, or $x = \frac{1}{2}$.

Consider this equation:

$$\left(9^{\frac{1}{2}}\right)^2 = 9$$

The equation above is true because the exponents are multiplied together: $\left(\frac{1}{2}\right)(2) = 1$. You also learned earlier that $\left(\sqrt{9}\right)^2 = 9$.

Put these two equations together:

If $\left(9^{\frac{1}{2}}\right)^2 = 9$ and $\left(\sqrt{9}\right)^2 = 9$,

then $9^{\frac{1}{2}} = \sqrt{9}$.

For expressions with positive bases, *a square root is equivalent to an exponent of* $\frac{1}{2}$.

Try this problem:

Simplify $\sqrt{7^{22}}$. $\qquad \left(7^{22}\right)^{1/2} \simeq 7^n$

You can approach the problem in one of two ways.

Option 1: Rewrite the square root as an exponent of $\frac{1}{2}$, then apply the power-raised-to-a-power rule (multiply exponents):

$$\sqrt{7^{22}} = \left(7^{22}\right)^{\frac{1}{2}} = 7^{\frac{22}{2}} = 7^{11}$$

Option 2: Rewrite what's inside the square root as a product of two identical numbers. The square root is therefore one of those numbers:

$$\sqrt{7^{22}} = \sqrt{7^{11} \times 7^{11}} = \sqrt{\left(7^{11}\right)^2} = 7^{11}$$

Notice that you get an exponent that is an integer. This tells you that the number 7^{22} is a perfect square, or the square of another integer: $7^{22} = \left(7^{11}\right)^2$. An integer raised to a positive, even power always produces a perfect square.

If you...	Then you...	Like this:
Take a square root of a positive number raised to a power	Rewrite the square root as an exponent of $\frac{1}{2}$, then multiply exponents OR Rewrite what's inside the root as a product of two identical numbers	$\sqrt{5^{12}} = \left(5^{12}\right)^{\frac{1}{2}}$ $= 5^6$ $\sqrt{5^{12}} = \sqrt{5^6 \times 5^6}$ $= 5^6$

Avoid changing the square root to an exponent of $\frac{1}{2}$ when you have variable expressions underneath the square root (or radical) sign, since the output depends on the sign of the variables.

Check Your Skills

13. If x is positive, simplify $\sqrt{x^6}$.

Answer can be found on page 110.

Cube Roots Undo Cubing

Cubing a number means raising it to the third power. Cube-rooting a number undoes that process. For example:

4^3	$=$	64	and	$\sqrt[3]{64}$	$=$	4
Four cubed	is	sixty-four,	and	the cube root of sixty-four	is	four

Many of the properties of square roots carry over to cube roots. You can approximate cube roots the same way:

$\sqrt[3]{66}$ is a little more than 4, but less than 5, because $\sqrt[3]{64} = 4$ and $\sqrt[3]{125} = 5$.

Like square-rooting, cube-rooting a positive number pushes it toward 1:

$$\sqrt[3]{17} < 17 \quad \text{but} \quad \sqrt[3]{0.17} > 0.17$$

The main difference in behavior between square roots and cube roots is that you *can* take the cube root of a negative number. You always get a negative number:

$$\sqrt[3]{-64} = -4 \quad \text{because } (-4)^3 = -64$$

As a fractional exponent, cube roots are equivalent to exponents of $\frac{1}{3}$, just as square roots are equivalent to exponents of $\frac{1}{2}$. Going further, fourth roots are equivalent to exponents of $\frac{1}{4}$, and so on.

Now you can deal with **fractional exponents**. Consider this example:

$$8^{\frac{2}{3}} =$$

Rewrite $\frac{2}{3}$ as $2 \times \frac{1}{3}$, making two successive exponents. This is the same as squaring first, then cube-rooting. For example:

$$8^{\frac{2}{3}} = 8^{2 \times \frac{1}{3}} = \left(8^2\right)^{\frac{1}{3}} = \sqrt[3]{8^2} = \sqrt[3]{64} = 4$$

You could also rewrite $\frac{2}{3}$ as $\frac{1}{3} \times 2$ instead, allowing you to take the cube root first and then square the result. Look at the numbers and decide which way you think is easier for this math:

$$8^{\frac{2}{3}} = 8^{\frac{1}{3} \times 2} = \left(8^{\frac{1}{3}}\right)^2 = \left(\sqrt[3]{8}\right)^2 = 2^2 = 4$$

Whichever path you choose, you don't have to write out every step of the math above. Talk yourself through the steps:

$\sqrt[3]{8^2}$ First, square the 8 to get 64, and then take the cube root of the 64 to get 4.

$\left(\sqrt[3]{8}\right)^2$ First, take the cube root of 8 to get 2, and then square the 2 to get 4.

If you...	Then you...	Like this:
Raise a number to a fractional power	Apply two exponents—the numerator as a power and the denominator as a fractional root, in whatever order seems easier to you	$125^{\frac{2}{3}} = \left(\sqrt[3]{125}\right)^2$ $= 5^2 = 25$

Check Your Skills

14. Simplify $64^{\frac{2}{3}}$.

Answer can be found on page 110.

Multiply Square Roots: Put Everything under the Root

Consider this example:

$$\sqrt{8} \times \sqrt{2} =$$

Earlier, you learned that $8^a 2^a = (8 \times 2)^a$. This principle holds true for fractional exponents as well:

$$\sqrt{8} \times \sqrt{2} = 8^{\frac{1}{2}} \times 2^{\frac{1}{2}} = (8 \times 2)^{\frac{1}{2}} = \sqrt{8 \times 2}$$

In other words, when you *multiply* separate square roots, you can just *put everything under the same radical sign*:

$$\sqrt{8} \times \sqrt{2} = \sqrt{8 \times 2} = \sqrt{16} = 4$$

This shortcut works for division, too. When you *divide* square roots, you can *put everything under the same radical sign*:

$$\frac{\sqrt{27}}{\sqrt{3}} = \sqrt{\frac{27}{3}} = \sqrt{9} = 3$$

As long as you're only multiplying and dividing, you can deal with more complicated expressions:

$$\frac{\sqrt{15} \times \sqrt{12}}{\sqrt{5}} = \sqrt{\frac{\overset{3}{\cancel{15}} \times 12}{\underset{1}{\cancel{5}}}} = \sqrt{36} = 6$$

Don't forget to simplify before you multiply!

One more thing: You can do the above when multiplying or dividing roots, but you *cannot* do the above when adding or subtracting roots. You'll learn more about this in a little bit.

If you...	Then you...	Like this:
Multiply or divide square roots	Combine everything under one radical sign, then simplify	$\sqrt{a} \times \sqrt{b} = \sqrt{ab}$ $\dfrac{\sqrt{a}}{\sqrt{b}} = \sqrt{\dfrac{a}{b}}$

Check Your Skills

Simplify the following expressions.

15. $\sqrt{20} \times \sqrt{5}$ $= 10$

16. $\dfrac{\sqrt{216}}{\sqrt{2} \times \sqrt{3}}$ $\sqrt{216} \rightarrow \sqrt{6 \cdot 6 \cdot 6} \quad \dfrac{6\sqrt{6}}{\sqrt{6}} \quad \boxed{6}$

Answers can be found on page 110.

Simplify Square Roots: Factor Out Squares

What does this product equal?

$$\sqrt{6} \times \sqrt{2} =$$

First, put everything under one radical:

$$\sqrt{6} \times \sqrt{2} = \sqrt{12}$$

You might think that you're done—after all, 12 is not a perfect square, so you won't get an integer out of $\sqrt{12}$. But what if you glance at the answers and $\sqrt{12}$ isn't there? It turns out that $\sqrt{12}$ is mathematically correct, but it can be simplified further.

Here's how: $12 = 4 \times 3$ and 4 is a perfect square, so it can be pulled out from under a square root sign:

$$\sqrt{12} = \sqrt{4 \times 3} = \sqrt{4} \times \sqrt{3} = 2\sqrt{3}$$

If the answer you get isn't a perfect square itself but does have a perfect square as a factor, then the GMAT will typically expect you to simplify that answer as far as it can go. Glance at the answers before you actually do that simplification, just to make sure—but expect that you will probably have to do so.

To simplify square roots, factor out squares.

If you...	Then you...	Like this:
Have the square root of a large number (or a root that doesn't match any answer choices)	Pull perfect-square factors out of the number under the radical sign	$\sqrt{50} = \sqrt{25 \times 2}$ $= \sqrt{25} \times \sqrt{2}$ $= 5\sqrt{2}$

Sometimes you can spot the perfect square factor, if you know your perfect squares. Consider this example:

$$\sqrt{360} =$$

If you have your perfect squares memorized, then 360 will likely make you think of the perfect square 36:

$$\sqrt{360} = \sqrt{36 \times 10} = \sqrt{36} \times \sqrt{10} = 6\sqrt{10}$$

What if you don't spot a perfect square? You can always *break the number down to primes*. This method will take longer, but it will work when you need it.

Consider $\sqrt{12}$ again. The prime factorization of 12 is $2 \times 2 \times 3$, or $2^2 \times 3$. Therefore, you can break it down:

$$\sqrt{12} = \sqrt{2^2 \times 3} = \sqrt{2^2} \times \sqrt{3} = 2\sqrt{3}$$

Each pair of prime factors under the radical (2×2, or 2^2) turns into a single copy as it emerges (becoming the 2 in $2\sqrt{3}$). In this exercise, it can be useful to write out the prime factorization without exponents so that you can spot the prime pairs quickly.

Take $\sqrt{360}$ again. Say you don't spot the perfect square factor (36). Write out the prime factorization of 360:

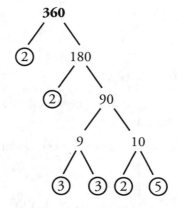

$$360 = 2 \times 2 \times 2 \times 3 \times 3 \times 5$$

Now pair off two 2's and two 3's, leaving an extra 2 and 5. Any pair of primes under a radical sign becomes a single copy of that prime outside the radical:

$$\sqrt{360} = \sqrt{2 \times 2} \times \sqrt{3 \times 3} \times \sqrt{2 \times 5} = 2 \times 3 \times \sqrt{2 \times 5} = 6\sqrt{10}$$

Check Your Skills

Simplify the following roots.

17. $\sqrt{96}$

18. $\sqrt{225}$

Answers can be found on page 110.

Add or Subtract under the Root: Pull Out Common Square Factors

Consider this example:

$$\sqrt{3^2 + 4^2} =$$

Don't fall into the trap. You *cannot* break this root into $\sqrt{3^2} + \sqrt{4^2}$. You can break up *products* (multiplication), not sums, under the square root. For instance, this is correct:

$$\sqrt{3^2 \times 4^2} = \sqrt{3^2} \times \sqrt{4^2} = 3 \times 4 = 12$$

But you can't do that when you're adding under the root. To evaluate $\sqrt{3^2 + 4^2}$, follow PEMDAS under the radical, *then* take the square root:

$$\sqrt{3^2 + 4^2} = \sqrt{9 + 16} = \sqrt{25} = 5$$

The same goes for subtraction:

$$\sqrt{13^2 - 5^2} = \sqrt{169 - 25} = \sqrt{144} = 12$$

Often, you have to crunch the numbers if they're small. However, when the numbers get large, the GMAT will give you a necessary shortcut: factoring out squares.

You'll need to find a square factor that is common to both terms under the radical. Consider this example:

$$\sqrt{3^{10} + 3^{11}} =$$

First, consider $3^{10} + 3^{11}$ by itself. What is the greatest factor that the two terms in the sum have in common? 3^{10}. Note that $3^{11} = 3^{10} \times 3$. Therefore, you can rewrite the equation as follows:

$$3^{10} + 3^{11} = 3^{10}(1 + 3) = 3^{10}(4)$$

Now plug that back into the square root:

$$\sqrt{3^{10} + 3^{11}} = \sqrt{3^{10}(1 + 3)} = \sqrt{3^{10}(4)} = \sqrt{3^{10}} \times \sqrt{4}$$

First, the easier part: $\sqrt{4} = 2$.

Next, since $3^{10} = \left(3^5\right)^2$, $\sqrt{3^{10}} = 3^5$.

The full simplification is: $\sqrt{3^{10}} \times \sqrt{4} = 3^5 \times 2$. The full answer also might be written in the form $3^5(2)$.

Alternatively, to simplify $\sqrt{3^{10}}$, apply the square root as an exponent of $\frac{1}{2}$:

$$\sqrt{3^{10}} = \left(3^{10}\right)^{\frac{1}{2}} = 3^{\frac{10}{2}} = 3^5$$

If you...	Then you...	Like this:
Add or subtract underneath the square root symbol	Factor out a square factor from the sum or difference OR *If they're small*, go ahead and crunch the numbers	$\sqrt{4^{14} + 4^{16}} = \sqrt{4^{14}\left(1 + 4^2\right)}$ $= \sqrt{4^{14}} \times \sqrt{1 + 16}$ $= 4^7\sqrt{17}$ OR $\sqrt{6^2 + 8^2} = \sqrt{36 + 64}$ $= \sqrt{100}$ $= 10$

3

Check Your Skills

19. $\sqrt{10^5 - 10^4} =$

Answer can be found on page 110.

$\sqrt{10^4(10-1)} \rightarrow 10^4(9)$

$10^4 \rightarrow 10^2\left(\sqrt{9}\right) = 300$

Check Your Skills Answer Key

1. $(-5)^3$, 7^0, 3^1, 2^2: In order to compare, expand each term.

$$(-5)^3 = -125$$
$$2^2 = 4$$
$$3^1 = 3$$
$$7^0 = 1$$

For the first term, since the base is a negative and the exponent is odd, the number must remain negative. All of the bases for the other terms are positive, so $(-5)^3$ must be the least term. Any number raised to the 0 power is equal to 1, so 7^0 is the next term. Finally, any number raised to the 1 power remains the same, so $2^2 = 4$ is greater than $3^1 = 3$.

2. b^{12}: The bases are the same, so add the exponents.

$$b^5 \times b^7 = b^{5+7} = b^{12}$$

3. y^4: The bases are the same, so subtract the bottom exponent from the top exponent.

$$\frac{y^6}{y^2} = y^{6-2} = y^4$$

4. d^{-1}: The bases are the same, so subtract the bottom exponent from the top exponent: $7 - 8 = -1$.

5. $\frac{1}{8}$: $2^{-3} = \frac{1}{2^3} = \frac{1}{8}$

6. 27: $\frac{1}{3^{-3}} = 3^3 = 27$

7. a^6: Simplify the parentheses first.

$$\frac{a^{15}}{a^0 \left(a^3\right)^3} = \frac{a^{15}}{a^0 a^9}$$

Then deal with all of the exponents.

$$\frac{a^{15}}{a^0 a^9} = \frac{a^{15}}{a^{0+9}} = a^{15-9} = a^6$$

8. 3^5: First, break the various numbers down to bases of 3.

$$\frac{3^4 9^2}{27} = \frac{3^4 \left(3^2\right)^2}{3^3}$$

Then, simplify the exponents.

$$\frac{3^4 \left(3^2\right)^2}{3^3} = \frac{3^4 3^4}{3^3} = 3^{4+4-3} = 3^5$$

9. $x^4 y^2 z^6$: Begin by applying the outside exponent to everything in the parentheses.

$$\left(\frac{x^2 y}{z^{-3}}\right)^2 = \frac{x^{(2)(2)} y^{(1)(2)}}{z^{(-3)(2)}}$$

Then, simplify all of the exponents.

$$\frac{x^{(2)(2)} y^{(1)(2)}}{z^{(-3)(2)}} = \frac{x^4 y^2}{z^{-6}} = x^4 y^2 z^6$$

10. **15**: First, break the bases down into common terms. In this case, the common primes are 3 and 5.

$$\frac{75^3 \times 45^3}{15^8} = \frac{\left(3 \times 5^2\right)^3 \times \left(3^2 \times 5\right)^3}{(3 \times 5)^8}$$

Next, apply the exponents outside of the parentheses to each element in the parentheses.

$$\frac{\left(3 \times 5^2\right)^3 \times \left(3^2 \times 5\right)^3}{(3 \times 5)^8} = \frac{3^3 \times 5^6 \times 3^6 \times 5^3}{3^8 \times 5^8}$$

Finally, simplify all of the exponents.

$$\frac{3^3 \times 5^6 \times 3^6 \times 5^3}{3^8 \times 5^8} = \frac{3^9 5^9}{3^8 5^8} = 3^1 5^1 = 15$$

11. **$5^3(29)$ or $125(29)$:**

$$5^5 + 5^4 - 5^3 = 5^3(5^2 + 5^1 - 1) = 5^3(25 + 5 - 1) = 5^3(29) = 125(29)$$

12. **27:** Any square root times itself equals the number underneath the square root symbol.

$$\sqrt{27} \times \sqrt{27} = \left(\sqrt{27}\right)^2 = 27$$

You can also think of it this way: If you square a square root, the two operations cancel out. Just cross off both the square and the square root symbol.

13. x^3: Since x is positive, x^3 is positive, too.

$$\sqrt{x^6} = \sqrt{x^3 \times x^3} = x^3$$

14. **16:** As always, think before you act. In this case, it's definitely easier to take the cube root before you square.

$$64^{\frac{2}{3}} = \left(\sqrt[3]{64}\right)^2 = 4^2 = 16$$

15. **10:** $\sqrt{20} \times \sqrt{5} = \sqrt{20 \times 5} = \sqrt{100} = 10$

16. **6:** Simplify before you multiply!

$$\frac{\sqrt{216}}{\sqrt{2} \times \sqrt{3}} = \sqrt{\frac{\overset{108}{\cancel{216}}}{\cancel{2} \times 3}} = \sqrt{\frac{108}{3}} = \sqrt{36} = 6$$

17. **4$\sqrt{6}$:** $\sqrt{96} = \sqrt{3 \times (2 \times 2) \times (2 \times 2) \times 2} = (2 \times 2)\sqrt{3 \times 2} = 4\sqrt{6}$

18. **15:** This is one of the common perfect squares; if you have it memorized, you don't have to do any work at all! The square root of 225 is 15. Here's the math.

$$\sqrt{225} = \sqrt{(3 \times 3) \times (5 \times 5)} = 3 \times 5 = 15$$

19. **300:** $\sqrt{10^5 - 10^4} = \sqrt{10^4(10 - 1)} = \sqrt{10^4 9} = 10^2 \times 3 = 300$

Chapter Review: Drill Sets

Drill 1

Simplify the following expressions by combining like terms. If the base is a number, leave the answer in exponential form (e.g., use 2^3, not 8).

1. $y^5 \times y^3$

2. $\left(a^3\right)^2$

3. $7^{-4} \times 7^3$

4. $\dfrac{(-3)^a}{(-3)^2}$

5. $\left(3^2\right)^{-3}$

6. $x^2 \times x^3 \times x^5$

7. $\left(5^2\right)^x$

Drill 2

Simplify the following expressions by combining like terms. If the base is a number, leave the answer in exponential form (e.g., use 2^3, not 8).

8. $3^4 \times 3^2 \times 3$

9. $y^7 \times y^8 \times y^{-6}$

10. $\dfrac{z^5 \times z^{-3}}{z^{-8}}$

11. $\dfrac{3^{2x} \times 3^{6x}}{3^{-3y}}$

12. $\left(y^2\right)^6 \times y^3$

13. $\left(m^y\right)^5 \times m^{y+5}$

14. $\dfrac{\left(2^x\right)^{-2} \times 2^3}{2^{2x}}$

Drill 3

Follow the directions for each question.

15. Compute the sum of $27^{\frac{1}{3}} + 9^{\frac{1}{2}} + \dfrac{3}{9^0}$.

16. Which of the following has the greatest value?

    (A) $-(5)^6$

    (B) 6^{-5}

    (C) $(-6)^5$

    (D) $(-5)^6$

    (E) 5^{-6}

17. Compute the sum of $6^{-3} - \left(\dfrac{1}{6}\right)^3 + 8^{\frac{2}{3}}$.

18. Which of the following is equal to $\left(\dfrac{4}{7}\right)^{-4}$?

    (A) $-\left(\dfrac{4}{7}\right)^4$

    (B) $\left(\dfrac{7}{4}\right)^4$

    (C) $\left(\dfrac{4}{7}\right)^{\frac{1}{4}}$

    (D) $\left(-\dfrac{4}{7}\right)^{\frac{1}{4}}$

    (E) $-\left(\dfrac{7}{4}\right)^4$

3

19. Which of the following has a value less than 1 ? (Select all that apply.)

    (A) $\dfrac{2^{-2}}{3^0}$

    (B) $\dfrac{3^{-2}}{4^{-2}}$

    (C) $\dfrac{(-3)^3}{(-5)^2}$

    (D) $\left(\dfrac{2}{3}\right)^{-2}$

    (E) $(-4)^3$

Drill 4

Simplify the following expressions by finding common bases. If the base is a number, leave the answer in exponential form (e.g., use 2^3, not 8).

20. $8^3 \times 2^6$

21. $\dfrac{36^3}{6^4}$

22. $25^4 \times 125^3$

23. $9^{-2} \times 27^2$

24. $2^{-7} \times 8^2$

Drill 5

Simplify the following expressions by pulling out as many common factors as possible.

25. $6^3 + 3^3$

    (A) 3^5

    (B) 3^9

    (C) $2(3^3)$

26. $81^2 - 9^3$

    (A) $3^7(2)$

    (B) $3^6(2^3)$

    (C) 3^{14}

27. $(-2)^3 - (5)^2 + (-4)^3$

28. $15^2 - 5^2$

    (A) $5^2(2)$

    (B) $5^2 2^3$

    (C) $5^2 3^2$

29. $4^3 + 4^3 + 4^3 + 4^3 + 3^2 + 3^2 + 3^2$

    (A) $4^4 + 3^3$

    (B) $4^{12} + 3^6$

    (C) $4^3(3^2)$

30. $\dfrac{3^{12} - 9^4}{27^2 + 9^4}$

    (A) $3^2(8)$

    (B) $\dfrac{5}{3^3}$

    (C) $3^3(20)$

Drill 6

Simplify the following expressions. All final answers should be integers.

31. $\sqrt{3} \times \sqrt{27}$

32. $\sqrt{2} \times \sqrt{18}$

33. $\dfrac{\sqrt{48}}{\sqrt{3}}$

34. $\sqrt{36} \times \sqrt{4}$

35. $\dfrac{\sqrt{54} \times \sqrt{3}}{\sqrt{2}}$

36. $\dfrac{\sqrt{640}}{\sqrt{2} \times \sqrt{5}}$

37. $\dfrac{\sqrt{48} \times \sqrt{7}}{\sqrt{21}}$

Drill 7

Simplify the following roots. Not every answer will be an integer.

38. $\sqrt{180}$

39. $\sqrt{490}$

40. $\sqrt{343}$

41. $\sqrt{135}$

42. $\sqrt{208}$

43. $\sqrt{432}$

Drill 8

Simplify the following roots. You will be able to eliminate the root completely in every question. Express answers as integers.

44. $\sqrt{35^2 - 21^2}$

45. $\sqrt{10\left(11^5 - 11^4\right)}$

46. $\sqrt{8^4 + 8^5}$

47. $\sqrt{2^9 + 2^7 - 2^6}$

48. $\sqrt{50^3 - 50^2}$

49. $\sqrt{\dfrac{10\left(13^4 + 13^2\right)}{17}}$

50. $\sqrt{5^7 - 5^5 + 5^4}$

3

Drill Sets Solutions

Drill 1

1. y^8: When the bases are multiplied, add the exponent.

$$y^5 \times y^3 = y^{(5+3)} = y^8$$

2. a^6: When raising a power to a power, multiply the exponents.

$$(a^3)^2 = a^{(3)(2)} = a^6$$
$$(a^3)^2 = a^{(3)(2)} = a^6$$

3. $7^{-1} = \frac{1}{7}$: When multiplying the bases, add the exponents.

$$7^{-4} \times 7^3 = 7^{(-4+3)} = 7^{-1} = \frac{1}{7}$$

4. $(-3)^{(a-2)}$: When dividing the bases, subtract the exponents.

$$\frac{(-3)^a}{(-3)^2} = (-3)^{(a-2)}$$

5. 3^{-6}: When raising a power to a power, multiply the exponents.

$$(3^2)^{-3} = 3^{(2)(-3)} = 3^{-6}$$

6. x^{10}: When multiplying the bases, add the exponents.

$$x^2 \times x^3 \times x^5 = x^{(2+3+5)} = x^{10}$$

7. 5^{2x}: When raising a power to a power, multiply the exponents.

$$(5^2)^x = 5^{(2)(x)} = 5^{2x}$$

Drill 2

8. 3^7:

$$3^4 \times 3^2 \times 3 = 3^{(4+2+1)} = 3^7$$

9. y^9:

$$y^7 \times y^8 \times y^{-6} = y^{(7+8+(-6))} = y^9$$

10. z^{10}: Multiply and divide in whatever order is most convenient. Break the process down to individual steps in order to minimize the risk of errors.

$$= \frac{z^5 \times z^{-3}}{z^{-8}} = \frac{z^{(5+(-3))}}{z^{-8}} = \frac{z^2}{z^{-8}} = z^{(2-(-8))} = z^{10}$$

11. 3^{8x+3y}:

$$\frac{3^{2x} \times 3^{6x}}{3^{-3y}} = \frac{3^{(2x+6x)}}{3^{-3y}} = \frac{3^{(8x)}}{3^{-3y}} = 3^{(8x-(-3y))} = 3^{8x+3y}$$

12. y^{15}:

$$\left(y^2\right)^6 \times y^3 = y^{12} \times y^3 = y^{(12+3)} = y^{15}$$

13. m^{6y+5}:

$$\left(m^y\right)^5 \times m^{y+5} = m^{5y} \times m^{y+5} = m^{5y+(y+5)} = m^{6y+5}$$

14. 2^{-4x+3}:

$$\frac{\left(2^x\right)^{-2} \times 2^3}{2^{2x}} = \frac{2^{-2x} \times 2^3}{2^{2x}} = \frac{2^{-2x+3}}{2^{2x}} = 2^{(-2x+3-2x)} = 2^{-4x+3}$$

Drill 3

15. **9:** If you are comfortable handling fractional exponents, you do not need to rewrite those forms using the radical signs shown in this solution; you can leave them in fraction form.

$$27^{\frac{1}{3}} + 9^{\frac{1}{2}} + \frac{3}{9^0} = \sqrt[3]{27} + \sqrt{9} + \frac{3}{1} = 3 + 3 + 3 = 9$$

16. **(D) $(-5)^6$:** The question asks for the answer with the greatest value. How can you compare efficiently? In general, don't calculate actual values unless you have to; without a calculator, this math would be messy. First, scan the answers. Can you tell whether some will be negative or positive?

Answer (A) will be negative because the negative sign is applied last. Answer (C) will be negative because any negative number raised to an odd power will remain negative. The other three answers will all be positive, so eliminate only (A) and (C) for now.

Answer (B) has a negative exponent, so the base 6 will turn into a fraction less than 1. Likewise, answer (E) will turn into a fraction less than 1.

Answer (D) is a negative number raised to an even power, so it will turn positive. The power itself is positive, so the final value will be greater than 1. Answer (D) is therefore greater than either (B) or (E). Eliminate those two. Answer (D) is the only one remaining.

3

17. **4:** The first two terms in the expression turn out to be the same. Because these terms are equal, when the second is subtracted from the first, they cancel out, leaving only the third term.

$$6^{-3} - \left(\frac{1}{6}\right)^3 + 8^{\frac{2}{3}} = \frac{1}{6^3} - \frac{1^3}{6^3} + \left(\sqrt[3]{8}\right)^2 = \frac{1}{6^3} - \frac{1}{6^3} + 2^2 = 4$$

18. **(B)** $\left(\frac{7}{4}\right)^4$: Make the answer choices work for you. Glance at them before you start manipulating the given statement. All of them have a positive exponent, so change the original term to the same format. To change a negative exponent to a positive, take the reciprocal or flip the fraction.

$$\left(\frac{4}{7}\right)^{-4} = \left(\frac{7}{4}\right)^4$$

19. **(A)** $\frac{2^{-2}}{3^0}$, **(C)** $\frac{(-3)^3}{(-5)^2}$, **and (E)** $(-4)^3$: The problem asks for values less than 1, so any expressions with negative values, the number 0 itself, or any values between 0 and 1 will fulfill the question.

(A) $\frac{2^{-2}}{3^0} = \frac{1}{3^0 \times 2^2} = \frac{1}{1 \times 4} = \frac{1}{4} < 1$. Correct.

Note: Dividing a smaller positive number by a greater positive number will result in a number less than 1. If you know this, you can stop at the second step.

(B) $\frac{3^{-2}}{4^{-2}} = \frac{4^2}{3^2} = \frac{16}{9} > 1$. Eliminate.

Note: Dividing a greater positive number by a smaller positive number will result in a number greater than 1. If you know this, you can stop at the second step.

(C) $\frac{(-3)^3}{(-5)^2} = \frac{-27}{25} = $ negative < 1. Correct.

(D) $\left(\frac{2}{3}\right)^{-2} = \left(\frac{3}{2}\right)^2 = \frac{3^2}{2^2} = \frac{9}{4} > 1$. Eliminate.

As with (B), dividing a greater positive number by a smaller positive number will result in a number greater than 1.

(E) $(-4)^3 = $ negative < 1. Correct.

Drill 4

20. **2^{15}:**

$$8^3 \times 2^6 = (2^3)^3 \times 2^6 = 2^9 \times 2^6 = 2^{9+6} = 2^{15}$$

21. **6^2:**

$$\frac{36^3}{6^4} = \frac{(6^2)^3}{6^4} = \frac{6^6}{6^4} = 6^{6-4} = 6^2$$

22. **5^{17}:**

$$25^4 \times 125^3 = (5^2)^4 \times (5^3)^3 = 5^8 \times 5^9 = 5^{17}$$

23. **3^2:**

$$9^{-2} \times 27^2 = (3^2)^{-2} \times (3^3)^2 = 3^{-4} \times 3^6 = 3^2$$

24. **2^{-1}:**

$$2^{-7} \times 8^2 = 2^{-7} \times (2^3)^2 = 2^{-7} \times 2^6 = 2^{-1}$$

Drill 5

25. **(A) 3^5:** Begin by breaking 6 down into its prime factors.

$$6^3 + 3^3 =$$
$$(2 \times 3)^3 + 3^3 =$$
$$(2^3)(3^3) + 3^3$$

Now each term contains 3^3. Factor it out.

$$\left(2^3\right)\left(3^3\right) + 3^3 =$$
$$3^3\left(2^3 + 1\right) =$$
$$3^3(9) =$$
$$3^3\left(3^2\right) = 3^5$$

26. **(B) $3^6(2^3)$:** Both bases are powers of 3. Rewrite the bases and pull out common factors.

$$81^2 - 27^2 =$$
$$\left(3^4\right)^2 - \left(3^3\right)^2 =$$
$$3^8 - 3^6 =$$
$$3^6\left(3^2 - 1\right) =$$
$$3^6(8) = 3^6\left(2^3\right)$$

27. **−97:** Before computing, eyeball the math. Can you tell whether the final value will be positive or negative? Raising a negative term to an odd exponent will keep the term negative. The only positive term in the expression is subtracted, so the result will be negative. Use this logic to check your work and avoid making a careless mistake.

$$(-2)^3 - (5)^2 + (-4)^3 =$$
$$(-2)(-2)(-2) - (5)(5) + (-4)(-4)(-4) =$$
$$(-8) - 25 + (-64) =$$
$$-33 - 64 = -97$$

28. **(B) $5^2 2^3$:** Begin by breaking 15 down into its prime factors.

$$15^2 - 5^2 =$$
$$(3 \times 5)^2 - 5^2 =$$
$$\left(3^2\right)\left(5^2\right) - 5^2$$

Now, both terms contain 5^2. Factor it out.

$$\left(3^2\right)\left(5^2\right) - 5^2 =$$
$$5^2\left(3^2 - 1\right) =$$
$$5^2\left(9 - 1\right) =$$
$$5^2 8$$

Compare that to the answers. What else can you manipulate?

$$5^2(8) = 5^2(2^3)$$

29. **(A) $4^4 + 3^3$:** Factor 4^3 out of the first four terms and factor 3^2 out of the last three terms.

$$4^3 + 4^3 + 4^3 + 4^3 + 3^2 + 3^2 + 3^2 =$$
$$4^3\left(1 + 1 + 1 + 1\right) + 3^2\left(1 + 1 + 1\right) =$$
$$4^3\left(4\right) + 3^2\left(3\right) =$$
$$4^{3+1} + 3^{2+1} =$$
$$4^4 + 3^3$$

30. **(A) $3^2(8)$:** Every base in the fraction is a multiple of 3. Begin by rewriting the bases as 3's.

$$\frac{3^{12} - 9^4}{27^2 + 9^4} = \frac{3^{12} - \left(3^2\right)^4}{\left(3^3\right)^2 + \left(3^2\right)^4} = \frac{3^{12} - 3^8}{3^6 + 3^8}$$

The terms in the numerator both contain 3^8, and the terms in the denominator both contain 3^6. Factor the numerator and denominator.

$$\frac{3^{12} - 3^8}{3^6 + 3^8} = \frac{3^8\left(3^4 - 1\right)}{3^6\left(1 + 3^2\right)} = \frac{3^8(80)}{3^6(10)}$$

Simplify the numerator and denominator.

$$\frac{3^8(80)}{3^6(10)} = 3^{8-6}(8) = 3^2(8)$$

Drill 6

31. **9:** Before you multiply larger numbers together, think about whether you can make the math easier by breaking the values down into primes. In this case, you can break down 27 into three 3's.

$$\sqrt{3} \times \sqrt{27} = \sqrt{3 \times 27} = \sqrt{3 \times 3^3} = \sqrt{3^4} = 3^2 = 9$$

When you break down 27, the term under the square root is more clearly a perfect square ($3^4 = 3^2 \times 3^2$).

32. **6:**

$$\sqrt{2} \times \sqrt{18} = \sqrt{2 \times 18} = \sqrt{36} = 6$$

33. **4:**

$$\frac{\sqrt{48}}{\sqrt{3}} = \sqrt{\frac{48}{3}} = \sqrt{16} = 4$$

34. **12:** Careful! You might get so used to combining automatically that you fail to notice that these two are already perfect squares. They can be simplified first and then multiplied.

$$\sqrt{36} \times \sqrt{4} = 6 \times 2 = 12$$

35. **9:** Simplify the numerator and denominator before you multiply, then break 27 down into 3's.

$$\frac{\sqrt{54} \times \sqrt{3}}{\sqrt{2}} = \sqrt{\frac{\overset{27}{\cancel{54}} \times 3}{\underset{1}{\cancel{2}}}} = \sqrt{3^4} = 3^2 = 9$$

36. **8:** Simplify before you multiply.

$$\frac{\sqrt{640}}{\sqrt{2} \times \sqrt{5}} = \sqrt{\frac{640}{2 \times 5}} = \sqrt{\frac{640}{10}} = \sqrt{64} = 8$$

37. **4:** Simplify before you multiply.

$$\frac{\sqrt{48} \times \sqrt{7}}{\sqrt{21}} = \sqrt{\frac{48 \times \overset{1}{\cancel{7}}}{\underset{3}{\cancel{21}}}} = \sqrt{\frac{\overset{16}{\cancel{48}} \times 1}{\underset{1}{\cancel{3}}}} = \sqrt{16} = 4$$

Alternatively, break larger terms into their prime factors, then simplify.

$$\frac{\sqrt{48} \times \sqrt{7}}{\sqrt{21}} = \sqrt{\frac{2 \times 2 \times 2 \times 2 \times \cancel{3} \times \cancel{7}}{\cancel{3} \times \cancel{7}}} = \sqrt{2^4} = 2^2 = 4$$

Drill 7

38. $6\sqrt{5}$: If needed, use a factor tree to break down large numbers.

$$\sqrt{180} = \sqrt{2^2 \times 3^2 \times 5} = 2 \times 3 \times \sqrt{5} = 6\sqrt{5}$$

39. $7\sqrt{10}$:

$$\sqrt{490} = \sqrt{2 \times 5 \times 7^2} = 7 \times \sqrt{2 \times 5} = 7\sqrt{10}$$

40. $7\sqrt{7}$:

$$\sqrt{343} = \sqrt{7^3} = \sqrt{7^2 \times 7} = 7\sqrt{7}$$

41. $3\sqrt{15}$:

$$\sqrt{135} = \sqrt{3^3 \times 5} = \sqrt{3^2 \times 3^1 \times 5} = 3\sqrt{15}$$

42. $4\sqrt{13}$:

$$\sqrt{208} = \sqrt{2^4 \times 13} = 2^2 \times \sqrt{13} = 4\sqrt{13}$$

43. $12\sqrt{3}$:

$$\sqrt{432} = \sqrt{2^4 \times 3^3} = 2^2 \times \sqrt{3^2 \times 3} = 2^2 \times 3 \times \sqrt{3} = 12\sqrt{3}$$

Drill 8

44. **28:** Pull out the greatest common factor of 35^2 and 21^2, namely 7^2.

$$\sqrt{\left(7^2 \times 5^2\right) - \left(7^2 \times 3^2\right)} = \sqrt{7^2 \left(5^2 - 3^2\right)} = \sqrt{7^2 \left(25 - 9\right)} = \sqrt{7^2 \left(16\right)}$$

Both 7^2 and 16 are perfect squares $(16 = 4^2)$. Therefore:

$$\sqrt{7^2 \left(16\right)} = \sqrt{7^2 \left(4^2\right)} = 7 \times 4 = 28$$

45. **1,210:** Pull out the greatest common factor of 11^4 and 11^5, namely 11^4.

$$\sqrt{10\left(11^5 - 11^4\right)} = \sqrt{10\left(11^4 \left(11 - 1\right)\right)} = \sqrt{10\left(11^4 \left(10\right)\right)}$$

$(10)(10)$ is the same as 10^2. The other term, 11^4, is also a perfect square $(11^4 = 11^2 \times 11^2)$. Pull the squares out of the square root.

$$\sqrt{\left(10^2\right)\left(11^4\right)} = (10)\left(11^2\right) = (10)(121) = 1{,}210$$

46. **192:** Pull out the greatest common factor of 8^4 and 8^5, namely 8^4.

$$\sqrt{8^4(1+8)} = \sqrt{8^4(9)} = \sqrt{8^4\left(3^2\right)}$$

Both 8^4 and 3^2 are perfect squares ($8^4 = 8^2 \times 8^2$). Therefore:

$$\sqrt{8^4\left(3^2\right)} = 8^2 \times 3 = 64 \times 3 = 192$$

47. **24:** Pull out the greatest common factor of 2^9, 2^7, and 2^6, namely 2^6.

$$\sqrt{2^6\left(2^3 + 2 - 1\right)} = \sqrt{2^6(8 + 2 - 1)} = \sqrt{2^6(9)} = \sqrt{2^6\left(3^2\right)}$$

Both 2^6 and 3^2 are perfect squares ($2^6 = 2^3 \times 2^3$). Therefore:

$$\sqrt{2^6\left(3^2\right)} = 2^3 \times 3 = 8 \times 3 = 24$$

48. **350:** Pull out the greatest common factor of 50^3 and 50^2, namely 50^2.

$$\sqrt{50^2(50 - 1)} = \sqrt{50^2(49)} = \sqrt{50^2\left(7^2\right)} = 50 \times 7 = 350$$

49. **130:** First focus on the numerator of the fraction. Pull out the greatest common factor of 13^4 and 13^2, namely 13^2.

$$\sqrt{\frac{10\left(13^2\left(13^2 + 1\right)\right)}{17}} = \sqrt{\frac{10\left(13^2(169 + 1)\right)}{17}} = \sqrt{\frac{10\left(13^2(170)\right)}{17}}$$

The denominator (17) divides evenly into 170, and the remaining terms are perfect squares:

$$\sqrt{\frac{10\left(13^2(170)\right)}{17}} = \sqrt{10\left(13^2(10)\right)} = \sqrt{\left(10^2\right)\left(13^2\right)} = (10)(13) = 130$$

50. **275:** Pull out the greatest common factor of 5^7, 5^5, and 5^4, namely 5^4.

$$\sqrt{5^4\left(5^3 - 5 + 1\right)} = \sqrt{5^4(125 - 5 + 1)} = \sqrt{5^4(121)} = \sqrt{5^4\left(11^2\right)}$$

Both 5^4 and 11^2 are perfect squares ($5^4 = 5^2 \times 5^2$). Therefore:

$$\sqrt{5^4\left(11^2\right)} = 5^2 \times 11 = 25 \times 11 = 275$$

CHAPTER 4

Fractions and Ratios

In This Chapter

In this chapter, you will learn to add, subtract, multiply, and divide fractions, as well as to manipulate fractions in more complex scenarios. You'll also learn how to move back and forth between fractions and ratios, as well as how to think about the connections between ratios and actual values.

CHAPTER 4 Fractions and Ratios

Fractions and ratios are closely related—but not exactly the same—and are common on the GMAT. It's also not uncommon to see something like $\frac{3}{5}$ in a problem but the number actually represents a ratio, not a fraction. It's important to know how to distinguish between the two types of values—and you'll learn how in this chapter.

Basics of Fractions: Part to Whole

A fraction expresses division.

The **numerator** on top is divided by the **denominator** on bottom:

Numerator

Fraction line $\longrightarrow \quad \frac{3}{4} \quad = \quad 3 \div 4$

Denominator

Three-fourths is three divided by four.

The result of the division is a number. If you punch "$3 \div 4 =$" into a calculator, you get the decimal 0.75. But you can also think of 0.75 as $\frac{3}{4}$, because $\frac{3}{4}$ and 0.75 are two different ways to write the same number. (You'll learn about decimals in the next chapter.)

Fractions express a **part-to-whole** relationship:

$$\frac{3}{4} = \frac{\text{part}}{\text{whole}}$$

3 pieces = part

4 pieces = whole

In the figure above, a circle represents a whole unit—a full pizza. The pizza has been divided into four equal parts, or fourths, because the denominator of the fraction is 4. In any fraction, the denominator tells you how many equal slices something has been broken into, in this case a pizza. In other words, the denominator indicates the size of a slice: Each slice is one-quarter of the pizza.

The numerator of the fraction is 3. In this case, the fraction is talking about three slices of the pizza. In any fraction, the numerator tells you how many of the slices you are talking about. Together, you are talking about three slices out of four total, or three parts to the total four. You can also say that you are talking about three-quarters of the pizza.

Since fractions express division, all the arithmetic rules of division apply. For instance, a negative divided by a positive gives you a negative, and so on:

$$\frac{-3}{4} = -3 \div 4 = -0.75 \qquad \frac{3}{-4} = 3 \div (-4) = -0.75$$

So $\frac{-3}{4}$ and $\frac{3}{-4}$ represent the same number. You can also write that number as $-\frac{3}{4}$. Just don't mix up the negative sign with the fraction bar.

PEMDAS also applies. The fraction bar means that you always *divide the entire numerator by the entire denominator*:

$$\frac{3x^2 + y}{2y^2 - z} = \left(3x^2 + y\right) \div \left(2y^2 - z\right)$$

The entire quantity $3x^2 + y$ is being divided by the entire quantity $2y^2 - z$.

If you rewrite a fraction, be ready to put parentheses around the numerator or denominator to preserve the correct order of operations.

Finally, if you divide by zero, your answer is considered "undefined" and the GMAT doesn't typically test this. So *a denominator can never equal zero*. If you have a variable expression in the denominator, the problem will tell you that the expression cannot equal zero. For example, if a problem contains the fraction $\frac{x}{y}$, the problem will tell you that $y \neq 0$ (that's a "does not equal" sign).

If the problem gives you $\frac{3x^2 + y}{2y^2 - z}$, then what cannot equal zero?

The entire denominator cannot equal zero. In other words:

$$2y^2 - z \neq 0 \quad \text{or} \quad 2y^2 \neq z$$

If the GMAT tells you that something does *not* equal something else (using the $\neq$ sign), the purpose is often to rule out dividing by zero somewhere in the problem.

To compare positive fractions with the same *denominator*, compare the numerators. The numerator tells you how many pieces you have. *The greater the numerator, the greater the fraction*, assuming positive numbers and the same denominators. A greater numerator indicates that you have more of the same-sized slices of pie:

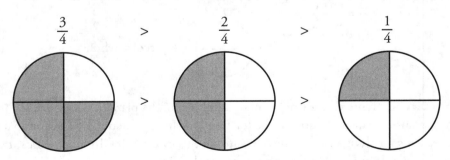

To compare positive fractions with the same *numerator*, compare the denominators. *The greater the denominator, the smaller the fraction.* Each slice of pie is smaller. So the same number of smaller slices represents a smaller portion of the pie:

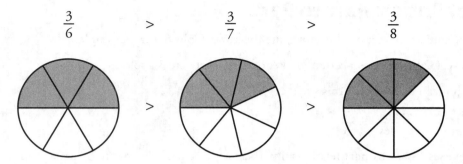

$$\frac{3}{6} \qquad > \qquad \frac{3}{7} \qquad > \qquad \frac{3}{8}$$

If the numerator and denominator are the same, then the fraction equals 1:

$$\frac{4}{4} \qquad = \qquad 4 \div 4 \qquad = \qquad 1$$

Four-   equals   four divided   and that equals   one
fourths   by four,

If the numerator is greater than the denominator (again, assume positive numbers), then you have more than one pizza:

$$\frac{5}{4} \qquad = \qquad 5 \div 4 \qquad = \qquad 1 \qquad + \qquad \frac{1}{4}$$

Five-   equals   five   and that   one   plus   one-
fourths   divided   equals   fourth
                    by four,

Another way to write $1 + \frac{1}{4}$ is $1\frac{1}{4}$ (read *one and one-fourth*). This is the only time in GMAT math when you put two things next to each other (1 and $\frac{1}{4}$) in order to *add* them. In all other circumstances, two things right next to each other means *multiplication*.

A **mixed number** such as $1\frac{1}{4}$ contains both an integer part, 1, and a fractional part, $\frac{1}{4}$. You can always rewrite a mixed number as a sum of the integer part and the fractional part; just split the integer and the fraction:

$$3\frac{3}{8} = 3 + \frac{3}{8}$$

In a positive **improper fraction** such as $\frac{5}{4}$, the numerator is greater than or equal to the denominator and the overall value is greater than or equal to 1. Improper fractions and mixed numbers are two different ways to express the same thing. Later, you'll learn how to convert between them.

A positive **proper fraction** such as $\frac{3}{4}$ has a value between 0 and 1. In a proper fraction, the numerator is smaller than the denominator.

Basics of Ratios: Part to Part

Ratios are very similar to fractions but not quite the same. Consider this example:

For every 2 bananas in a certain basket of fruit, there are 3 apples.

This relationship can be rewritten this way:

$$\frac{\text{Number of bananas}}{\text{Number of apples}} = \frac{2}{3}$$

This ratio expresses a **part-to-part** relationship: There are 2 "parts" bananas to 3 "parts" apples. Even though the ratio can be written as a fraction, as shown above, it is still a part-to-part relationship.

Ratios are more commonly shown in part-to-part relationships, but you can also transform this same information into a *part-to-whole relationship*. There are 5 parts total, because 2 parts + 3 parts = 5 parts, so the whole is 5. There are 2 bananas for every 5 pieces of fruit, and there are 3 apples for every 5 pieces of fruit:

$$\frac{\text{Number of bananas}}{\text{Total pieces of fruit}} = \frac{2}{5} \qquad \frac{\text{Number of apples}}{\text{Total pieces of fruit}} = \frac{3}{5}$$

The part-to-whole form can be called either a fraction or a ratio. You can choose to move back and forth between part-to-whole and part-to-part relationships, depending on how you think a particular problem would be easier to solve.

The language and symbols of ratios are peculiar. Note the following equivalent expressions:

For every 2 bananas, there are 3 apples.
There are 2 bananas for every 3 apples.
The ratio of bananas to apples is 2 to 3.
The ratio of bananas to apples is 2 : 3.
The ratio of bananas to apples is $\frac{2}{3}$.
The number of bananas in the basket is $\frac{2}{3}$ the number of apples in the basket.

All of those sentences convey the same information. The last one is the most indirect. Why is it a part-to-part and not a part-to-whole?

The sentence can be rewritten in this way: For every two bananas in the basket, there are 3 apples. That's a part-to-part relationship, so the "fraction" in the sentence is really a ratio.

Add Fractions with the Same Denominator: Add the Numerators

Addition will mostly take place with part-to-whole relationships such as fractions; you'll learn more about operations common to part-to-part relationships later.

The numerator of a fraction tells you how many slices of the pizza you have. So when you add fractions, you add the numerators. You just have to make sure that the slices are the same size—in other words, that the denominators are equal. For example:

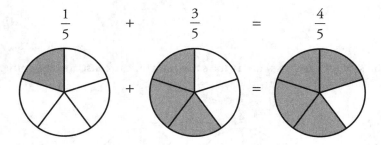

In words, one-fifth plus three-fifths equals four-fifths. The *fifth* is the size of the slice, so the denominator (5) doesn't change.

Since $4 = 1 + 3$, you can write the fraction with $1 + 3$ in the numerator:

$$\frac{1}{5} + \frac{3}{5} = \frac{1+3}{5} = \frac{4}{5}$$

The same process applies with subtraction. Subtract the numerators and leave the denominator the same:

$$\frac{9}{14} - \frac{4}{14} = \frac{9-4}{14} = \frac{5}{14}$$

If variables are involved, add or subtract the same way. Just make sure that the denominators in the original fractions are equal. It doesn't matter how complicated they are. For example:

$$\frac{3a}{b} + \frac{4a}{b} = \frac{3a+4a}{b} = \frac{7a}{b} \qquad \frac{5x^2}{z+w} - \frac{2x^2}{z+w} = \frac{5x^2 - 2x^2}{z+w} = \frac{3x^2}{z+w}$$

If you can't simplify the numerator, leave it as a sum or a difference. The denominator stays the same, because it just tells you the *size* of the slices you're adding or subtracting. For example:

$$\frac{x}{y} + \frac{z}{y} = \frac{x+z}{y} \qquad \frac{3n}{2w^3} - \frac{5m}{2w^3} = \frac{3n-5m}{2w^3}$$

Here's another rule for your list:

If you...	Then you...	Like this:
Add or subtract fractions that have the same denominator	Add or subtract the numerators, leaving the denominator alone	$\frac{2}{7} + \frac{3}{7} = \frac{2+3}{7} = \frac{5}{7}$

Check Your Skills

1. $\dfrac{3x}{yz^2} + \dfrac{7x}{yz^2} =$

Answer can be found on page 158.

Add Fractions with Different Denominators: Find a Common Denominator

Consider this example:

$$\frac{1}{4} + \frac{3}{8} =$$

The denominators (the sizes of the slices) aren't the same, so you can't just add the numerators this time:

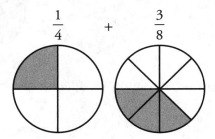

To add these fractions exactly, manipulate the fractions so that the slices are the same size. In other words, the fractions need to have a **common denominator**—that is, the *same* denominator. Once they have the same denominator, you can add the numerators.

But wait! This is a multiple-choice test, so first see whether you can eyeball the answer. The first fraction is $\frac{1}{4}$ and the second is a bit bigger than $\frac{1}{4}$, so the sum has to be more than $\frac{1}{2}$. In addition, both starting fractions are less than $\frac{1}{2}$, so the sum has to be less than 1.

The sum, therefore, has to be between $\frac{1}{2}$ and 1. Depending on the details of the specific problem, that knowledge might be enough to answer the question.

If not, find common denominators to add the two fractions. Since a fourth of a pizza is twice as big as an eighth, take the fourth in the first circle and cut it in two:

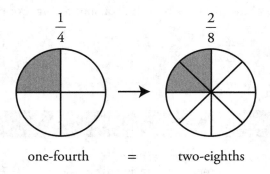

You have the same amount of pizza—the shaded area hasn't changed in size. So one-fourth $\left(\frac{1}{4}\right)$ equals two-eighths $\left(\frac{2}{8}\right)$.

When you cut the fourth in two, you end up with twice as many slices. So the numerator is doubled. But you're breaking the whole circle into twice as many pieces, so the denominator is doubled as well. If you double both the numerator and the denominator, the fraction's value stays exactly the same—this is why you are allowed to make this move:

$$\frac{1}{4} = \frac{1 \times 2}{4 \times 2} = \frac{2}{8}$$

Without changing the value of $\frac{1}{4}$, you have renamed it $\frac{2}{8}$. Now you can add it to $\frac{3}{8}$:

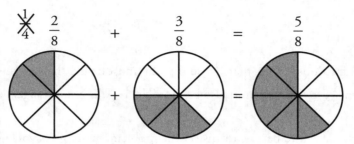

Here's the math all in one line:

$$\frac{1}{4} + \frac{3}{8} = \frac{1 \times 2}{4 \times 2} + \frac{3}{8} = \frac{2}{8} + \frac{3}{8} = \frac{5}{8}$$

To add fractions with different denominators, find a common denominator. That is, rename the fractions so that they have the same denominator. Then add the new numerators. The same holds true for subtraction.

How do you rename a fraction without changing its value? Multiply the top and bottom by the same number. For example:

$$\frac{1}{4} = \frac{1 \times 2}{4 \times 2} = \frac{2}{8} \qquad \frac{3}{4} = \frac{3 \times 25}{4 \times 25} = \frac{75}{100} \qquad \frac{5}{9} = \frac{5 \times 7}{9 \times 7} = \frac{35}{63}$$

Here's why this works. Doubling the top and doubling the bottom of a fraction is the same as multiplying the fraction by $\frac{2}{2}$. (More on fraction multiplication later.)

Notice that $\frac{2}{2}$ is equal to 1. And multiplying a number by 1 doesn't change the number. So when you

multiply by $\frac{2}{2}$, you aren't really changing the number, you're just changing its appearance, or renaming it:

$$\frac{1}{4} = \frac{1}{4} \times \frac{2}{2} = \frac{1 \times 2}{4 \times 2} = \frac{2}{8}$$

You can rename fractions that have variables in them, too. You can even multiply the top and bottom by the same variable:

$$\frac{x}{y} = \frac{x}{y} \times \frac{2}{2} = \frac{x \times 2}{y \times 2} = \frac{2x}{2y} \qquad\qquad \frac{a}{2} = \frac{a}{2} \times \frac{b}{b} = \frac{a \times b}{2 \times b} = \frac{ab}{2b}$$

Just make sure the expression on the bottom can never equal zero, of course. Here's another rule for your list:

If you...	Then you...	Like this:
Want to give a fraction a different denominator but keep the value the same	Multiply the top and bottom of the fraction by the same number	$\frac{1}{4} = \frac{1 \times 2}{4 \times 2} = \frac{2}{8}$

Try this problem:

$$\frac{1}{4} + \frac{1}{3} =$$

First, eyeball this. Both fractions are less than $\frac{1}{2}$, so the sum must be less than 1. The first fraction is $\frac{1}{4}$ and the second is larger than $\frac{1}{4}$, so the sum must also be more than $\frac{1}{2}$. The sum is again somewhere between $\frac{1}{2}$ and 1.

Even if that isn't enough for you to be able to answer the problem, it still gives you a good sense (we call this *number sense*) of what you should expect when you calculate the exact value. Developing your number sense will help you to make fewer mistakes on the test.

What should the common denominator of these fractions be? It needs to be both a multiple of 4 *and* a multiple of 3. That is, it should be a multiple that 3 and 4 have in common. The easiest multiple to pick is usually the **least common multiple** (LCM) of 3 and 4. If you need a refresher on that concept, return to Chapter 2.

The least common multiple of 4 and 3 is 12. Rename the two fractions so that they each have a denominator of 12:

$$\frac{1}{4} = \frac{1 \times 3}{4 \times 3} = \frac{3}{12} \qquad\qquad \frac{1}{3} = \frac{1 \times 4}{3 \times 4} = \frac{4}{12}$$

Once you have a common denominator, add the numerators:

$$\frac{1}{4} + \frac{1}{3} = \frac{1 \times 3}{4 \times 3} + \frac{1 \times 4}{3 \times 4} = \frac{3}{12} + \frac{4}{12} = \frac{7}{12}$$

The process works the same if you subtract fractions or even have more than two fractions. Try this problem:

$$\frac{5}{6} + \frac{2}{9} - \frac{3}{4} =$$

First, find the common denominator by finding the least common multiple. All three denominators (6, 9, and 4) are composed of 2's and 3's:

$$6 = 2 \times 3 \qquad\qquad 9 = 3 \times 3 \qquad\qquad 4 = 2 \times 2$$

The LCM will contain two 2's (because there are two 2's in 4) and two 3's (because there are two 3's in 9):

$$2 \times 2 \times 3 \times 3 = 36$$

To make each of the denominators equal 36, multiply the fractions by $\frac{6}{6}$, $\frac{4}{4}$, and $\frac{9}{9}$, respectively:

$$\frac{5}{6} = \frac{5 \times 6}{6 \times 6} = \frac{30}{36} \qquad \frac{2}{9} = \frac{2 \times 4}{9 \times 4} = \frac{8}{36} \qquad \frac{3}{4} = \frac{3 \times 9}{4 \times 9} = \frac{27}{36}$$

Now that the denominators are all the same, add and subtract normally:

$$\frac{5}{6} + \frac{2}{9} - \frac{3}{4} = \frac{30}{36} + \frac{8}{36} - \frac{27}{36} = \frac{30 + 8 - 27}{36} = \frac{11}{36}$$

The process works even if you have variables. Try adding these two fractions:

$$\frac{2}{x} + \frac{3}{2x} =$$

First, find the common denominator by finding the least common multiple of x and $2x$. The LCM is $2x$. So give the first fraction a denominator of $2x$, then add:

$$\frac{2}{x} + \frac{3}{2x} = \frac{2 \times 2}{x \times 2} + \frac{3}{2x} = \frac{4}{2x} + \frac{3}{2x} = \frac{4 + 3}{2x} = \frac{7}{2x}$$

Now that you know how to do the math in the traditional way, it's time to learn a useful shortcut: the **Double-Cross**. (In general, you'll be better able to remember how to implement a shortcut when you first understand the textbook approach.)

Here's an example that you solved earlier:

$$\frac{1}{4} + \frac{1}{3} = \frac{1 \times 3}{4 \times 3} + \frac{1 \times 4}{3 \times 4} = \frac{3}{12} + \frac{4}{12} = \frac{7}{12}$$

This time, you're going to take some different steps. First, draw three arrows on the problem and rewrite the addition sign just above the top arrows:

The arrows mean multiplication. Multiply along each arrow, and place the product where the arrow points:

The new denominator, 12, is the result of multiplying along the bottom, as shown above.

To get the new numerator, add the two new numbers on top:

The result is the same whether you use the shortcut or do the math in the traditional way; the shortcut just compresses several steps.

The double-cross will always work, and it is usually faster, but there is one circumstance in which the traditional approach is faster.

Take a look at this example:

$$\frac{1}{3} + \frac{1}{6}$$

In this case, 3 and 6 share a common denominator of 6, so you only need to change one fraction: $\frac{1}{3}$. Since this is the case, it's faster to find common denominators:

$$\frac{1}{3} + \frac{1}{6} = \frac{2}{6} + \frac{1}{6} = \frac{3}{6} = \frac{1}{2}$$

Is this one easier to solve the traditional way or via the double-cross? Try it:

$$\frac{5}{6} + \frac{3}{8}$$

Traditional	The Double-Cross
Find a common denominator. The smallest one is 24: $$\frac{5 \times 4}{6 \times 4} + \frac{3 \times 3}{8 \times 3}$$	Draw your arrows and multiply: $$\overset{40}{\underset{6}{}}\,\overset{}{5} + \overset{18}{\underset{8}{}}\,\overset{}{3} = \frac{}{48}$$
Add the two fractions: $$\frac{20}{24} + \frac{9}{24} = \frac{29}{24}$$	Add the numerators. Note that 58 and 48 are both divisible by 2, so simplify as a last step: $$\frac{58}{48} = \frac{29 \times \cancel{2}}{24 \times \cancel{2}} = \frac{29}{24}$$

As usual, the traditional method has more work up front (finding a common denominator), but this time the shortcut had an extra step at the end: You have to simplify. Why?

This will happen when the two denominators share factors, as 6 and 8 do. The least common denominator, 24, is smaller than multiplying those two numbers together: $6 \times 8 = 48$. If you choose the double-cross in this circumstance, be aware that you will need to simplify at the end (you'll learn various methods for doing so later in this chapter). If the numbers are large enough to be cumbersome, then you may want to use the traditional method, even though you do have to find common denominators.

If you...	Then you...	Like this:
Add or subtract fractions with different denominators	Put the fractions in terms of a common denominator, then add or subtract OR Use the double-cross method	$$\frac{1}{3} - \frac{1}{6} = \frac{1 \times 2}{3 \times 2} - \frac{1}{6}$$ $$= \frac{2}{6} - \frac{1}{6} = \frac{1}{6}$$ OR $$\overset{7}{}\,\frac{1}{3} - \overset{3}{}\,\frac{1}{7} = \frac{4}{21}$$

Check Your Skills

2. $\dfrac{1}{2} + \dfrac{3}{4} =$

3. $\dfrac{2}{3} - \dfrac{3}{8} =$

Answers can be found on page 158.

Compare Fractions: Use the Double-Cross

The double-cross is very versatile; it can also be used to compare two fractions. Consider this example:

Which is greater, $\dfrac{3}{5}$ or $\dfrac{4}{7}$?

Use a modified version of the double-cross. Multiply along the top two crossed arrows, but ignore the denominator:

1. Set the two fractions up near each other.

2. Multiply "up" the arrows. Put the resulting number at the top of each respective arrow.

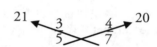

3. Compare the numbers. The side with the bigger number is the bigger fraction.
    $21 > 20$, so $\dfrac{3}{5}$ is greater than $\dfrac{4}{7}$.

The full process for the double-cross results in a common denominator, so you're just shortcutting the full process. Since the denominator will be the same, all you need to do is compare the numerators.

If you...	Then you...	Like this:
Want to compare fractions	Do the two top calculations of the double-cross	⟨28⟩ ↖ 4 3 ↗ 27 9 ╳ 7

Check Your Skills

For each of the following pairs of fractions, decide which fraction is greater.

4. $\dfrac{5}{8}, \dfrac{4}{7}$

5. $\dfrac{3}{10}, \dfrac{3}{13}$

Answers can be found on page 158.

From Improper Fraction to Mixed Number: Break Up the Numerator

The number $\frac{13}{4}$ is an improper fraction because $13 > 4$. How would you write $\frac{13}{4}$ as a mixed number?

First, estimate; that's often enough for to solve the rest of the problem. The fraction $\frac{13}{4}$ is pretty close to $\frac{12}{4}$, which equals 3. So the given fraction is a bit more than 3.

If you do need the exact number, what's left over after you take out that 3?

$$\frac{13}{4} = \frac{12}{4} + \frac{1}{4} = 3\frac{1}{4}$$

The $\frac{12}{4}$ equals 3 and then there's an extra $\frac{1}{4}$ left over. As a mixed number, $\frac{13}{4}$ equals $3\frac{1}{4}$.

To convert an improper fraction to a mixed number, break up the numerator to get an integer and a "left-over" fraction.

To find the integer, look for the greatest multiple of the denominator that is less than or equal to the numerator. In the case given above, 12 is the greatest multiple of 4 that is still less than 13, and since $12 = 4 \times 3$, the number 3 is the integer. Then figure out what else you need to add to get the original numerator: $12 + 1 = 13$. Take that number (1, in this case) and place it over the original denominator: $\frac{1}{4}$.

Here's another way to understand this process. Fraction addition can be done both forward and in reverse:

Forward: $\frac{2}{7} + \frac{4}{7} = \frac{2+4}{7} = \frac{6}{7}$ Reverse: $\frac{6}{7} = \frac{2+4}{7} = \frac{2}{7} + \frac{4}{7}$

In other words, you can *rewrite a numerator as a sum, then split the fraction.*

If you...	Then you...	Like this:
Want to convert an improper fraction to a mixed number	Rewrite the numerator as a sum, then split the fraction	$\frac{13}{4} = \frac{12+1}{4}$ $= \frac{12}{4} + \frac{1}{4}$ $= 3\frac{1}{4}$

Check Your Skills

Change the following improper fractions to mixed numbers.

6. $\frac{11}{6}$

7. $\frac{100}{11}$

Answers can be found on page 159.

From Mixed Number to Improper Fraction: Add

You can rewrite $5\frac{2}{3}$ as an improper fraction.

First, rewrite the mixed number as a sum: $5\frac{2}{3} = 5 + \frac{2}{3}$.

Now add these two numbers together by rewriting 5 as a fraction. You can always write any integer as a fraction by putting it over 1.

$$5 = \frac{5}{1} \qquad \text{This is true because } 5 \div 1 = 5.$$

So $5\frac{2}{3} = 5 + \frac{2}{3} = \frac{5}{1} + \frac{2}{3}$. At this point, you're adding fractions with different denominators, so find a common denominator.

The least common multiple of 1 and 3 is 3, so convert $\frac{5}{1}$ to a fraction with a 3 in its denominator:

$$\frac{5}{1} = \frac{5 \times 3}{1 \times 3} = \frac{15}{3}$$

Finally, complete the addition. Here are the steps from start to finish:

$$5\frac{2}{3} = 5 + \frac{2}{3} = \frac{5}{1} + \frac{2}{3} = \frac{15}{3} + \frac{2}{3} = \frac{15 + 2}{3} = \frac{17}{3}$$

This is essentially the reverse of the process to convert an improper fraction to a mixed number.

If you'd like, memorize this shortcut: The new numerator is $3 \times 5 + 2 = 17$. That is: (denominator) (integer part of the mixed number) + numerator = the new numerator. That formula always works! The denominator, 3, stays the same, so the complete fraction is $\frac{3 \times 5 + 2}{3} = \frac{17}{3}$.

If you understand how this shortcut is equivalent to the addition process above, you'll be better able to remember the shortcut. You start with 5, then multiply it by the denominator 3 to get a new numerator (with a common denominator) for just that first part of the fraction. Then you still need to add the numerator of the other fraction—the 2. Bingo! You've got the new numerator for the overall answer.

If you...	Then you...	Like this:
Want to convert a mixed number to an improper fraction	Convert the integer to a fraction over 1, then add it to the fractional part OR Use the shortcut: $\dfrac{(\text{denom})(\text{integer}) + \text{num}}{\text{denom}}$	$7\frac{3}{8} = \frac{7}{1} + \frac{3}{8}$ $= \frac{56}{8} + \frac{3}{8}$ $= \frac{59}{8}$ $\dfrac{8 \times 7 + 3}{8} = \dfrac{56 + 3}{8} = \dfrac{59}{8}$

Check Your Skills

Change the following mixed numbers to improper fractions.

8. $3\dfrac{3}{4}$

9. $6\dfrac{3}{4}$

Answers can be found on page 159.

Simplify a Fraction: Cancel Common Factors on the Top and Bottom

Consider this problem:

$$\frac{5}{9}+\frac{1}{9}=$$

(A) $\dfrac{4}{9}$

(B) $\dfrac{6}{18}$

(C) $\dfrac{2}{3}$

You know how to add fractions with the same denominator: $\dfrac{5}{9}+\dfrac{1}{9}=\dfrac{5+1}{9}=\dfrac{6}{9}$. This is mathematically correct. But $\dfrac{6}{9}$ is not among the answer choices. Try **simplifying** or **reducing** the fraction to its lowest terms.

To simplify a fraction, cancel out common factors from the numerator and denominator:

$$\frac{6}{9}=\frac{2\times 3}{3\times 3}\qquad \text{Since 3 is a common factor on the top and bottom, cancel it.}$$

$$\frac{6}{9}=\frac{2\times \cancel{3}}{3\times \cancel{3}}=\frac{2}{3}$$

Earlier, you learned that you can multiply the top and bottom of a fraction by the same number without changing the value of the fraction. The fraction stays the same because you are multiplying the whole fraction by the equivalent of 1. For example:

$$\frac{2}{3}=\frac{2\times 3}{3\times 3}=\frac{6}{9}$$

To simplify a fraction, you're dividing the fraction by the equivalent of 1; this also leaves the value unchanged. As you divide away the $\dfrac{3}{3}$ (which equals 1), the look of the fraction changes from $\dfrac{6}{9}$ to $\dfrac{2}{3}$, but the value of the fraction is the same.

This process works with both numbers and variables. Try reducing the following fraction:

$$\frac{18x^2}{60x}=$$

Start canceling common factors on the top and bottom. You can do so in any order. If you want, you can even break any numbers all the way down to primes, then cancel. The math below uses parentheses to indicate all the multiplication:

$$\frac{18x^2}{60x} = \frac{(2)(3)(3)(x)(x)}{(2)(2)(3)(5)(x)}$$

The top and the bottom each contain a 2, a 3, and an x. These are the common factors to cancel. Here's one way to do so:

$$\frac{18x^2}{60x} = \frac{(\cancel{2})(3)(\cancel{3})(x)(\cancel{x})}{(\cancel{2})(2)(\cancel{3})(5)(\cancel{x})} = \frac{3x}{10}$$

Here's another way to do so:

$$\frac{18x^{\cancel{2}}}{60\cancel{x}} = \frac{18x}{60} = \frac{(\cancel{6})(3x)}{(\cancel{6})(10)} = \frac{3x}{10}$$

If you feel comfortable canceling all at once, feel free to do so—but if this opens you up to careless mistakes, write the math out more fully. Here's a more compressed version:

$$\frac{\overset{3}{\cancel{18}}\, x\, \cancel{x}}{\underset{10}{\cancel{60}}\, \cancel{x}} = \frac{3x}{10}$$

When you are given larger numbers, break them down in whatever way seems easiest to you. For example, here's one way to simplify this fraction:

$$\frac{120}{16} = \frac{4 \times 30}{4 \times 4} = \frac{30}{4} = \frac{15}{2}$$

But you could also simplify this way:

$$\frac{120}{16} = \frac{\overset{3}{\cancel{12}} \times \overset{5}{\cancel{10}}}{\cancel{2} \times \cancel{2} \times \cancel{2} \times 2} = \frac{3 \times 5}{2} = \frac{15}{2}$$

If you...	Then you...	Like this:
Want to simplify a fraction	Cancel out common factors from the top and bottom	$\frac{14}{35} = \frac{2 \times \cancel{7}}{5 \times \cancel{7}} = \frac{2}{5}$

Check Your Skills

Solve the following problems.

10. Simplify $\frac{25}{40}$.

11. What is 24 divided by 16 ?

Answers can be found on page 159.

Ratios vs. Actual Numbers

It's time for another part-to-part lesson. Imagine again that you have a ratio of 2 bananas to 3 apples in your fruit basket. The ratio, by itself, does not indicate how many actual bananas and apples there really are.

In this case, there are at least 2 bananas and at least 3 apples in this basket of fruit—but there may be more. There could also be 4 bananas and 6 apples, or 6 bananas and 9 apples, or 20 bananas and 30 apples, and so on. In other words, *a ratio tells you the relationship between two things, but not necessarily the actual numbers of those two things*.

Notice what those pairs of numbers have in common? The part-to-part relationship must simplify to $\frac{2}{3}$. For example, $\frac{4}{6} = \frac{2}{3}$ and $\frac{6}{9} = \frac{2}{3}$. A ratio is similar to a fraction in its simplest form—you cannot cancel out any more common factors in the top and bottom. There are many different possible combinations of actual numbers that will simplify to $\frac{2}{3}$.

By the way, don't use single-letter abbreviations for the units *bananas* and *apples*. Instead, write something like 2 ban : 3 ap or $\frac{\text{ban}}{\text{ap}} = \frac{2}{3}$. *Never* write 2B to mean 2 bananas or 3A to mean 3 apples because you may confuse those labels with variables. Use single letters (such as B and A) to represent variables, not units. The expression 2B would mean "2 times the number of bananas."

Next, if you know that the ratio of bananas to apples is 2 to 3 and you are given one of the real numbers, then you can figure out the other real number. For instance, if there are 15 apples, then how many bananas are there?

Sketch out the information in a table:

	Part (bananas)	Part (apples)	Whole (total)
Ratio	2	3	$2 + 3 = 5$
Actual		15	

The 15 apples represent 3 parts of the ratio. Use that knowledge to calculate a new number: **the unknown multiplier**. The unknown multiplier is the number by which you multiply the *ratio* to get to the *actual* number. In this case, the unknown multiplier is 5, because $3 \times 5 = 15$:

	Part (bananas)	Part (apples)	Whole (total)
Ratio	2	3	5
× Unknown Multiplier		× 5	
= **Actual**		15	

The unknown multiplier is always the same for all parts of a ratio. Once you know what it is, apply it to everything in the ratio:

	Part (bananas)	Part (apples)	Whole (total)
Ratio	2	3	5
× Unknown Multiplier	× 5	←— × 5 —→	× 5
= **Actual**	10	15	25

Now you also know the actual number of bananas (10) and the actual total number of pieces of fruit (25). If you represent the numbers of bananas and apples as a fraction, that fraction reduces to the original ratio:

$$\frac{\overset{2}{\cancel{10}}}{\underset{3}{\cancel{15}}} = \frac{2}{3}$$

When you don't yet know what the unknown multiplier is, call it y, and write out an algebraic representation of the actual numbers:

	Part (bananas)	Part (apples)	Whole (total)
Ratio	2	3	5
×	y	y	y
Actual	$2y$	$3y$	$5y$

If there are 12 bananas, how many apples are there?

$12 = 2y =$ the actual number of bananas, so $y = 6$. The actual number of apples, therefore, is $3y = 3(6) = 18$.

Take a look at the table again. The actual number of bananas is always $2y$, and the actual number of apples is always $3y$. This formulation *guarantees* the ratio of 2 to 3:

$$\frac{\text{Number of bananas}}{\text{Number of apples}} = \frac{2y}{3y} = \frac{2\cancel{y}}{3\cancel{y}} = \frac{2}{3}$$

The unknown multiplier y is a common factor that cancels out to leave $\frac{2}{3}$, every single time.

Since the actual values will always simplify down to the same ratio, you can always set up an algebraic representation to solve. For example:

> The ratio of apples to bananas in a certain display is 4 to 7. If there are 63 bananas, how many apples are there?

The ratio of apples to bananas is $4 : 7$. The actual number of apples is $4y$, and the actual number of bananas is $7y$, where y represents the unknown multiplier. The problem states that there are 63 actual bananas, so set up an equation to solve:

$$7y = 63$$
$$y = 9$$

Now plug in the unknown multiplier to solve for the number of apples: $4y = 4(9) = 36$. There are 36 apples.

Alternatively, you can set up the table:

	Apples	Bananas	Total
R	4	7	
×	× 9	× 9	
A	36	63	

You don't need to fill out every cell in the table. Fill out only the parts you need to answer the question.

For example:

> The ratio of apples to bananas in a certain display is 4 to 7. If there are 28 apples, how many total pieces of fruit are there?

Again, you can set up an equation or use the table to solve—use what you find easier.

This time, you need the ratio total, which is equal to the sum of the individual parts of the ratio: $4 + 7 = 11$. Using y as the unknown multiplier, the actual number of apples is $4y$ and the actual number of pieces of fruit is $11y$. Therefore:

$$4y = 28$$
$$y = 7$$

Total pieces of fruit $= 11y = 11 \times 7 = 77$

Here's the table approach:

	Ap	Ban	Tot
R	4	7	11
×	× 7		× 7
A	28		77

If there are 28 apples, then there are 77 total pieces of fruit.

If you...	Then you...	Like this:			
Are told that "the ratio of sharks to dolphins is 3 to 13"	Write each quantity in terms of the unknown multiplier OR Create the table	Sharks = 3x Dolphins = 13x			
			Sh	Dol	Tot
		R	3	13	
		×			
		A	3x	13x	

Check Your Skills

12. The ratio of blue marbles to white marbles in a bag is 3 : 5. If there are 15 white marbles in the bag, how many blue marbles are in the bag?

Answer can be found on page 159.

Multiply Fractions: Simplify First, Then Multiply

What do you need to do to find $\frac{1}{2}$ of 6 ?

One-half of 6 is 3. When you take $\frac{1}{2}$ of 6, you divide 6 into two equal parts (since the denominator of $\frac{1}{2}$ is 2). Then you keep one of those two parts (since the numerator of $\frac{1}{2}$ is 1). If your starting point is 6, then that one part equals 3:

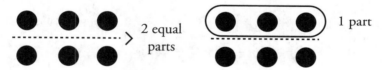

One-half *of* 6 is the same thing as one-half *times* 6. It's also the same thing as 6 divided by 2. Either way, you get 3:

$$\frac{1}{2} \times 6 = 3 \qquad\qquad \frac{6}{2} = 6 \div 2 = 3$$

Try this problem:

What is $\frac{1}{2}$ of $\frac{3}{4}$?

Since *of* means *multiply*, the question is asking this: What is $\frac{1}{2} \times \frac{3}{4}$?

To multiply two fractions, multiply the tops together and multiply the bottoms together:

$$\frac{1}{2} \times \frac{3}{4} = \frac{1 \times 3}{2 \times 4} = \frac{3}{8}$$

The rule works for integers, too. Just put the integer over 1:

$$\frac{1}{2} \times 6 = \frac{1}{2} \times \frac{6}{1} = \frac{6}{2} = 3$$

When you're asked to multiply two fractions, don't multiply immediately. Whenever possible, *cancel factors before you multiply out.* Simplify before you multiply. Take a look at these numbers; can you simplify first?

$$\frac{33}{7} \times \frac{14}{3} =$$

The terribly long way to do this multiplication is to multiply the tops, then multiply the bottoms, then reduce the numerator and denominator. You'd first have to multiply 33×14 (ugh) and then take whatever that big number is and divide it by $7 \times 3 = 21$ (double ugh).

But you *can* cancel factors before multiplying! First, when multiplying, you're allowed to combine everything into one big fraction:

$$\frac{33}{7} \times \frac{14}{3} = \frac{33 \times 14}{7 \times 3}$$

Next, break the larger numbers into smaller factors and simplify:

$$\frac{33 \times 14}{7 \times 3} = \frac{11 \times \cancel{3} \times \cancel{7} \times 2}{\cancel{7} \times \cancel{3}} = \frac{11 \times 2}{1 \times 1} = 22$$

You are always allowed to cancel across the fractions as long as the two fractions are multiplied or divided. In this circumstance, a factor on top of the first fraction can cancel with a factor on the bottom of the second fraction. (Do *not* cancel across the fractions, however, if the two fractions are added or subtracted.)

Here is another way to write out the same problem. First, cancel out a 3 from the 33 and 3. Then, cancel out a 7 from the 7 and 14:

$$\frac{33}{7} \times \frac{14}{3} = \frac{\overset{11}{\cancel{33}}}{7} \times \frac{14}{\cancel{3}} = \frac{\overset{11}{\cancel{33}}}{\cancel{7}} \times \frac{\overset{2}{\cancel{14}}}{\cancel{3}} = 11 \times 2 = 22$$

When doing multiplication or division, if a factor cancels to 1, you don't need to write it down. If *all* of the factors in the denominator cancel to 1, the answer will not be a fraction. Technically, the bottom of the calculation shown above is $1 \times 1 = 1$, but the answer $\frac{22}{1}$ simplifies to 22.

A negative sign can make fraction multiplication a little trickier. Again, a negative sign can appear anywhere in a fraction:

$$-\frac{2}{3} = \frac{-2}{3} = \frac{2}{-3}$$

When you multiply a fraction by -1, you put a negative sign in the fraction. Where you put it is up to you, though the most common place to put the negative sign is out in front of the fraction. For example:

$$-1 \times \frac{3}{5} = -\frac{3}{5} = \frac{-3}{5} = \frac{3}{-5}$$

In general, think of it as multiplying *either* the numerator *or* the denominator by -1:

$$-1 \times \frac{8}{7} = -\frac{8}{7} \text{ OR } \frac{-8}{7} \text{ OR } \frac{8}{-7}$$

If the fraction is already negative and you multiply it by -1, then cancel out both negatives, because $-1 \times -1 = 1$:

$$-1 \times -\frac{8}{7} = -1 \times -1 \times \frac{8}{7} = \frac{8}{7}$$

$$-1 \times \frac{-8}{7} = \frac{-1 \times (-8)}{7} = \frac{8}{7}$$

If you...	Then you...	Like this:
Multiply two fractions	Multiply tops and multiply bottoms, canceling common factors first	$\frac{20}{9} \times \frac{6}{5} = \frac{\overset{4}{\cancel{20}}}{\underset{3}{\cancel{9}}} \times \frac{\overset{2}{\cancel{6}}}{\cancel{5}}$ $= \frac{4 \times 2}{3} = \frac{8}{3}$

You can use the concept of multiplying by a fraction to multiply any number by 5 or divide any number by 5—without a calculator and very quickly. For example:

What is 42×5 ?

First, the number 5 could be written in fraction form as $\frac{10}{2}$. So when you're asked to multiply by 5, you can instead multiply by 10 and then divide by 2:

$$42 \times \frac{10}{2} = \frac{420}{2} = 210$$

And since multiplication and division can be done in any order, you could divide by 2 first, then multiply by 10:

$$42 \times \frac{10}{2} = 21 \times 10 = 210$$

Do the math in whatever order seems easier to you.

If you...	Then you...	Like this:
Multiply by 5	Multiply by $\frac{10}{2}$	$67 \times 5 = 67 \times \frac{10}{2}$ $= 33.5 \times 10 = 335$

Check Your Skills

Solve the following problems. Simplify all fractions.

13. What is $\dfrac{3}{10} \times \dfrac{6}{7}$?
14. What is 91×5?

Answers can be found on page 160.

Square a Proper Fraction: It Gets Smaller

What do you need to do to find $\left(\dfrac{1}{2}\right)^2$?

Now that you can multiply fractions, you can apply exponents:

$$\left(\dfrac{1}{2}\right)^2 = \dfrac{1}{2} \times \dfrac{1}{2} = \dfrac{1}{4}$$

This is interesting: $\dfrac{1}{4}$ is *less* than $\dfrac{1}{2}$.

When you square a number greater than 1, it gets bigger. When you square a negative number, it also gets bigger (because it becomes positive). But when you square a proper fraction (i.e., a fraction between 0 and 1), it gets *smaller*.

The same holds true for greater powers (cubes, etc.). *If you square, cube, or apply any greater power to a fraction between 0 and 1, the number will get smaller.*

In general, if you multiply *any* positive number by a proper fraction, the result is smaller than the original number. You are taking a *fraction* of that number. For instance, if you take $\dfrac{1}{2}$ of 3, you will get $\dfrac{1}{2} \times 3 = \dfrac{3}{2}$. The answer is smaller than the starting number, 3.

By the way, what happens if you square 0 or 1?

You get the same number! $0^2 = 0$ and $1^2 = 1$.

If you...	Then you...	Like this:
Square a proper fraction (between 0 and 1)	Get a smaller number	$\left(\dfrac{1}{3}\right)^2 = \dfrac{1}{9}$ $\dfrac{1}{3} > \dfrac{1}{9}$
Square a number greater than 1 or square a negative number	Get a greater number	$2^2 = 4$ $(-3)^2 = 9$
Square 0 or 1	Get the same number	$0^2 = 0$ $1^2 = 1$

Take a Reciprocal: Flip the Fraction

The **reciprocal** of an integer is "1 over" that number. For example, the reciprocal of 5 is 1 over 5, or $\frac{1}{5}$.

Any number times its reciprocal equals 1:

$$5 \times \frac{1}{5} = \frac{5}{1} \times \frac{1}{5} = \frac{5}{5} = 1 \qquad \text{(You could also cancel all the factors as you multiply.)}$$

Consider this example:

What is the reciprocal of $\frac{2}{3}$?

To find the reciprocal of a fraction, *flip the fraction*. The reciprocal of $\frac{2}{3}$ is $\frac{3}{2}$, since the product of $\frac{2}{3}$ and $\frac{3}{2}$ is 1:

$$\frac{2}{3} \times \frac{3}{2} = \frac{6}{6} = 1$$

If you write an integer as a fraction over 1, then the flipping rule works for integers as well. The integer 9 is $\frac{9}{1}$, and the reciprocal of 9 is $\frac{1}{9}$.

Keep track of negative signs. The reciprocal of a negative fraction will also be negative, since the two fractions need to multiply to positive 1. For example:

$$\frac{-5}{6} \times \frac{6}{-5} = \frac{-30}{-30} = 1$$

The reciprocal of $-\frac{5}{6}$ is $-\frac{6}{5}$.

If you...	Then you...	Like this:
Want the reciprocal of a fraction	Flip the fraction	Fraction　　Reciprocal $\frac{4}{7} \longrightarrow \frac{7}{4}$ $\frac{4}{7} \times \frac{7}{4} = 1$

Divide by a Fraction: Multiply by the Reciprocal

What do you need to do to find $6 \div 2$?

Interestingly, $6 \div 2$ gives the same result as $6 \times \frac{1}{2}$:

$$6 \div 2 = 6 \times \frac{1}{2} = 3 \qquad \text{Dividing by 2 is the same as multiplying by } \frac{1}{2}.$$

The numbers 2 and $\frac{1}{2}$ are reciprocals of each other. This pattern generalizes: *Dividing by a number is the same as multiplying by its reciprocal.* Try this problem:

What is $\frac{5}{6} \div \frac{4}{7}$?

First, find the reciprocal of the second fraction (the one you're dividing *by*). Then multiply the first fraction by that reciprocal:

$$\frac{5}{6} \div \frac{4}{7} = \frac{5}{6} \times \frac{7}{4} = \frac{35}{24}$$

Sometimes, you see a double-decker fraction. This is a fancy way of showing one fraction divided by another. The longer fraction bar is the primary division; the fraction under that bar is the one to flip:

$$\frac{\frac{5}{6}}{\frac{4}{7}} = \frac{5}{6} \div \frac{4}{7} = \frac{5}{6} \times \frac{7}{4} = \frac{35}{24}$$

This works with variables as well. Flip the bottom fraction and multiply:

$$\frac{\frac{3}{x}}{\frac{5}{x}} = \frac{3}{x} \times \frac{x}{5} = \frac{3}{\cancel{x}} \times \frac{\cancel{x}}{5} = \frac{3}{5}$$

As always, dividing by 0 is forbidden, so the problem above would specify that x cannot equal 0.

Remember that trick about multiplying by 5? What if you want to divide by 5 instead? Here's how:

What is $42 \div 5$?

Dividing by 5 is the same thing as dividing by $\frac{10}{2}$. To divide by a fraction, multiply by its reciprocal: $\frac{2}{10}$. So when you are asked to divide by 5, you can instead multiply the number by 2 and then divide by 10:

$$42 \times \frac{2}{10} = \frac{84}{10} = 8.4$$

You can again do the math in whatever order you like:

$$42 \times \frac{2}{10} = 4.2 \times 2 = 8.4$$

If you...	Then you...	Like this:
Divide something by a fraction	Multiply by that fraction's reciprocal	$\frac{3}{2} \div \frac{7}{11} = \frac{3}{2} \times \frac{11}{7}$
Divide by 5	Multiply by $\frac{2}{10}$	$63 \div 5 = 63 \times \frac{2}{10}$ $= \frac{126}{10} = 12.6$

Check Your Skills

Evaluate and simplify the following expressions.

15. $\frac{1}{6} \div \frac{1}{11}$

16. What is $183 \div 5$?

Answers can be found on page 160.

Addition and Subtraction: Pull Out a Common Factor

If a fraction contains addition or subtraction in the numerator or denominator, you may be able to pull out a common factor and cancel it out.

The fraction bar always tells you to *divide the entire numerator by the entire denominator*. To respect PEMDAS, think of the fraction bar as a grouping symbol, like parentheses. For example:

$$\frac{2x^2 - 6}{2y^2 - 4z} = \left(2x^2 - 6\right) \div \left(2y^2 - 4z\right)$$

That math looks messy. Here's a cleaner example first:

$$\frac{9x - 6}{3x}$$

The entire quantity $9x - 6$ is divided by $3x$. In other words, you have $(9x - 6) \div 3x$.

To simplify $\frac{9x - 6}{3x}$, *find a common factor of every term in the numerator and every term in the denominator.* That is, find a common factor that you can pull out of both the $9x$ *and* the 6 in the numerator, as well as the $3x$ in the denominator.

What factor does $3x$ have in common with the quantity $9x - 6$? Notice that x is not a common factor, because you can't pull it out of the *entire* numerator; the 6 does not contain an x. But you can pull a 3 out because both 9 and 6 have 3 as a factor.

In the numerator, pull out a 3:

$$\frac{9x - 6}{3x} = \frac{3(3x - 2)}{3x}$$

Now, the numerator and denominator have a common factor. Cancel it out of the top and bottom:

$$\frac{9x - 6}{3x} = \frac{\cancel{3}(3x - 2)}{\cancel{3}x} = \frac{3x - 2}{x}$$

The common factor could include a variable. For example:

$$\frac{9y^2 - 6y}{12y} = \frac{\cancel{3y}(3y - 2)}{\cancel{3y}(4)} = \frac{3y - 2}{4}$$

If you feel comfortable simplifying without explicitly pulling out the common factor first, you can do so. Remember to cancel from each separate term, $9y^2$, $6y$, and $12y$:

$$\frac{9y^2 - 6y}{12y} = \frac{\overset{3}{\cancel{9}}y^{\cancel{2}} - \overset{2}{\cancel{6}}\cancel{y}}{\underset{4}{\cancel{12}}\cancel{y}} = \frac{3y - 2}{4}$$

If you find yourself making too many careless mistakes that way, though, then pull out the common factor before you cancel.

Finally, you can pull out a common factor in both the numerator and the denominator. Here is the complicated example from the beginning of this section:

$$\frac{2x^2 - 6}{2y^2 - 4z} = \frac{\cancel{2}\left(x^2 - 3\right)}{\cancel{2}\left(y^2 - 2z\right)} = \frac{x^2 - 3}{y^2 - 2z}$$

If you...	Then you...	Like this:
Have addition or subtraction in the numerator (or denominator, not shown)	Pull out a factor from the entire numerator and cancel that factor with the same one in the denominator	$\dfrac{5x + 10y}{25y} = \dfrac{\cancel{5}(x + 2y)}{\cancel{5}(5y)}$ $= \dfrac{x + 2y}{5y}$

Check Your Skills

Simplify the following expression.

17. $\dfrac{4x^2 + 20xy}{12x}$

$\dfrac{4x(x + 5y)}{4x(3)} \rightarrow \dfrac{x + 5y}{3}$

Answer can be found on page 160.

Addition in the Numerator: Split into Two Fractions

After you've canceled common factors, you still might not see your answer among the answer choices. In that case, you can try one more thing. Remember this math when changing an improper fraction to a mixed number?

$$\frac{13}{4} = \frac{12 + 1}{4} = \frac{12}{4} + \frac{1}{4}$$

If you have a sum in the numerator, you can rewrite the fraction as the sum of two fractions. The same is true if you have a difference—that is, subtraction. (Note: You cannot do this if the addition or subtraction is in the denominator. You'll learn why in the next section.)

Consider this example, with two possible sets of answer choices:

$$\frac{9x - 6}{3x} =$$

First set:

(A) $3x - 6$

(B) $\dfrac{3x - 2}{x}$

(C) $\dfrac{3x - 6}{x}$

Second set:

(A) $3x - 6$

(B) $3 - \dfrac{2}{x}$

(C) $3 - \dfrac{6}{x}$

On any Problem Solving problem, always glance at the answers before starting to solve. The form of the answers will often provide clues about the best way to solve.

In the first set of answers, answers (B) and (C) are all in one big fraction, so splitting out the numerator into two fractions is not the way to go. In the second set of answers, however, none are in the one-big-fraction form, so here, you do want to split out the numerator into two fractions.

Either way, the first step is still to cancel common factors from the numerator and denominator:

$$\frac{9x - 6}{3x} = \frac{\cancel{3}(3x - 2)}{\cancel{3}x} = \frac{3x - 2}{x}$$

If you were given the first set of answers, you'd be done now—the answer is (B). But if you were given the second set, go further by splitting the fraction into two fractions:

$$\frac{3x - 2}{x} = \frac{3x}{x} - \frac{2}{x}$$

Now, simplify the first fraction further by canceling the common factor of x on the top and bottom. Here's the full math:

$$\frac{9x - 6}{3x} = \frac{\cancel{3}(3x - 2)}{\cancel{3}x} = \frac{3x - 2}{x} = \frac{3\cancel{x}}{\cancel{x}} - \frac{2}{x} = 3 - \frac{2}{x}$$

This matches answer (B) of the second set. Is $3 - \frac{2}{x}$ simpler than $\frac{3x - 2}{x}$? Not really; it's just a different form. The form you want is the form that matches the particular answer choices you happened to be given for that problem—so train yourself to glance at the answers *before* you start solving and *while* you're solving. The form of the answers will guide your work.

Try this problem involving square roots:

$$\frac{10\sqrt{2} + \sqrt{6}}{2\sqrt{2}} =$$

(A) $\dfrac{5 + \sqrt{6}}{2}$

(B) $5 + \dfrac{\sqrt{6}}{2}$

(C) $5 + \dfrac{\sqrt{3}}{2}$

In this case, it's hard to spot a common factor in the numerator that will cancel with one in the denominator. So, if you're stuck, try splitting the fraction in two:

$$\frac{10\sqrt{2} + \sqrt{6}}{2\sqrt{2}} = \frac{10\sqrt{2}}{2\sqrt{2}} + \frac{\sqrt{6}}{2\sqrt{2}}$$

Now deal with the two fractions separately. Cancel a $\sqrt{2}$ out of the top and bottom of the first fraction:

$$\frac{10\cancel{\sqrt{2}}}{2\cancel{\sqrt{2}}} = \frac{10}{2} = 5$$

The second fraction, $\dfrac{\sqrt{6}}{2\sqrt{2}}$, is trickier. Cast your mind back to the Exponents and Roots chapter. Better yet, go browse right now to look for something you could use in this situation; if you can find it yourself, you'll remember it better later.

When you divide roots, you can combine the numbers under one square root sign. For example:

$$\frac{\sqrt{9}}{\sqrt{3}} = \sqrt{\frac{9}{3}} = \sqrt{3}$$

Apply that rule to the slightly more complex fraction in the given problem. Separate out the extra 2 on the bottom. Introduce a factor of 1 on top as a temporary placeholder:

$$\frac{\sqrt{6}}{2\sqrt{2}} = \frac{1 \times \sqrt{6}}{2 \times \sqrt{2}} = \frac{1}{2} \times \sqrt{\frac{6}{2}} = \frac{1}{2} \times \sqrt{3} = \frac{\sqrt{3}}{2}$$

Put the two parts together: $5 + \dfrac{\sqrt{3}}{2}$. The answer is (C).

If you...	Then you...	Like this:
Have addition or subtraction in the numerator	Might split the fraction into two fractions	$\dfrac{a+b}{c} = \dfrac{a}{c} + \dfrac{b}{c}$

Check Your Skills

18. $\dfrac{x+y}{xy}$ is equivalent to which of the following for all nonzero values of x and y?

(A) $\dfrac{1}{x} + \dfrac{1}{y}$

(B) $\dfrac{1+y}{y}$

(C) $\dfrac{x+1}{x}$

Answer can be found on page 160.

Addition in the Denominator: Pull Out a Common Factor

To simplify a fraction with addition or subtraction in the *denominator*, you can do one of the same things as before, but not both. You can *pull out a common factor from the denominator, and cancel with a factor in the numerator.* Do not, however, split into two fractions. For example:

$$\frac{4x}{8x - 12} =$$

You can factor a 4 out of $8x - 12$ and cancel it with the 4 in the numerator:

$$\frac{4x}{8x - 12} = \frac{4x}{4(2x - 3)} = \frac{\cancel{4}x}{\cancel{4}(2x - 3)} = \frac{x}{2x - 3}$$

That's all legal so far. But you *cannot* go any further. *Never split a fraction in two because of addition or subtraction in the denominator.* Consider this example:

Is $\dfrac{1}{3 + 4}$ equal to $\dfrac{1}{3} + \dfrac{1}{4}$?

Try it! The fraction $\dfrac{1}{3 + 4} = \dfrac{1}{7}$. But the fraction $\dfrac{1}{3} + \dfrac{1}{4} = \dfrac{7}{12}$. Those are not the same value.

Do not split $\dfrac{x}{2x - 3}$ into anything else. That's as far as you can go.

If you...	Then you...	Like this:
Have addition or subtraction in the denominator	Pull out a factor from the entire denominator and cancel that factor with one in the numerator... but *never* split the fraction in two!	$\dfrac{3y}{y^2 + xy} = \dfrac{3\cancel{y}}{\cancel{y}(y + x)}$ $= \dfrac{3}{y + x}$

Check Your Skills

19. $\dfrac{5a^3}{15ab^2 - 5a^3}$ is equivalent to which of the following?

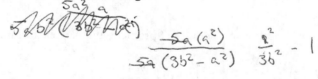

(A) $\dfrac{a^2}{3b^2} - 1$

(B) $\dfrac{a^2}{3b^2 - a^2}$

(C) $\dfrac{1}{15ab^2}$

Answer can be found on page 160.

Add, Subtract, Multiply, Divide Ugly Fractions: Add Parentheses

Complicated fractions, such as $\dfrac{4x}{8x - 12}$, can be a headache, but they follow the same rules of addition, subtraction, multiplication, and division as do all other fractions:

- **Addition:** Use the double-cross, or find a common denominator, then add numerators.
- **Subtraction:** Use the double-cross, or find a common denominator, then subtract numerators.
- **Multiplication:** Cancel common factors, then multiply tops and multiply bottoms.
- **Division:** Flip, then multiply.

With complicated fractions, the most important point to remember is this: *Treat the numerators and denominators as if they have parentheses around them.* This preserves the order of operations (PEMDAS).

Consider this sum:

$$\frac{1}{y+1} + \frac{2}{y} =$$

[handwritten:] $\dfrac{2(y+1) + y}{y(y+1)} = \dfrac{3y+2}{y^2+y}$

The same principle of addition holds. Do these fractions have the same denominator? No. So use the double-cross to add:

$$\underset{y+1}{\overset{(1)(y)}{1}} \overset{+}{\underset{}{}} \underset{y}{\overset{(2)(y+1)}{2}} = \frac{y + (2y+2)}{y(y+1)} = \frac{3y+2}{y(y+1)}$$

You could also write the answer as $\dfrac{3y+2}{y^2 + y}$.

Consider this product:

$$\left(\frac{2w+4}{z^3 + z}\right)\left(\frac{z}{2}\right) =$$

You could just multiply the tops and multiply the bottoms, but don't forget to cancel common factors wherever possible *before* you multiply. Start by pulling out factors from the ugly fraction on the left:

$$\frac{2w+4}{z^3 + z} = \frac{2(w+2)}{z\left(z^2 + 1\right)}$$

Now plug that new form back into the original problem and cancel common factors:

$$\left(\frac{2(w+2)}{z\left(z^2+1\right)}\right)\left(\frac{z}{2}\right) = \left(\frac{\cancel{2}(w+2)}{\cancel{z}\left(z^2+1\right)}\right)\left(\frac{\cancel{z}}{\cancel{2}}\right) = \frac{w+2}{z^2+1}$$

If you...	Then you...	Like this:
Add, subtract, multiply, or divide fractions with complicated numerators and/or denominators	Throw parentheses around those numerators and/or denominators, then proceed normally—find common denominators, cancel common factors, etc.	$\dfrac{3}{m+2} - \dfrac{2}{m} = \underset{m+2}{\overset{(3)(m)}{3}} \overset{-}{} \underset{m}{\overset{(2)(m+2)}{2}}$ $= \dfrac{3m - (2m+4)}{m(m+2)}$ $= \dfrac{m-4}{m(m+2)}$

Check Your Skills

20. $\dfrac{x+1}{x-1} - \dfrac{3}{4} =$

[handwritten:] $\dfrac{4x+4 - (3(x-1))}{4x-4} \longrightarrow \dfrac{x+7}{4x-4}$

Answer can be found on page 161.

Fractions within Fractions: Work Your Way Out

Remember double-decker fractions in fraction division?

$$\frac{\frac{5}{6}}{\frac{4}{7}} = \frac{5}{6} \div \frac{4}{7} = \frac{5}{6} \times \frac{7}{4} = \frac{35}{24}$$

When you see a fraction within a fraction, first glance at the answers. How far apart are they? This gives you an idea of how aggressively you can estimate. Can you tell whether the answer will be positive or negative? If positive, is it greater than one or less than one?

Try this problem:

$$\frac{1}{1+\frac{1}{3}} =$$

Forget about the entire expression for a moment. Just focus on the deepest level: the fraction within the fraction. In this case, the denominator is $1 + \frac{1}{3}$. This value is a bit greater than 1.

Dividing 1 (the numerator) by something greater than 1 will give you a positive number less than 1. The answer, then, has to be between 0 and 1. Depending on the mix of answers, that could be enough to solve.

If you do need to find a more precise number, again start with whatever part has the fraction within the fraction:

$$1 + \frac{1}{3} = \frac{3}{3} + \frac{1}{3} = \frac{4}{3}$$

Now plug that back into the original expression:

$$\frac{1}{1+\frac{1}{3}} = \frac{1}{\frac{4}{3}}$$

Dividing by a fraction is the same as multiplying by the reciprocal of that fraction. Here's how that works:

$$\frac{1}{\frac{4}{3}} = \frac{1}{1} \div \frac{4}{3} = \frac{1}{1} \times \frac{3}{4} = \frac{3}{4}$$

That's the answer (and notice that it is, indeed, between 0 and 1). In fact, if the numerator of the fraction is plain 1, then you can take a great shortcut. Literally just flip the fraction in the denominator and you're done:

$$\frac{1}{\frac{4}{3}} = \frac{3}{4}$$

Here's what happens if the numerator is not a plain 1:

$$\frac{3}{\frac{1}{4}} = 3 \div \frac{1}{4} = \frac{3}{1} \times \frac{4}{1} = 12$$

As before, you can turn the problem into multiplication by flipping the fraction from the denominator.

Try another one, this time a three-level problem:

$$\cfrac{1}{2+\cfrac{1}{3+\cfrac{1}{4}}} =$$

Estimate first. At the deepest level, $3 + \frac{1}{4}$ is a little more than 3. The fraction $\cfrac{1}{3+\frac{1}{4}}$ is approximately $\frac{1}{3}$.

The next level is approximately $2 + \frac{1}{3}$, or a bit more than 2. The overall fraction, then, is $\cfrac{1}{\text{a bit more than 2}}$, or a bit *less* than $\frac{1}{2}$. So the correct answer should be between 0 and $\frac{1}{2}$.

Why *less* than $\frac{1}{2}$? The value $\cfrac{1}{\text{a bit more than 2}}$ is somewhere between $\frac{1}{2}$ and $\frac{1}{3}$.

To do the full math, start at the deepest level: $3 + \frac{1}{4}$. Turn this into a mixed fraction: $3 + \frac{1}{4} = \frac{12}{4} + \frac{1}{4} = \frac{13}{4}$.

Now move up a level:

$$\cfrac{1}{2+\cfrac{1}{3+\cfrac{1}{4}}} = \cfrac{1}{2+\cfrac{1}{\cfrac{13}{4}}}$$

The next part to tackle is again that fraction-within-a-fraction portion. Since it's a plain 1 over a fraction, you can just flip the fraction:

$$\cfrac{1}{2+\cfrac{1}{3+\cfrac{1}{4}}} = \cfrac{1}{2+\cfrac{1}{\cfrac{13}{4}}} = \cfrac{1}{2+\cfrac{4}{13}}$$

Next, add the two terms in the bottom part:

$$2 + \frac{4}{13} = \frac{26}{13} + \frac{4}{13} = \frac{30}{13}$$

Finally, replace that in the original fraction. You've almost reached the surface:

$$\cfrac{1}{2+\cfrac{1}{3+\cfrac{1}{4}}} = \cfrac{1}{2+\cfrac{1}{\cfrac{13}{4}}} = \cfrac{1}{2+\cfrac{4}{13}} = \cfrac{1}{\cfrac{30}{13}}$$

Finally, you have another 1 divided by a fraction. Take the reciprocal:

$$\cfrac{1}{\cfrac{30}{13}} = \frac{13}{30}$$

That was a lot of steps! If you can do that problem, then you can tackle any fraction within a fraction that you might see on the real test. (But estimate first; that might be enough to get you to the correct answer without doing the long version of the math.)

If you...	Then you...	Like this:
Encounter a fraction within a fraction	Estimate first; if that's not enough, work your way out from the deepest level inside	$$\cfrac{1}{y+\cfrac{1}{2-\dfrac{3}{y}}}$$ Focus here

Check Your Skills

21. $\dfrac{1+\dfrac{3}{4}}{2} =$ $\dfrac{7}{4}$ $\dfrac{7}{8}$

Answer can be found on page 161.

4

Check Your Skills Answer Key

1. $\dfrac{10x}{yz^2}$: The denominator stays the same. Add the numerators.

$$\frac{3x}{yz^2} + \frac{7x}{yz^2} = \frac{3x + 7x}{yz^2} = \frac{10x}{yz^2}$$

2. $\dfrac{5}{4}$ **or** $1\dfrac{1}{4}$: The denominator 4 is already a multiple of the denominator 2. Because only one of the fractions needs to change, use the traditional method to solve this one.

$$\frac{1}{2} + \frac{3}{4} = \frac{1 \times 2}{2 \times 2} + \frac{3}{4} = \frac{2}{4} + \frac{3}{4} = \frac{2 + 3}{4} = \frac{5}{4}$$

3. $\dfrac{7}{24}$: This time, the 3 and the 8 don't share any factors, so the double-cross will likely be more efficient.

$$^{16}\diagdown\frac{2}{3}\underset{\diagdown}{\overset{-}{\diagup}}\frac{3}{8}^{9}\diagup = \frac{7}{24}$$

Here is the traditional approach:

$$\frac{2}{3} - \frac{3}{8} = \frac{2 \times 8}{3 \times 8} - \frac{3 \times 3}{8 \times 3} = \frac{16}{24} - \frac{9}{24} = \frac{16 - 9}{24} = \frac{7}{24}$$

4. $\dfrac{5}{8}$: Use the double-cross.

$$\frac{5}{8} \qquad \frac{4}{7}$$

$$\text{(35)} \longleftarrow \frac{5}{8} \diagup\diagdown \frac{4}{7} \longrightarrow 32$$

The greater number, 35, is next to the greater fraction, $\dfrac{5}{8}$.

5. $\dfrac{3}{10}$: The numerators of the two fractions are the same, but the denominator of $\dfrac{3}{10}$ is smaller, so $\dfrac{3}{10} > \dfrac{3}{13}$. Alternatively, use the double-cross.

$$^{39}\diagdown\frac{3}{10}\diagdown\diagup\frac{3}{13}^{30}\diagup$$

39 is greater than 30, so $\dfrac{3}{10}$ is the greater fraction.

6. $1\dfrac{5}{6}$: Split the numerator into two parts and simplify.

$$\frac{11}{6} = \frac{6+5}{6} = \frac{6}{6} + \frac{5}{6} = 1 + \frac{5}{6} = 1\frac{5}{6}$$

7. $9\dfrac{1}{11}$: Think of multiples of 11. How close can you get to 100 without going over? $11 \times 9 = 99$, with 1 left over.

$$\frac{100}{11} = \frac{99+1}{11} = \frac{99}{11} + \frac{1}{11} = 9 + \frac{1}{11} = 9\frac{1}{11}$$

8. $\dfrac{15}{4}$: The shortcut is $\dfrac{4 \times 3 + 3}{4} = \dfrac{12+3}{4} = \dfrac{15}{4}$. Here's the full path:

$$3\frac{3}{4} = 3 + \frac{3}{4} = \frac{3}{1} \times \frac{4}{4} + \frac{3}{4} = \frac{12}{4} + \frac{3}{4} = \frac{15}{4}$$

9. $\dfrac{27}{4}$: The shortcut is $\dfrac{6 \times 4 + 3}{4} = \dfrac{24+3}{4} = \dfrac{27}{4}$. Here's the full path:

$$6\frac{3}{4} = 6 + \frac{3}{4} = \frac{6}{1} \times \frac{4}{4} + \frac{3}{4} = \frac{24}{4} + \frac{3}{4} = \frac{27}{4}$$

10. $\dfrac{5}{8}$: Find common factors and cancel them.

$$\frac{25}{40} = \frac{5 \times 5}{8 \times 5} = \frac{5 \times \cancel{5}}{8 \times \cancel{5}} = \frac{5}{8}$$

11. $\dfrac{3}{2}$: First, write the question in fraction form. Then, simplify.

$$\frac{24}{16} = \frac{3 \times 8}{2 \times 8} = \frac{3 \times \cancel{8}}{2 \times \cancel{8}} = \frac{3}{2}$$

12. **9:** The ratio of blue : white is 3 : 5. There are 15 white marbles, so set the 15 equal to $5x$ and solve for the unknown multiplier.

$$5x = 15$$
$$x = 3$$

Therefore, the number of blue marbles is $3x$, which is $3(3) = 9$.

Alternatively, use the table to solve.

	B	W	Total
R	3	5	
×	× 3	× 3	
A	9	15	

13. $\dfrac{9}{35}$: Simplify before you multiply.

$$\frac{3}{10} \times \frac{6}{7} = \frac{3}{\cancel{10}_5} \times \frac{\cancel{6}^3}{7} = \frac{9}{35}$$

14. **455:** Multiply by 10 over 2.

$$91 \times 5 = 91 \times \frac{10}{2} = 45.5 \times 10 = 455$$

You could also do this: $(90 + 1)5 = (90)5 + 1(5) = 450 + 5 = 455$.

15. $\dfrac{11}{6}$: To divide by a fraction, multiply by its reciprocal.

$$\frac{1}{6} \div \frac{1}{11} = \frac{1}{6} \times \frac{11}{1} = \frac{11}{6}$$

16. **36.6:** To divide by 5, multiply by $\dfrac{2}{10}$.

$$183 \div 5 = 183 \times \frac{2}{10} = \frac{366}{10} = 36.6$$

To multiply 183 by 2 without a calculator, break 183 into $180 + 3$. Multiply each part by 2: $(180)2 = 360$ and $(3)2 = 6$. Add up the two parts: $360 + 6 = 366$.

17. $\dfrac{x+5y}{3}$: Pull out a common term and cancel.

$$\frac{4x^2 + 20xy}{12x} = \frac{\cancel{4x}\,(x + 5y)}{\cancel{4x}\,(3)} = \frac{x + 5y}{3}$$

18. **(A)** $\dfrac{1}{x} + \dfrac{1}{y}$: Split the fraction into two and simplify.

$$\frac{x + y}{xy} = \frac{\cancel{x}}{\cancel{x}\,y} + \frac{\cancel{y}}{x\,\cancel{y}} = \frac{1}{y} + \frac{1}{x}$$

19. **(B)** $\dfrac{a^2}{3b^2 - a^2}$: Don't split the denominator!

$$\frac{5a^3}{15ab^2 - 5a^3} = \frac{\cancel{5}\,a^{\cancel{3}^2}}{\cancel{5}\,\cancel{a}\,\left(3b^2 - a^2\right)} = \frac{a^2}{3b^2 - a^2}$$

20. $\frac{x+7}{4x-4}$: Find a common denominator, then simplify.

$$\frac{x+1}{x-1} - \frac{3}{4} = \frac{x+1}{x-1} - \frac{3}{4} = \frac{(4x+4)-(3x-3)}{4(x-1)} = \frac{x+7}{4x-4}$$

21. $\frac{7}{8}$: Eyeball this first. The denominator is 2 and the numerator is a bit less than 2, so the answer should be a bit less than 1. If you had multiple-choice answers, you could cross off any that are greater than 1 right now. To do the math, work your way out from the fraction-within-a-fraction.

$$\frac{1+\frac{3}{4}}{2} = \frac{\frac{4}{4}+\frac{3}{4}}{2} = \frac{\frac{7}{4}}{2} = \frac{7}{4} \times \frac{1}{2} = \frac{7}{8}$$

Chapter Review: Drill Sets

Drill 1

1. Fill in the missing information in the table below.

1 : 2	=	3 : __	=	__ : 14	=	__ : 22
1 : __	=	4 : 20	=	__ : 25	=	15 : __
3 : __	=	__ : 8	=	__ : 36	=	33 : 44
__ : 7	=	20 : __	=	40 : 56	=	60 : __
4 : 11	=	__ : 22	=	36 : __	=	__ : 132

Drill 2

For each of the following pairs of fractions, decide which fraction is greater.

2. $\frac{1}{4}, \frac{3}{4}$

3. $\frac{1}{5}, \frac{1}{6}$

4. $\frac{53}{52}, \frac{85}{86}$

5. $\frac{7}{9}, \frac{6}{10}$

6. $\frac{700}{360}, \frac{590}{290}$

Drill 3

7. If there are 20 birds and 6 dogs in a park, which of the following is the ratio of dogs to birds?

    (A) 3 : 13

    (B) 3 : 10

    (C) 10 : 3

8. In a class of 24 students made up of only juniors and seniors, 12 are juniors. Which of the following is the ratio of juniors to seniors in the class?

    (A) 1 : 1

    (B) 1 : 2

    (C) 2 : 1

9. If there are 45 red marbles and 35 green marbles in a bag, which of the following is the ratio of green to red marbles?

    (A) 9 : 16

    (B) 7 : 9

    (C) 9 : 7

10. There are 21 trout and 24 catfish in a pond. There are no other fish in the pond. What is the ratio of catfish to the total number of fish in the pond?

    (A) 7 : 15

    (B) 8 : 15

    (C) 7 : 8

Drill 4

Add or subtract the following fractions. Put answers in their most simplified form. For any problems with variables in the denominator, assume that the denominator does not equal zero.

11. $\frac{4}{9} + \frac{8}{11}$

12. $\frac{20}{12} - \frac{5}{3}$

13. $\frac{a}{12} - \frac{b}{6} - \frac{b}{4}$

14. $\sqrt{\frac{7}{5}} - \sqrt{\frac{5}{7}}$

    (A) $\frac{2}{\sqrt{35}}$

    (B) $\sqrt{35}$

    (C) $\frac{\sqrt{2}}{\sqrt{35}}$

15. $\dfrac{x^2z}{yz} + \dfrac{x^2z}{xy} - \dfrac{3xz}{y}$

(A) $\dfrac{x^2 - 4xz}{y}$

(B) $\dfrac{x(x + 2z)}{y}$

(C) $\dfrac{2x^2z + 3xz}{xyz}$

16. $\dfrac{24}{3\sqrt{2}} - \dfrac{4}{\sqrt{2}}$

(A) $2\sqrt{2}$

(B) 4

(C) $8\sqrt{2}$

Drill 5

Convert: If given an improper fraction, convert to a mixed number; if given a mixed number, convert to an improper fraction.

17. $\dfrac{47}{15}$

18. $\dfrac{72}{12}$

19. $6\dfrac{3}{7}$

Drill 6

Simplify the following expressions. For any problems with variables in the denominator, assume that the denominator does not equal zero.

20. $\dfrac{2\sqrt{18}}{15}$

21. $\dfrac{17^2 \times 22}{11 \times 34}$

22. $\dfrac{2r\sqrt{54}}{r^2s\sqrt{12}}$

23. $\dfrac{6x^8yz^5}{46x^6y^2z^3}$

24. If $a > 0$, $\dfrac{3ab^2\sqrt{50}}{\sqrt{18a^2}} =$

Drill 7

Multiply or divide the following fractions. Put any resulting fractions in their most simplified form. For any problems with variables in the denominator, assume that the denominator does not equal zero.

25. $\dfrac{6}{25} \div \dfrac{9}{10}$

26. $\dfrac{3}{11} \div \dfrac{3}{11}$

27. $\dfrac{\sqrt{12}}{5} \times \dfrac{\sqrt{60}}{2^4} \times \dfrac{\sqrt{45}}{3^2}$

28. $\dfrac{\sqrt{18}}{\sqrt{4}} \div \dfrac{\sqrt{9}}{\sqrt{18}}$

29. $\dfrac{xy^3z^4}{x^3y^4z^2} \div \dfrac{x^6y^3z}{x^3y^5z^2}$

30. $\dfrac{12^2}{9^2} \div \dfrac{6^3}{3^5}$

Drill 8

Simplify the following fractions. For any problems with variables in the denominator, assume that the denominator does not equal zero.

31. $\dfrac{9a + 4b}{3ab}$

(A) $\dfrac{13}{3}$

(B) $\dfrac{3a + 4b}{ab}$

(C) $\dfrac{3}{b} + \dfrac{4}{3a}$

32. $\dfrac{6a}{33a + 21ab}$

33. $\dfrac{8x^2 + 40x}{32x - 24x^2}$

Drill 9

Simplify the following expressions. Simplify any fractions fully, but it is not necessary to convert improper fractions into mixed fractions. For any problems with variables in the denominator, assume that the denominator does not equal zero.

34. $\dfrac{3+4}{1+2} - \dfrac{1+2}{3+4}$

35. $\dfrac{-t+1}{t-2} \times \dfrac{-t}{2}$

36. $\dfrac{b+6}{6} - \dfrac{3+b}{6}$

37. $\dfrac{3x^2+3y}{40} + \dfrac{x^2+y}{8}$

40. $\dfrac{8}{2-\frac{2}{3}}$

   (A) 6

   (B) 4

   (C) 3

41. $\dfrac{1}{1-\frac{2}{y+1}}$

   (A) $\dfrac{y+3}{y+1}$

   (B) $\dfrac{y}{y-1}$

   (C) $\dfrac{y+1}{y-1}$

Drill 10

Match the following expressions to their simplified forms. For any problems with variables in the denominator, assume that the denominator does not equal zero.

38. $\dfrac{x+3}{15} \times \dfrac{10}{x+3}$

   (A) $\dfrac{2}{3}$

   (B) $2x+6$

   (C) $\dfrac{2(x+3)}{3}$

39. $\dfrac{(n+2n)}{n^4} \times \dfrac{(2n)^2}{(15n-5n)}$

   (A) $\dfrac{2n^2+4n^3}{15n^4-5n^5}$

   (B) $\dfrac{6}{5n^2}$

   (C) $\dfrac{n^7}{n^{16}}$

Drill Sets Solutions

Drill 1

1. If the table does not give you the most basic form of the ratio, find that basic form first. Use it to find the other answers.

1 : 2	=	3 : **6**	=	**7** : 14	=	**11** : 22
1 : **5**	=	4 : 20	=	**5** : 25	=	15 : **75**
3 : **4**	=	**6** : 8	=	**27** : 36	=	33 : 44
5 : 7	=	20 : **28**	=	40 : 56	=	60 : **84**
4 : 11	=	**8** : 22	=	36 : **99**	=	**48** : 132

Drill 2

2. $\frac{3}{4}$: When denominators are the same and the fractions are positive, the *greater* numerator is the *greater* fraction. The numerator of $\frac{3}{4}$ is greater, so $\frac{3}{4} > \frac{1}{4}$.

3. $\frac{1}{5}$: When numerators are the same and the fractions are positive, the *smaller* denominator is the *greater* fraction. The denominator of $\frac{1}{5}$ is smaller, so $\frac{1}{5} > \frac{1}{6}$.

4. $\frac{53}{52}$: Finding a common denominator or using the double-cross would both involve ugly multiplication. Instead, estimate by using some things you know about fractions. In the first fraction, $\frac{53}{52}$, the numerator is greater than the denominator, so the fraction is greater than 1. In the second fraction, $\frac{85}{86}$, the denominator is greater than the numerator, so the fraction is less than 1. Therefore, $\frac{53}{52} > \frac{85}{86}$.

5. $\frac{7}{9}$: Use what you know about fractions to compare the relative size of these. The first fraction, $\frac{7}{9}$, has both a greater numerator and a smaller denominator than the second fraction. Therefore, $\frac{7}{9} > \frac{6}{10}$. Alternatively, eyeball their relative values. Both are less than 1, but not by the same amount. Compare each to an easier fraction, namely, $\frac{1}{2}$. The first fraction, $\frac{6}{10}$, is a little greater than $\frac{5}{10} = \frac{1}{2}$. The other, $\frac{7}{9}$, is further away from $\frac{1}{2}$, since $\frac{4.5}{9} = \frac{1}{2}$, so $\frac{7}{9}$ is the greater number.

6. $\frac{590}{290}$: These numbers are ugly, so try simplifying and estimating. First, simplify each fraction by dropping the extra zeros. Now compare $\frac{70}{36}$ and $\frac{59}{29}$. The first fraction is greater than 1 but less than 2, because 70 is less than twice 36 ($2 \times 36 = 72$). The second fraction is greater than 2, because 59 is more than twice 29 ($2 \times 29 = 58$). So $\frac{590}{290} > \frac{700}{360}$.

Drill 3

7. **(B) 3 : 10:** The information was given in the order birds : dogs, but the question asks about dogs : birds. Make sure to put your answer in the correct order. If there are 6 dogs and 20 birds in the park, the ratio of dogs to birds is 6 : 20. Divide by 2 to simplify this ratio to its base form, 3 : 10.

8. **(A) 1 : 1:** If 12 of the 24 students in the class are juniors, then the remainder, $24 - 12 = 12$, are seniors. Therefore, the ratio of juniors to seniors in the class is 12 : 12. Divide by 12 to simplify the ratio to its base form, 1 : 1.

9. **(B) 7 : 9:** The question first presents the information in the order of red marbles to green, but the question asks about the ratio of green to red. There are 35 green to 45 red, or 35 : 45. Divide each number by 5 to simplify the ratio to its base form, 7 : 9.

10. **(B) 8 : 15:** If there are 21 trout and 24 catfish, then there are $21 + 24 = 45$ total fish in the pond. The ratio of catfish to the total is therefore 24 : 45. Simplify the ratio to its base form. The number 3 is a factor of both numbers, so divide by 3 to get 8 : 15.

Drill 4

11. $\dfrac{116}{99}$: The denominators don't share any factors, so use the double-cross.

$$\frac{4}{9} + \frac{8}{11} = \frac{\overset{44}{\nwarrow} 4 + 8 \overset{72}{\nearrow}}{9 \times 11} = \frac{116}{99}$$

12. **0:** Reduce the first fraction before looking for a common denominator. Remove a factor of 4 from the numerator and the denominator.

$$\frac{\overset{5}{\cancel{20}}}{\underset{3}{\cancel{12}}} - \frac{5}{3} = \frac{5}{3} - \frac{5}{3} = 0$$

13. $\dfrac{a - 5b}{12}$: Find a common denominator and subtract. Because both 6 and 4 are factors of 12, 12 is the least common denominator.

$$\frac{a}{12} - \frac{b}{6} - \frac{b}{4} = \frac{a}{12} - \frac{2b}{12} - \frac{3b}{12} = \frac{a - 2b - 3b}{12} = \frac{a - 5b}{12}$$

14. **(A)** $\dfrac{2}{\sqrt{35}}$: First, break out the square root signs.

$$\sqrt{\frac{7}{5}} - \sqrt{\frac{5}{7}} = \frac{\sqrt{7}}{\sqrt{5}} - \frac{\sqrt{5}}{\sqrt{7}}$$

Then use the double-cross.

$$\frac{\overset{\sqrt{(7)(7)}}{\nwarrow} \quad - \quad \overset{\sqrt{(5)(5)}}{\nearrow}}{\dfrac{\sqrt{7}}{\sqrt{5}} - \dfrac{\sqrt{5}}{\sqrt{7}}} \longrightarrow \sqrt{(5)(7)}$$

Now, you can simplify. $\sqrt{(7)(7)}$ reduces to 7 and $\sqrt{(5)(5)}$ reduces to 5.

$$\frac{7 - 5}{\sqrt{35}} = \frac{2}{\sqrt{35}}$$

15. **(B)** $\dfrac{x(x+2z)}{y}$: Start by simplifying the individual fractions.

$$\frac{x^2z}{yz} - \frac{x^2z}{xy} + \frac{3xz}{y} = \frac{x^2\cancel{z}}{y\cancel{z}} - \frac{x^{\cancel{2}}z}{\cancel{x}y} + \frac{3xz}{y} = \frac{x^2}{y} - \frac{xz}{y} + \frac{3xz}{y}$$

Now, you have common denominators, so you can add and subtract.

$$\frac{x^2}{y} - \frac{xz}{y} + \frac{3xz}{y} = \frac{x^2 - xz + 3xz}{y} = \frac{x^2 + 2xz}{y}$$

Glance at the answers. No match yet, but can you see a way to manipulate what you have to match one of the answers? Pull out a common factor on top.

$$\frac{x^2 + 2xz}{y} = \frac{x(x+2z)}{y}$$

16. **(A) $2\sqrt{2}$:** Glance at the answers. They aren't fractions, so there must be some way to eliminate the denominators. When you see a root on the bottom of a fraction, you can get rid of it by multiplying by that root over itself. In this case, the root is $\sqrt{2}$, so multiply the fraction by $\dfrac{\sqrt{2}}{\sqrt{2}}$.

$$\frac{24}{3\sqrt{2}} \times \frac{\sqrt{2}}{\sqrt{2}} - \frac{4}{\sqrt{2}} \times \frac{\sqrt{2}}{\sqrt{2}} = \frac{24 \times \sqrt{2}}{3 \times 2} - \frac{4 \times \sqrt{2}}{2} = \frac{24\sqrt{2}}{6} - \frac{4\sqrt{2}}{2}$$

Simplify the terms before combining the fractions.

$$\frac{24\sqrt{2}}{6} - \frac{4\sqrt{2}}{2} = 4\sqrt{2} - 2\sqrt{2} = 2\sqrt{2}$$

Drill 5

17. $3\dfrac{2}{15}$:

$$\frac{47}{15} = \frac{45+2}{15} = \frac{45}{15} + \frac{2}{15} = 3 + \frac{2}{15} = 3\frac{2}{15}$$

18. **6:** This one simplifies to an integer.

$$\frac{72}{12} = 6$$

19. $\dfrac{45}{7}$: Via the shortcut, the numerator is $(7)(6) + 3 = 45$, and the denominator remains 7. Here's the full calculation.

$$6\frac{3}{7} = 6 + \frac{3}{7} = \frac{6 \times 7}{1 \times 7} + \frac{3}{7} = \frac{42}{7} + \frac{3}{7} = \frac{45}{7}$$

Drill 6

20. $\dfrac{2\sqrt{2}}{5}$: Begin by simplifying the square root in the numerator. When simplifying a square root, always look for factors that are perfect squares; in this example, $18 = 2 \times 9 = 2 \times 3^2$. The 3^2 can be removed from the square root to become 3.

$$\frac{2\sqrt{18}}{15} = \frac{2\sqrt{2 \times 3 \times 3}}{15} = \frac{2 \times \cancel{3} \times \sqrt{2}}{\cancel{3} \times 5} = \frac{2\sqrt{2}}{5}$$

21. **17**: Eliminate any common factors. To find all common factors, rewrite larger numbers as their prime factors.

$$\frac{17^2 \times 22}{11 \times 34} = \frac{17 \times \cancel{17} \times \cancel{2} \times \cancel{11}}{\cancel{11} \times \cancel{2} \times \cancel{17}} = 17$$

22. $\dfrac{3\sqrt{2}}{rs}$: To begin, simplify the square roots in the numerator and denominator by looking for pairs of factors (that is, perfect squares).

$$\sqrt{54} = \sqrt{2 \times 3 \times 3 \times 3} = 3\sqrt{2 \times 3}$$
$$\sqrt{12} = \sqrt{2 \times 2 \times 3} = 2\sqrt{3}$$

Because the numbers remaining inside the square roots have a factor of 3 in common, it's useful to leave the numerator broken into the root of 2 and the root of 3.

$$3\sqrt{2 \times 3} = 3\sqrt{2}\sqrt{3}$$

Therefore:

$$\frac{2r\sqrt{54}}{r^2s\sqrt{12}} = \frac{2r\left(3\sqrt{2}\sqrt{3}\right)}{r^2s\left(2\sqrt{3}\right)} = \frac{\cancel{2}\,\cancel{r}\left(3\sqrt{2}\,\cancel{\sqrt{3}}\right)}{r^{\cancel{2}}s\left(\cancel{2}\,\cancel{\sqrt{3}}\right)} = \frac{3\sqrt{2}}{rs}$$

23. $\dfrac{3x^2z^2}{23y}$: There are two good ways to simplify a fraction with variables raised to powers. One approach is to use exponent rules to rewrite the expression so that the cancellations are more clear.

$$\frac{6x^8yz^5}{46x^6y^2z^3} = \frac{\left(\overset{3}{\cancel{6}}\right)\left(\cancel{x^6}\right)\left(x^2\right)\left(\cancel{y}\right)\left(\cancel{z^3}\right)\left(z^2\right)}{\left(\underset{23}{\cancel{46}}\right)\left(\cancel{x^6}\right)\left(\cancel{y}\right)\left(y\right)\left(\cancel{z^3}\right)} = \frac{3x^2z^2}{23y}$$

Alternatively, use exponent rules to simplify directly.

$$\frac{6}{46} \times \frac{x^8}{x^6} \times \frac{y}{y^2} \times \frac{z^5}{z^3} = \frac{3}{23}x^2y^{-1}z^2$$

To combine these into one fraction, leave x^2 and z^2 in the numerator but place y in the denominator because $y^{-1} = \dfrac{1}{y}$.

$$\frac{3}{23}x^2y^{-1}z^2 = \frac{3x^2z^2}{23y}$$

24. **$5b^2$:** Begin by simplifying the square roots, then pull any squares out.

$$\sqrt{50} = \sqrt{2 \times 25} = 5\sqrt{2}$$
$$\sqrt{18a^2} = \sqrt{2 \times 9 \times a^2} = 3a\sqrt{2}$$

Cancel common terms on the top and bottom.

$$\frac{3ab^2\sqrt{50}}{\sqrt{18a^2}} = \frac{\cancel{3}\,\cancel{a}b^2 \times 5\cancel{\sqrt{2}}}{\cancel{3}\,\cancel{a}\,\cancel{\sqrt{2}}} = 5b^2$$

Drill 7

25. $\frac{4}{15}$**:** Flip the second fraction to multiply.

$$\frac{6}{25} \div \frac{9}{10} = \frac{6}{25} \times \frac{10}{9} = \frac{\overset{2}{\cancel{6}}}{\underset{5}{\cancel{25}}} \times \frac{\overset{2}{\cancel{10}}}{\underset{3}{\cancel{9}}} = \frac{4}{15}$$

26. **1:** Flip the second fraction to multiply.

$$\frac{3}{11} \div \frac{3}{11} = \frac{3}{11} \times \frac{11}{3} = \frac{\cancel{3} \times \cancel{11}}{\cancel{11} \times \cancel{3}} = 1$$

27. $\frac{1}{4}$**:** Combine everything into one fraction, with the entire numerator under one square root sign, and factor the larger numbers.

$$\frac{\sqrt{12}}{5} \times \frac{\sqrt{60}}{2^4} \times \frac{\sqrt{45}}{3^2} = \frac{\sqrt{12 \times 60 \times 45}}{5 \times 2^4 \times 3^2} = \frac{\sqrt{4 \times 3 \times 4 \times 15 \times 3 \times 15}}{5 \times 2^2 \times 2^2 \times 3^2}$$

Simplify the square roots by pulling out perfect squares, then simplify the fraction.

$$\frac{\sqrt{3^2 \times 4^2 \times 15^2}}{5 \times 4 \times 4 \times 3 \times 3} = \frac{3 \times 4 \times 15}{5 \times 4 \times 4 \times 3 \times 3} = \frac{\cancel{3} \times \cancel{4} \times \cancel{15}}{\cancel{5} \times \cancel{4} \times 4 \times \cancel{3} \times \cancel{3}} = \frac{1}{4}$$

28. **3:** To divide by a fraction, multiply by its reciprocal.

$$\frac{\sqrt{18}}{\sqrt{4}} \div \frac{\sqrt{9}}{\sqrt{18}} = \frac{\sqrt{18}}{\sqrt{4}} \times \frac{\sqrt{18}}{\sqrt{9}}$$

Feel free to simplify the perfect squares as soon as you spot them.

$$\frac{\sqrt{18}}{\sqrt{4}} \times \frac{\sqrt{18}}{\sqrt{9}} = \frac{\sqrt{18^2}}{2 \times 3} = \frac{18}{6} = 3$$

29. $\dfrac{yz^3}{x^5}$: To divide, multiply by the reciprocal.

$$\frac{xy^3z^4}{x^3y^4z^2} \div \frac{x^6y^3z}{x^3y^5z^2} = \frac{xy^3z^4}{x^3y^4z^2} \times \frac{x^3y^5z^2}{x^6y^3z}$$

If you'd like, first rearrange to place the like terms together (to minimize careless mistakes). Cancel any terms that match exactly.

$$\frac{(x)\cancel{(x^3)}}{\cancel{(x^3)}(x^6)} \times \frac{\cancel{(y^3)}(y^5)}{(y^4)\cancel{(y^3)}} \times \frac{(z^4)\cancel{(z^2)}}{\cancel{(z^2)}(z)} = \frac{x}{x^6} \times \frac{y^5}{y^4} \times \frac{z^4}{z}$$

For the remaining terms, use exponent rules to simplify the like variables.

$$\frac{x}{x^6} \times \frac{y^5}{y^4} \times \frac{z^4}{z} = x^{(1-6)}y^{(5-4)}z^{(4-1)} = x^{-5}y^1z^3 = \frac{yz^3}{x^5}$$

30. **2:** Begin by multiplying by the reciprocal.

$$\frac{12^2}{9^2} \div \frac{6^3}{3^5} = \frac{12^2}{9^2} \times \frac{3^5}{6^3}$$

Break the larger numbers down into numbers with similar bases, then use exponent rules to simplify.

$$\frac{12^2}{9^2} \times \frac{3^5}{6^3} = \frac{(2 \times 6)^2}{\left(3^2\right)^2} \times \frac{3^5}{6^3} = \frac{2^2 \times 6^2}{3^4} \times \frac{3^5}{6^3} = \frac{2^2}{1} \times \frac{3^5}{3^4} \times \frac{6^2}{6^3} = \frac{2^2 3^1}{6^1} = \frac{12}{6} = 2$$

Drill 8

31. (C) $\dfrac{3}{b} + \dfrac{4}{3a}$: Split this fraction into two fractions with a common denominator of $3ab$, and then simplify further.

$$\frac{9a + 4b}{3ab} = \frac{9a}{3ab} + \frac{4b}{3ab} = \frac{\overset{3}{\cancel{9}}\,\cancel{a}}{\cancel{3}\,\cancel{a}b} + \frac{4\cancel{b}}{3a\cancel{b}} = \frac{3}{b} + \frac{4}{3a}$$

32. $\dfrac{2}{11 + 7b}$: Be careful when dealing with addition or subtraction in the denominator; you can't split these into two separate fractions. Instead, look for a common term that you can cancel.

$$\frac{6a}{33a + 21ab} = \frac{3 \times 2 \times a}{3a(11 + 7b)} = \frac{\cancel{3a} \times 2}{\cancel{3a}\,(11 + 7b)} = \frac{2}{11 + 7b}$$

33. $\dfrac{x+5}{4-3x}$: Every term has the common factor $8x$. Pull this out and cancel.

$$\frac{8x^2 + 40x}{32x - 24x^2} = \frac{\cancel{8x}\,(x + 5)}{\cancel{8x}\,(4 - 3x)} = \frac{x + 5}{4 - 3x}$$

Drill 9

34. $\frac{40}{21}$: Begin by simplifying each fraction.

$$\frac{3+4}{1+2} - \frac{1+2}{3+4} = \frac{7}{3} - \frac{3}{7}$$

Use the double-cross method to subtract.

$$\overset{49}{\nwarrow} 7 \overset{-}{\underset{3}{\times}} 3 \overset{9}{\nearrow} = \frac{40}{21}$$

35. $\frac{t^2 - t}{2t - 4}$:

$$\frac{-t+1}{t-2} \times \frac{-t}{2} = \frac{(-t+1) \times (-t)}{(t-2) \times 2} = \frac{t^2 - t}{2t - 4}$$

36. $\frac{1}{2}$: When subtracting fractions with more than one term in the numerator, put the subtracted term in parentheses to remind yourself to distribute the negative sign.

$$\frac{b+6}{6} - \frac{3+b}{6} = \frac{b+6-(3+b)}{6} = \frac{b+6-3-b}{6} = \frac{3}{6} = \frac{1}{2}$$

37. $\frac{x^2 + y}{5}$:

$$\frac{3x^2 + 3y}{40} + \frac{x^2 + y}{8} = \frac{3x^2 + 3y}{40} + \frac{5x^2 + 5y}{40} = \frac{\left(3x^2 + 3y\right) + \left(5x^2 + 5y\right)}{40} = \frac{8x^2 + 8y}{40}$$

Don't stop yet! All three terms are divisible by 8.

$$\frac{8x^2 + 8y}{40} = \frac{\cancel{8}x^2 + \cancel{8}y}{\underset{5}{\cancel{40}}} = \frac{x^2 + y}{5}$$

Drill 10

38. **(A)** $\frac{2}{3}$: Simplify before you multiply.

$$\frac{x+3}{15} \times \frac{10}{x+3} = \frac{\cancel{(x+3)}}{\underset{3}{\cancel{15}}} \times \frac{\overset{2}{\cancel{10}}}{\cancel{(x+3)}} = \frac{2}{3}$$

39. **(B)** $\dfrac{6}{5n^2}$: First, combine the like terms in the numerator of the first fraction and the denominator of the second fraction.

$$\frac{(n+2n)}{n^4} \times \frac{(2n)^2}{(15n-5n)} = \frac{3n}{n^4} \times \frac{4n^2}{10n}$$

If you like, rearrange to place the numbers and variables near each other (to minimize careless mistakes), then simplify.

$$\frac{(3)(4)}{(10)} \times \frac{(n)(n^2)}{(n^4)(n)} = \frac{(3)\left(\overset{2}{\cancel{4}}\right)}{\left(\underset{5}{\cancel{10}}\right)} \times \frac{(\cancel{n})(n^2)}{(n^4)(\cancel{n})} = \frac{6}{5} \times n^{2-4} = \frac{6}{5} \times n^{-2} = \frac{6}{5n^2}$$

40. **(A) 6:** Begin by simplifying the denominator.

$$\frac{8}{2 - \dfrac{2}{3}} = \frac{8}{\dfrac{6}{3} - \dfrac{2}{3}} = \frac{8}{\dfrac{4}{3}}$$

Dividing by $\dfrac{4}{3}$ is the same as multiplying by $\dfrac{3}{4}$.

$$\frac{8}{\dfrac{4}{3}} = 8 \times \frac{3}{4} = \overset{2}{\cancel{8}} \times \frac{3}{\cancel{4}} = 6$$

41. **(C)** $\dfrac{y+1}{y-1}$: Work from the inside out. First, combine the terms in the denominator.

$$\frac{1}{1 - \dfrac{2}{y+1}} = \frac{1}{\dfrac{y+1}{y+1} - \dfrac{2}{y+1}} = \frac{1}{\dfrac{(y+1)-(2)}{y+1}} = \frac{1}{\dfrac{y-1}{y+1}}$$

A numerator of 1 divided by a fraction is equal to the reciprocal of that fraction.

$$\frac{1}{\dfrac{y-1}{y+1}} = \frac{y+1}{y-1}$$

Fractions, Decimals, Percents, and Ratios

In This Chapter

- Four Ways to Express Parts of a Whole

- From Decimal to Percent: Move the Point Two Places Right

- From Decimal or Percent to Fraction: Put Number over Whole and Simplify

- From Fraction to Decimal or Percent: Memorize or Use Long Division

- Multiply a Decimal by a Power of 10: Shift the Point

- Add or Subtract Decimals: Line Up the Points

- Multiply Two Decimals: Ignore Points at First

- Multiply a Decimal and a Big Number: Trade Decimal Places

- Divide Two Decimals: Move Points in the Same Direction

- 20% of $55 = 0.2 \times $55

- Percent of a Percent Of: Multiply Twice

- Ratios and Percents: Convert Fractions to Percents

- Non-Integer Ratios

In this chapter, you will learn how and when to convert among fractions, decimals, percents, and ratios. You'll also learn how to add, subtract, multiply, and divide decimals. Finally, you'll learn how to manipulate percentages, find percent change, and solve percent problems using ratio principles.

CHAPTER 5 Fractions, Decimals, Percents, and Ratios

Fractions, decimals, percents, and ratios are so closely related that you will often be asked to move from one form to another while solving a problem. Other times, you may choose to switch to another form because it's easier to perform a certain calculation in that form.

Four Ways to Express Parts of a Whole

In the prior chapter, you learned that a fraction is one way to express a *part-to-whole* relationship. There are in fact three common ways to express such a relationship—and even a fourth, when you count ratios.

Say you have the shaded part of this orange. You can express how much you have in terms of a part-to-whole relationship in these ways:

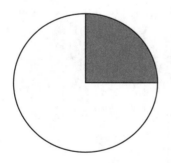

1. You have $\frac{1}{4}$ of the orange.   **Fraction**

2. You have 0.25 of the orange.   **Decimal**

3. You have 25% of the orange.   **Percent**

In general, a part-to-whole relationship is the standard relationship expressed by a fraction, a decimal, or a percent.

You can also say that the **ratio** of your piece to the whole orange is 1 part to 4 parts, or 1 : 4. The more common usage of a ratio is *part-to-part*, though, so pay attention to the details of the problem; it will give you the information necessary to determine whether it's a part-to-part or part-to-whole relationship. (An example of a part-to-part relationship is this: You have 1 part of the orange and someone else has 3 parts, for a part-to-part relationship of 1 : 3. In this usage, the whole is the sum of the parts: $1 + 3 = 4$.)

The main difference between the forms is how you think about the whole:

$$\frac{1}{4} = 1 \text{ out of 4 pieces of the whole}$$

$$0.25 = 0.25 \text{ of the whole}$$

$$25\% = 25 \text{ out of 100 pieces of the whole}$$

$$1 : 4 = (\text{"1 to 4"}) \; 1 \text{ out of 4 pieces of the whole}$$

In other words, what is each form "out of"? What is the whole that you are dividing by?

- Fractions are *out of* the denominator (4 in this case).
- Decimals are *out of* 1 (the whole). You've already done the division.
- Percents are *out of* 100. *Percent* literally means *per hundred*, or divided by 100.
- Part-to-whole ratios are *out of* the second term in the ratio (4 in this case). (Again, only use this form when the problem tells you to.)

Which form is most useful depends on the problem at hand. You might say any of the following:

- The container is $\frac{1}{2}$ full.
- The container is filled to 0.5 of its capacity.
- The container is 50% full.
- The ratio of the contents of the container to its total capacity is 1 to 2.

By the way, the part can be greater than the whole:

- I ate $\frac{5}{4}$ boxes of cereal. (I ate more than one box.)
- I ate 1.25 boxes of cereal.
- I ate 125% of one box of cereal.
- The ratio of what I ate to a whole box of cereal was 5 to 4.

From Decimal to Percent: Move the Point Two Places Right

Decimals are out of 1. Percents are out of 100. So to convert a decimal to a percent, move the decimal point to the *right two places*. Add zeros if necessary:

$$0.53 = 53\% \qquad 0.4 = 0.40 = 40\% \qquad 0.03 = 3\% \qquad 1.7 = 1.70 = 170\%$$

A percent might still contain a visible decimal point when you're done:

$$0.4057 = 40.57\% \qquad 0.002 = 0.2\% \qquad 0.0005 = 0.05\%$$

Note that, when converting from a decimal to a percent, the new number seems greater than the original. In reality, the two are equal; they represent the exact same number. But the way you write the number makes it *look* greater in percent form and lesser in decimal form. This can help you to remember to move the decimal point to the *right* when converting to the percent form.

To convert a percent to a decimal, *go in reverse*. That is, move the decimal point two places to the *left*. If the decimal point isn't visible, it's actually just before the percent sign. Add zeros if necessary as you move left. For example:

$$39\% = 39.\% = 0.39 \qquad 8\% = 0.08 \qquad 225\% = 2.25$$

$$13.4\% = 0.134 \qquad 0.7\% = 0.007 \qquad 0.001\% = 0.00001$$

Remember: The decimal form of the number *looks* like it's less than the percent form. That can help you to remember to move the decimal point to the *left* when converting to decimal form.

If you . . .	Then you . . .	Like this:
Want to convert a decimal to a percent	Move the decimal point two places to the right to make the number *seem* greater (though it isn't really)	$0.036 = 3.6\%$
Want to convert a percent to a decimal	Move the decimal point two places to the left to make the number *seem* smaller (though it isn't really)	$41.2\% = 0.412$

Check Your Skills

1.   Convert 0.035 to a percent.

Answer can be found on page 202.

From Decimal or Percent to Fraction: Put Number over Whole and Simplify

The decimal 0.25 is twenty-five one-hundredths. Put the decimal over its whole, 1, and then manipulate the fraction until no decimals are left:

$$0.25 = \frac{0.25}{1}$$

$$= \frac{0.5}{2} \quad \text{Multiply by } \frac{2}{2}. \text{ Still a decimal.}$$

$$= \frac{1}{4} \quad \text{Multiply by } \frac{2}{2} \text{ again. No more decimals.}$$

You can write out the math as shown above—place the decimal over 1 and then figure out what to multiply into the numerator and denominator to get the decimal to disappear. (Always multiply the numerator and denominator by the same value.)

If you feel comfortable with powers of 10, you can also directly put a power of 10 (10, 100, 1,000, etc.) in the denominator of the fraction. Which power of 10? Put as many zeros in your power of 10 as you have digits to the right of the decimal point. For example:

0.3	$=$	$\dfrac{3}{10}$
Zero point three	is	three-tenths, or three over ten
0.25	$=$	$\dfrac{25}{100}$
Zero point two five	is	twenty-five one-hundredths
0.007	$=$	$\dfrac{7}{1,000}$
Zero point zero zero seven	is	seven one-thousandths

As with any fractions, cancel common terms to simplify:

$$0.4 = \frac{4}{10} = \frac{\overset{2}{\cancel{4}}}{\underset{5}{\cancel{10}}} = \frac{2}{5} \qquad 0.75 = \frac{75}{100} = \frac{\overset{3}{\cancel{75}}}{\underset{4}{\cancel{100}}} = \frac{3}{4}$$

Take a look at this example:

$$0.0102 = \frac{102}{10,000}$$

To convert the decimal to a fraction, first count the number of decimal places to the right of the decimal point; in this case, there are four. The denominator, then, will be a 1 plus four zeros: 10,000.

For the numerator, go back to the original decimal and move the decimal point all the way to the right. The number is now 00102 or just 102. In other words, drop both the decimal point and any "leading zeros" (zeros that appear to the left of the first nonzero number) and just use the resulting number.

Then, simplify the fraction:

$$\frac{102}{10,000} = \frac{\overset{51}{\cancel{102}}}{\underset{5,000}{\cancel{10,000}}} = \frac{51}{5,000}$$

As always, look for opportunities to estimate. Depending upon the details of the problem, it may be close enough to do this:

$$0.0102 = \frac{102}{10,000} \approx \frac{\cancel{100}}{\cancel{10,000}} = \frac{1}{100}$$

You might even be able to do this:

$$0.0102 \approx 0.01 = \frac{1}{100}$$

To convert a percent to a fraction, write the given number over the whole, 100, every time. Remember that *percent* literally means *per hundred*. For example:

$$45\% = \frac{45}{100} = \frac{\overset{9}{\cancel{45}}}{\underset{20}{\cancel{100}}} = \frac{9}{20} \qquad 8\% = \frac{8}{100} = \frac{\overset{2}{\cancel{8}}}{\underset{25}{\cancel{100}}} = \frac{2}{25}$$

Alternatively, you can first convert the percent to a decimal by moving the decimal place and then follow the process to convert a decimal to a fraction:

$$2.5\% = 0.025 = \frac{25}{1,000} = \frac{\overset{1}{\cancel{25}}}{\underset{40}{\cancel{1,000}}} = \frac{1}{40}$$

If you don't convert to a decimal first, be sure to write the fraction over 100, no matter the form of the percent:

$$2.5\% = \frac{2.5}{100} = \frac{5}{200} = \frac{1}{40}$$

You'll learn how to divide decimals later in this chapter.

If you...	Then you...	Like this:
Want to convert a decimal to a fraction	Drop the decimal point and any leading zeros; put that number over the appropriate power of 10, then simplify	$0.036 = \frac{36}{1,000}$ $= \frac{9}{250}$
Want to convert a percent to a fraction	Write the percent over 100, then simplify OR Convert first to a decimal, then follow the process for converting decimals to fractions	$4\% = \frac{4}{100}$ $= \frac{1}{25}$ $3.6\% = 0.036$ $0.036 = \frac{36}{1,000}$ $= \frac{9}{250}$

Check Your Skills

2. Convert 0.375 to a fraction and simplify.
3. Convert 24% to a fraction and simplify.

Answers can be found on page 202.

From Fraction to Decimal or Percent: Memorize or Use Long Division

A fraction represents division. The decimal equivalent is the result of that division.

On the GMAT, most cases in which you might need to *convert a fraction to a decimal* will involve common fractions that you can memorize. For instance, $\frac{1}{4}$ is a common fraction. Memorize the fact that $\frac{1}{4} = 0.25 = 25\%$, and you won't have to do any converting at all. At the end of this section, you'll find more common fraction–decimal–percent equivalents to memorize.

You can also use long division, if necessary—though check first to see whether the problem allows you to estimate. Most of the time, if the fraction is not a common one to memorize, then you can estimate; in other words, the test rarely makes you do long division. If you do perform long division, keep track of the decimal point on top (in the solution), but don't worry about keeping the decimals down below:

$\frac{1}{4} = ?$ Divide 1 by 4. $\frac{5}{8} = ?$ Divide 5 by 8.

$$
\begin{array}{r}
0.25 \\
4\overline{)1.00} \\
\underline{-8} \\
20 \\
\underline{-20} \\
0
\end{array}
\qquad \frac{1}{4} = 0.25
\qquad\qquad
\begin{array}{r}
0.625 \\
8\overline{)5.000} \\
\underline{-48} \\
20 \\
\underline{-16} \\
40 \\
\underline{-40} \\
0
\end{array}
\qquad \frac{5}{8} = 0.625
$$

In some cases, the decimal never ends because the long division never ends. You get a repeating decimal:

$$\frac{1}{3} = 0.333... = 0.\overline{3}$$

$$
\begin{array}{r}
0.33... \\
3\overline{)1.000} \\
\underline{-\;9} \\
10 \\
\underline{-\;9} \\
10 \\
...
\end{array}
$$

If the denominator contains only factors of 2 and/or 5, then the decimal will end (this is called a *terminating decimal* because it terminates, or ends). In this case, multiply the numerator and denominator by the same number to turn the denominator into a power of 10. For example:

$$\frac{1}{4} = \frac{1 \times 25}{4 \times 25} = \frac{25}{100} = 0.25 \qquad \frac{1}{20} = \frac{1 \times 5}{20 \times 5} = \frac{5}{100} = 0.05$$

However, if the denominator contains factors other than 2's and 5's, and the fraction is not a common one that you have memorized, then you would have to use long division to convert to an exact percent. This is cumbersome, so it's important to check whether you can estimate before you bother to divide.

To convert a fraction to a percent, first convert to a decimal, then convert the decimal to a percent:

$$\frac{1}{2} = \frac{1 \times 5}{2 \times 5} = \frac{5}{10} = 0.5 = 50\%$$

If you...	Then you...	Like this:
Want to convert a fraction to a decimal	Memorize the common conversions	$\frac{1}{10} = 0.1$
	OR	
	Multiply the top and bottom by the same number such that the denominator becomes a power of 10, but only if the denominator contains only 2's and 5's as factors	$\frac{1}{50} = \frac{1 \times 2}{50 \times 2}$ $= \frac{2}{100}$ $= 0.02$
	OR	
	Do long division	$\frac{7}{8} \rightarrow 8)\overline{7.000}^{\,0.875}$

Memorize the following common conversions. Flash cards are a great tool to help; grab some index cards and start writing.

Tenths and Fifths

Fraction	Decimal	Percent
$\dfrac{1}{10}$	0.1	10%
$\dfrac{2}{10} = \dfrac{1}{5}$	0.2	20%
$\dfrac{3}{10}$	0.3	30%
$\dfrac{4}{10} = \dfrac{2}{5}$	0.4	40%
$\dfrac{5}{10} = \dfrac{1}{2}$	0.5	50%
$\dfrac{6}{10} = \dfrac{3}{5}$	0.6	60%
$\dfrac{7}{10}$	0.7	70%
$\dfrac{8}{10} = \dfrac{4}{5}$	0.8	80%
$\dfrac{9}{10}$	0.9	90%
$\dfrac{10}{10} = \dfrac{5}{5} = 1$	1.0	100%
$\dfrac{11}{10}$	1.1	110%
$\dfrac{12}{10} = \dfrac{6}{5}$	1.2	120%

Eighths and Fourths

Fraction	Decimal	Percent
$\frac{1}{8}$	0.125	12.5%
$\frac{2}{8} = \frac{1}{4}$	0.25	25%
$\frac{3}{8}$	0.375	37.5%
$\frac{4}{8} = \frac{2}{4} = \frac{1}{2}$	0.5	50%
$\frac{5}{8}$	0.625	62.5%
$\frac{6}{8} = \frac{3}{4}$	0.75	75%
$\frac{7}{8}$	0.875	87.5%
$\frac{8}{8} = \frac{4}{4} = 1$	1	100%
$\frac{10}{8} = \frac{5}{4}$	1.25	125%
$\frac{12}{8} = \frac{6}{4} = \frac{3}{2}$	1.5	150%

For the set above, note that $\frac{1}{8}$ is half of $\frac{2}{8}$. You may already know that $\frac{2}{8} = \frac{1}{4} = 25\%$. Use this to help memorize the fact that $\frac{1}{8}$ is half that, or 12.5%. Every eighth fraction increases by another 12.5%.

Thirds

Fraction	Decimal	Percent
$\frac{1}{3}$	0.3333...	33.33...%
$\frac{2}{3}$	0.6666...	66.66...%
$\frac{3}{3} = 1$	1	100%

Just a couple more!

Fraction	Decimal	Percent
$\dfrac{1}{100}$	0.01	1%
$\dfrac{1}{20}$	0.05	5%

Check Your Skills

4. Change $\dfrac{3}{5}$ to a decimal.

5. Convert $\dfrac{3}{8}$ to a percent.

Answers can be found on page 202.

Multiply a Decimal by a Power of 10: Shift the Point

Decimals are tenths, hundredths, thousandths, and so on. One-tenth is a power of 10, namely 10^{-1}. One-hundredth is also a power of 10, namely 10^{-2}:

$$0.1 = \frac{0.1}{1} = \frac{1}{10} = \frac{1}{10^1} = 10^{-1}$$

$$0.01 = \frac{0.01}{1} = \frac{1}{100} = \frac{1}{10^2} = 10^{-2}$$

You can write any decimal as a fraction with a power of 10 in the denominator, or as a product involving a power of 10. For example, $0.03 = \dfrac{3}{100} = \dfrac{3}{10^2}$. The power of 10 determines where the decimal point falls.

In this example, the power of 10 is 2 so, in the decimal form, the decimal point is two places to the left of 3.

As a result, if you multiply or divide a decimal by a power of 10, you move the decimal point to the right or to the left.

If you multiply by 10 itself, which can also be written 10^1, then you shift the decimal point one place to the right:

$$0.004 \times 10 = 0.04$$

The 10 cancels with one power of 10 in the denominator:

$$\frac{4}{1,000} \times 10 = \frac{4}{1,00\cancel{0}} \times \cancel{10} = \frac{4}{100}$$

You can also see it in terms of exponents. The additional 10 increases the overall exponent from -3 to -2:

$$4 \times 10^{-3} \times 10^1 = 4 \times 10^{-2}$$

If you multiply by 100, or 10^2, you shift the decimal point two places to the right:

$$0.004 \times 100 = 0.4 \qquad \text{That is, } \frac{4}{1,000} \times 100 = \frac{4}{10} \qquad \text{Or } 4 \times 10^{-3} \times 10^2 = 4 \times 10^{-1}$$

When you multiply by a power of 10, the exponent of that power is the number of places you move the decimal. If the power of 10 is positive, move the decimal to the *right* to *increase* the number:

$$43.8723 \times 10^3 = 43{,}872.3 \qquad \text{Move the decimal three places to the right.}$$

If you divide by a power of 10, move to the *left* to *decrease* the number:

$$782.95 \div 10 = 78.295 \qquad \text{Move the decimal 1 place to the left.}$$
$$57{,}234 \div 10^4 = 5.7234 \qquad \text{Move the decimal 4 places to the left.}$$

If you're asked to multiply by a negative power of 10, flip the power to positive and divide instead. Move the decimal to the left to decrease the number, since you're dividing:

$$4 \times 10^{-3} = 4 \div 10^3 = 0.004 \qquad \text{Move three places to the left.}$$

Likewise, if you're asked to divide by a negative power of 10, change the power to positive and multiply instead. Move the decimal to the right to increase the number, since you're multiplying:

$$62 \div 10^{-2} = 62 \times 10^2 = 6{,}200 \qquad \text{Move two places to the right.}$$

All of these procedures work the same way for repeating decimals:

$$\frac{1}{3} \times 10 = 0.333... \times 10 = 3.33... \qquad \text{Move one place to the right.}$$

If you...	Then you...	Like this:
Multiply a decimal by a positive power of 10	Move the decimal point right a number of places, corresponding to the exponent of the 10	$0.007 \times 10^2 = 0.7$ $7 \div 10^{-2} = 7 \times 10^2 = 700$
Divide a decimal by a positive power of 10	Move the decimal point left a number of places, corresponding to the exponent of the 10	$6 \div 10^3 = 0.006$ $6 \times 10^{-2} = 6 \div 10^2 = 0.06$

Check Your Skills

Multiply. Give each answer as a single value.

6. 32.753×10^2

7. $43{,}681 \times 10^{-4}$

Answers can be found on pages 202–203.

Add or Subtract Decimals: Line Up the Points

When you add or subtract decimals, write the decimals vertically, with the decimal points lined up:

$$0.3 + 0.65 = \qquad\qquad\qquad 0.65 - 0.5 =$$

$$
\begin{array}{r}
0.30 \\
+ \ 0.65 \\
\hline
0.95
\end{array}
\qquad\qquad
\begin{array}{r}
0.65 \\
- \ 0.50 \\
\hline
0.15
\end{array}
$$

You can add zeros on the right to help you line up. For instance, turn 0.5 into 0.50 before you subtract it from 0.65.

If you...	Then you...	Like this:
Add or subtract decimals	Line up the decimal points vertically	$\begin{array}{r} 4.035 \\ +0.120 \\ \hline 4.155 \end{array}$

Check Your Skills

Add or subtract. Give each answer as a single value.

8. $3.128 + 0.045$

9. $1.8746 - 0.313$

Answers can be found on page 203.

Multiply Two Decimals: Ignore Points at First

Consider this example:

$0.25 \times 0.5 =$

First, multiply the numbers together as if they were integers. In other words, ignore the decimal points:

$25 \times 5 = 125$

This number is too big to be the actual answer; it needs to be adjusted. In the original problem, count all the digits to the right of the decimal points:

0.25 has two digits to the right.   0.5 has one digit to the right.

There were a total of three digits originally to the right, so move the decimal point of the answer three places to the *left*, in order to compensate. In other words, since there were three digits to the right of the decimal originally, the result should also have three digits to the right of the decimal:

125 becomes 0.125.   Therefore, $0.25 \times 0.5 = 0.125$.

Here's why this process works:

$$0.25 = 25 \times 10^{-2} \qquad\qquad 0.5 = 5 \times 10^{-1}$$

$$0.25 \times 0.5 = (25 \times 10^{-2}) \times (5 \times 10^{-1}) = 125 \times 10^{-3} = 0.125$$

The powers of 10 tell you where to put the decimal point. Here is another example:

$3.5 \times 20 =$	There is one digit to the right of the decimal point.
$35 \times 20 = 700$	Ignore the decimals and multiply.
$3.5 \times 20 = 70.0$	Account for the one decimal place in 3.5 by moving the final decimal
$= 70$	point one place to the left.

Count the zeros to the right of the decimal point as well. For example:

$0.01 \times 0.05 =$	There are four digits (including zeros) to the right of the decimal points.
$1 \times 5 = 5$	Ignore the decimals and multiply.
$0.01 \times 0.05 = 0.0005$	Account for the original four decimal places by moving the decimal point four places to the left.

If you...	Then you...	Like this:
Multiply two decimals	Ignore the decimal points, multiply integers, then place the decimal point by counting the original number of decimals	$0.2 \times 0.5 = ?$ $2 \times 5 = 10$ 0.10 $0.2 \times 0.5 = 0.1$

Check Your Skills

Multiply. Give each answer as a single value.

10. 0.6×1.1

11. 0.004×0.032

Answers can be found on page 203.

Multiply a Decimal and a Big Number: Trade Decimal Places

Now consider this example:

$4,000,000 \times 0.0003 =$

When doing multiplication and one number is very big and the other one is very small, you can trade powers of 10 from the big one (4,000,000) to the small one (0.0003). In other words, move one decimal point *left* and the other one *right*. Just make sure that you move the same number of places.

This multiplication would be easier if you had no decimals at all. To make this happen, move the decimal in 0.0003 to the right four places to get 3. To compensate, move the decimal in 4,000,000 to the left four places. That makes that number more manageable, too:

$4,000,000 \times 0.0003 = 4,000,000 \times 0.0003 = 400 \times 3 = 1,200$

Here's why this math is valid:

$4,000,000 \times 0.0003 = (4 \times 10^6) \times (3 \times 10^{-4}) = 12 \times 10^{6-4} = 12 \times 10^2 = 1,200$

If you...	Then you...	Like this:
Multiply a small decimal and a big number	Trade decimal places from the big number to the small one	$50,000 \times 0.007 =$ $50 \times 7 = 350$

Check Your Skills

Multiply. Give the answer as a single value.

12. $520,000 \times 0.0004$

Answer can be found on page 203.

187

Divide Two Decimals: Move Points in the Same Direction

When you divide decimals, first write the division as a fraction if it isn't in that form already. For example:

$$\frac{300}{0.05} =$$

Since this is a fraction, you can move the decimals in the *same* direction on the top and bottom. Multiply by the necessary power of 10 or just move the decimals directly, whatever you're most comfortable doing:

$$\frac{300}{0.05} = \frac{300 \times 100}{0.05 \times 100} = \frac{300}{0.05} = \frac{30,000}{5} = 6,000$$

In this case, turn 0.05 into 5 by moving its decimal two places to the right. Then do the same thing on top. Add zeros as necessary.

Try this problem:

$$\frac{3.39}{0.003} =$$

The 3.39 only needs two moves to get rid of the decimal, while 0.003 needs three moves. Now what? Go with the greater number of moves, and add zeros to the other number to compensate:

$$\frac{3.39}{0.003} = \frac{3.39}{0.003} = \frac{3,390}{3} = 1,130$$

If you...	Then you...	Like this:
Divide two decimals	Move the decimal points in the same direction to eliminate decimals as far as you can	$\frac{0.0020}{0.0004} = \frac{20}{4} = 5$

Check Your Skills

13. Simplify $\frac{0.00084}{0.00007}$.

Answer can be found on page 203.

20% of $55 = 0.2 × $55

In everyday life, percents are the most common way of expressing part-to-whole relationships. You often see signs advertising "25% off," but you don't see as many signs advertising "$\frac{1}{4}$ off" or "0.75 of the original price." So your intuition about percents is probably already pretty good, and that's useful on the GMAT.

However, percents are not necessarily the most useful form for actual *computation*. If you need to crunch numbers, think about what form would be easiest to use (fractions, decimals, or percents) for the specific math that has to happen.

Consider this example:

30% of $60 =

The word *of* means *times* in math terms. In other words, *of* indicates multiplication.

One way to calculate a percent is to benchmark using easy-to-calculate percentages. For instance, it's easy to find 10% of a number: Move the decimal one place to the left. So start by taking 10% of 60, or $(0.1)(60) = 6$. Since 30% is three times as much as 10%, multiply: $6 \times 3 = 18$.

Alternatively, you could convert 30% to a decimal, trade decimal places, and then multiply:

$$30\% \text{ of } \$60 = 0.30 \times \$60 = 3 \times \$6 = \$18$$

You even have a third option. Use the fraction form of 30%:

$$30\% \text{ of } \$60 = \frac{30}{100} \times \$60 = \frac{3\cancel{0}}{10\cancel{0}} \times \$60 = \frac{3}{1\cancel{0}} \times \$6\cancel{0} = \$18$$

Which path did you find the easiest? Do similar math that way in future.

A problem could also be written out in sentence form:

What is 20% of $55 ?

Before you do anything, think about this question logically. You have $55 and you're going to give someone 20% of it. Approximately how much is that?

The answer should be less than $55, first of all, and it should be a decent amount less, since 20% isn't all that much of the total. If you calculate an answer that turns out to be greater than $55 (or less than $55 but close to it), you'll know to check your work.

In the Word Problems chapter, you'll learn more about translating words into math. For now, know these translations:

What	can be translated as	y	(some variable)
is	can be translated as	$=$	(the equals sign)
of	can be translated as	$\times$	(multiplication)

Translate the full question to math as follows:

What	is	20%	of	$55 ?
y	$=$	0.20	$\times$	$55

Now crunch the numbers on the right. Note that when there are zeros at the end of a decimal, you don't count those zeros when counting decimals:

$0.20 \times \$55 =$	For purposes of counting the decimal point, do *not* include the 0 after the 2.
$0.2 \times \$55 =$	Compute $2 \times \$55 = \$110$, then move the decimal point.
$0.2 \times \$55 = \$11$	

Alternatively, translate 20% to a fraction rather than to a decimal (or use your memorized fraction equivalent):

$$20\% = \frac{20}{100} = \frac{1}{5}$$

$$\frac{1}{5} \times \$55 = \frac{1}{\cancel{5}} \times \overset{11}{\cancel{\$55}} = \$11$$

Or use the benchmark 10% and then multiply by 2 to get 20%:

$$(10\%)(55) = (0.1)(55) = 5.5$$

$$5.5 \times 2 = 11$$

The translation gets a little tougher when you encounter the phrase "what percent." For example:

What percent of 125 is 25 ?

Use a variable to represent the word *what*. Remember that the word *percent* by itself means *divided by 100*, so it can be translated as $\frac{}{100}$.

As a result, "what percent" can be translated as $\frac{y}{100}$.

Here's the translation:

What percent	of	125	is	25 ?
$\frac{y}{100}$	$\times$	125	$=$	25

Find the answer by solving for *y*. Solving equations for a variable will be covered in depth in chapter 6, *Equations*. Here's how to solve this question:

$$\frac{y}{100} \times 125 = 25$$

$$\frac{y}{\underset{4}{\cancel{100}}} \times \overset{5}{\cancel{125}} = 25$$

$$\frac{y}{4} \times 5 = 25$$

$$\frac{y}{4} = 25$$

$$y = 20$$

One note: If you are given the variable *x*, use something other than $\times$ (multiplication sign) to indicate multiplication so that you don't mix up $\times$ and the variable *x* on your paper. Parentheses are a good alternative:

$$\left(\frac{x}{100}\right)125 = 25$$

Here's a last example:

16 is 2% of what number?

This one is definitely weirder than the earlier ones. Should the answer be greater or smaller than 16?

The number 16 represents just 2% of some other number, so that other number must be greater than 16. And since 2% is a very small percentage, that other number should be a decent amount greater than 16. Now, do the math and see whether your answer makes sense with that logic.

Translate word by word. Change 2% either to 0.02 or to $\frac{2}{100} = \frac{1}{50}$:

16	is	2%	of	what number?
16	=	0.02	×	y

Now solve for y:

$$16 = (0.02)\,y$$

$$\frac{16}{0.02} = y$$

$$\frac{16}{0.02} = y$$

$$\frac{1,600}{2} = y$$

$$800 = y$$

Does that make sense? Yes, 800 is a decent amount greater than 16.

If you...	Then you...	Like this:
See "30% of"	Convert 30% into a decimal or fraction, then multiply OR Use benchmarks: Find 10%, then multiply by 3 to reach 30%	30% of 200 = 0.30 × 200 = 60 10% of 200 = (0.1)(200) = 20 30% = 20 × 3 = 60
See "what percent of"	Turn *what percent* into $\frac{y}{100}$, then multiply	What percent of 200 is 60 ? $\left(\frac{y}{100}\right)200 = 60$

Check Your Skills

Solve the following problem.

14. 21 is 30% of what number?

Answer can be found on page 204.

$$\frac{210}{3} = 70$$

Percent Change: Divide Change in Value by Original Value

A story problem might ask about percent change. For example:

> You have $200 in a bank account. You deposit an additional $30 in that account. By what percent did the value of the bank account increase?

Whenever some amount changes and you care about percents, set up this equation:

Original + Change = New

This equation holds true in two ways. First, it holds true for the actual amounts or values, which in this case are in dollars. This is unsurprising. Here is the first equation:

Original value	+	Change in value	=	New value
$200	+	$30	=	$230

This equation also holds true for percents, as long as you mean *percents of the original value*. For example:

Original percent (% of original)	+	Change percent (% of original)	=	New percent (% of original)
100%	+	?	=	?

The original percent is always 100%, since the original value is always 100% of itself.

The change percent is better known as the **percent change**.

You had $200 in your bank account. You added $30. What percent of $200 is $30?

One method is the benchmark approach. How can you get to the value of 30 using some combination of "easy" small percentages of the original number?

200 = 100%	The starting number is always 100% of itself.
20 = 10%	20 (or 10%) is close to 30, but not quite there…
10 = 5%	…add 10 (or 5% more) to get to 30!
	20 + 10 = 30, which represents 10% + 5% = 15%.

The percent change is 15%.

To do this kind of back-of-the-envelope calculation in the future, break down the starting 100% figure into more manageable pieces: 50%, 10%, 5%, or 1%. Then add (or subtract) what you need in order to solve for the percent that you were asked to find.

Or work algebraically. Turn *what percent* into $\frac{y}{100}$, translate the rest, and solve for y:

$$\left(\frac{y}{100}\right)200 = 30$$

$$\left(\frac{y}{\cancel{100}}\right)2\cancel{00} = 30$$

$$2y = 30$$

$$y = 15$$

Therefore, 15% is the percent change.

So the original percent is 100% and the percent change is 15%. What is the new percent? Add them up!

Original percent (% of original)		Change percent (% of original)		New percent (% of original)
	$+$		$=$	
100%	$+$	15%	$=$	115%

If the problem asks you for the percent change, the answer is 15%. If the problem asks you what percent of the original price the new price represents, the answer is 115%.

If you want to use an algebraic approach, use this shortcut to calculate the percent change more directly. The percent change equals the change in value divided by the original value:

$$\text{Percent change (as \% of original)} = \frac{\text{Change in value}}{\text{Original value}}$$

$$\frac{\text{Change in value}}{\text{Original value}} = \frac{\$30}{\$200} = \frac{\overset{15}{\cancel{\$30}}}{\underset{100}{\cancel{\$200}}} = \frac{15}{100} = 15\%$$

The additional $30 corresponds to a 15% change in the value of the account. Take note of the way that the math was simplified. You're looking for a percent, so you want the denominator to be 100 (since *per cent* means *of 100*). Don't simplify to $\frac{3}{20}$, since you'd just have to do more math to figure out what percent that is.

You can also directly calculate the *new percent*. For example:

> You have $200 in a bank account. You deposit an additional $30 in that account. The new balance in the bank account is what percent of the original balance?

Here's the formula:

$$\text{New percent (as \% of original)} = \frac{\text{New value}}{\text{Original value}}$$

$$\frac{\text{New value}}{\text{Original value}} = \frac{\$230}{\$200} = \frac{115}{100} = 115\%$$

As before, make the denominator 100; in this case, divide by 2 and you're there.

If the value of the account *decreases*, the equations still hold true. You just have a negative change. In other words, subtract the change this time. Consider this example:

> You have $200 in a bank account. You make a withdrawal that reduces the value of the account by 40%. How much money remains in the account?

First, use your number sense. You have less than $200 left. Do you have less than $100 left?

In order to have exactly $100 left, you would have to have withdrawn 50% of the value. The problem says you only withdrew 40%, though, so you have a little more than $100 left in the account. Do the math and see whether your answer fits that logic.

Solve for the new percent first. A 40% decrease from the original is a negative 40% change:

Original percent (% of original)	+	Change percent (% of original)	=	New percent (% of original)
100%	+	−40%	=	60%

If you take out 40%, what's left is 60% of the original. Now find that new value:

$$\text{New percent (as \% of original)} = \frac{\text{New value}}{\text{Original value}}$$

$$60\% = \frac{\text{New value}}{\$200}$$

$$\left(\frac{60}{100}\right)200 = y$$

$$\left(\frac{60}{100}\right)2\cancel{00} = y$$

$$120 = y$$

Therefore, $120 remains in the account.

Alternatively, use the benchmark approach. How can you get to 60% using values for some combination of 50%, 10%, 5%, and 1% of the original number?

60% = 50% + 10%	
200 = 100%	The original value is always 100% of itself.
100 = 50%	50% is half of the original value.
20 = 10%	To find 10%, move the decimal one place to the left.

Therefore, 60% = 100 + 20 = 120, so $120 remains in the account.

If you...	Then you...	Like this:
Need to find percent change of an original percent	Use the formula: $\%\ \text{Change} = \dfrac{\text{Change in value}}{\text{Original value}}$ OR Use benchmarks to find the percentage	Original = $200; $30 added: $\dfrac{\$30}{\$200} = 15\%$ Original = $200; $30 added: $\begin{aligned} 200 &= 100\% \\ 20 &= 10\% \\ 10 &= 5\% \\ \hline 30 &= 15\% \end{aligned}$
Need to find a new percent of an original percent	Use the formula: $\text{New}\ \% = \dfrac{\text{New value}}{\text{Original value}}$	Original = $200; $30 added: $\dfrac{230}{200} = 115\%$

If you learn the following shortcuts, you'll make your job easier on the GMAT.

A percent INCREASE of...	...is the same as this NEW percent...	...which is the same as multiplying the ORIGINAL VALUE by...
10%	110%	1.1
20%	120%	1.2, or $\dfrac{6}{5}$
25%	125%	1.25, or $\dfrac{5}{4}$
50%	150%	1.5, or $\dfrac{3}{2}$
100%	200%	2

A percent DECREASE of...	...is the same as this NEW percent...	...which is the same as multiplying the ORIGINAL VALUE by...
10%	90%	0.9
20%	80%	0.8, or $\dfrac{4}{5}$
25%	75%	0.75, or $\dfrac{3}{4}$
50%	50%	0.5, or $\dfrac{1}{2}$
75%	25%	0.25, or $\dfrac{1}{4}$

Percent more than is just like *percent increase*. You do exactly the same math. Consider this example:

$230 is what percent more than $200 ?

Think of $230 as the new value and $200 as the original value. Again, 230 is 15% more than the starting point of $200.

Likewise, *percent less than* is just like *percent decrease*. Consider this example:

$120 is what percent less than $200 ?

Think of $120 as the new value and $200 as the original value. Again, you'll get a 40% decrease or difference.

Which number you call the *original value* matters. The original value is always after the word *than*. It's the value you're comparing the other value *to*. For example:

$110 is what percent *more than* $100 ?

Therefore, $100 is the original value. You're comparing the new number $110 to a starting point of $100, not the other way around.

Finally, watch out for the language *percent OF* versus *percent MORE THAN*. These two expressions don't mean the same thing. Consider these examples:

30 is what percent *more than* 20 ?

30 is what percent *of* 20 ?

The first question (*more than*) is asking you to find only the *increase* from the original number, 20, to the new number, 30. What percentage of 20 do you need to take and add to 20 in order to get to 30? Count the difference: $30 - 20 = 10$.

What percent of the original number does 10 represent?

$$20 = 100\%$$
$$10 = 50\%$$

In other words, 30 is 50 percent *more than* 20. Double-check the math to see that this works, using the percent change formula, since *more than* is a signal to use percent change:

$$\% \text{ change} = \frac{10}{20} = \frac{50}{100} = 50\%$$

The math above works the same way for *less than* problems, except you'll be subtracting the change rather than adding it.

The second question (*30 is what percent of 20*) is different. The word *of* signals that you need to count the original 20 as part of the percent calculation along with the newly added 10. Since 20 is 100% of 20, you have to have at least 100% already. Since you still have to add more to get up to 30, the answer should be greater than 100%. What additional percentage do you need to add to get from 20 to 30?

$$20 = 100\%$$
$$10 = 50\%$$
$$20 + 10 = 30$$
$$100\% + 50\% = 150\%$$

In other words, 30 is 150 percent *of* 20. Translate the equation to double-check the math:

$$30 = \frac{150}{100} \times 20$$

$$30 = \frac{15\cancel{0}}{1\cancel{0}\cancel{0}} \times 2\cancel{0}$$

$$30 = 30$$

By the way, did you happen to notice that the answer to the second one (150%) is exactly 100% more than the first one (50%)? This isn't a coincidence! It will always be the case that the *percent of* question will be 100% more than the *percent more than* question for the exact same original and new numbers.

Recall how the *percent of* question wants you to count the original number in the calculation but the *percent more than* question does not? The original number is always 100% of itself, so the *percent of* calculation will always include this 100% and the *percent more than* question will not.

Keep *percent of* and *percent more than* or *percent less than* distinct; they mean different things!

If you...	Then you...	Like this:
Need to find a *percent more than* or *percent less than*	Treat the problem like a percent increase or a percent decrease	$230 is what percent more than $200 ? $\frac{\$30}{\$200} = 15\%$
Need to find a *percent of*	Translate the equation and solve OR Use benchmarks; set the starting number to 100%	50 is what percent of 25 ? $50 = \frac{y}{100}(25)$ $50 = \frac{y}{4}$ $200 = y$ $25 = 100\%$ $50 = 200\%$

Check Your Skills

15. What is the percent decrease from 90 to 72 ?

Answer can be found on page 204.

Percent of a Percent Of: Multiply Twice

You may be asked to take a percent of a percent of a number. For example:

What is 120% of 150% of 30 ?

These are *percents of*, so turn the percents into decimal or fractional equivalents and multiply 30 by *both* of those equivalents. As always, simplify before you multiply.

But wait! Before you do math, eyeball this. Both 120% and 150% are greater than 100%, so the answer should be greater than the starting value of 30.

Okay, now do the math:

$$120\% = \frac{6}{5} \qquad 150\% = \frac{3}{2}$$

$$120\% \text{ of } 150\% \text{ of } 30 = \frac{6}{5} \times \frac{3}{2} \times 30 = \frac{\cancel{6}^{3}}{5} \times \frac{3}{\cancel{2}} \times 30 = \frac{\cancel{6}}{\cancel{5}} \times \frac{3}{\cancel{2}} \times \cancel{30}^{6} = 9 \times 6 = 54$$

Therefore, 120% of 150% of 30 equals 54. Percent changes often come one after the other. *When you have successive percent changes, multiply the original value by each new percent* (converted to a suitable fraction or decimal).

Try this problem:

> The price of a share, originally $50, increases 10% on Monday and then increases a further 20% on Tuesday. What is the total change in the price of a share, in dollars?

A percent increase of 10% is equivalent to a new percent of 110%, or multiplying the original value by 1.1 or $\frac{11}{10}$.

A percent increase of 20% is equivalent to a new percent of 120%, or multiplying the original value by 1.2 or $\frac{6}{5}$.

Compute the new value by multiplying the original value by each of these factors:

$$\$50 \times \frac{11}{10} \times \frac{6}{5} = \cancel{\$50} \times \frac{11}{\cancel{10}} \times \frac{6}{\cancel{5}} = 11 \times 6 = \$66$$

The change in value is $66 − $50 = $16.

Notice that that change is *not* 10% + 20% = 30% of $50 (that would equal $15, not $16). Never *add* successive percents (in this case, 10% and 20%) if you need to find the exact number. If you add the percents, the answer could be approximately right ($15 is *close* to $16), but it will not be the exact number. (If estimation is good enough, you can add the percents *just* to estimate—but then cross off any answer that exactly matches your estimation.)

If you...	Then you...	Like this:
Have successive percent changes	Multiply the original value by the new percents for *each* percent change	$50 is increased by 10%, and the result is increased by 20% $$\$50\left(\frac{11}{10}\right)\left(\frac{6}{5}\right) = \$66$$

Check Your Skills

16. What is 80% of 75% of 120 ?

Answer can be found on page 204.

Ratios and Percents: Convert Fractions to Percents

You can also use ratio concepts to solve percent problems, and vice versa. You can create ratios between any two components of a ratio, whether part-to-part or part-to-whole:

> Cars : Trucks = 2 : 3 (part-to-part)
> Cars : Total Vehicles = 2 : 5 (part-to-whole)
> Trucks : Total Vehicles = 3 : 5 (part-to-whole)

Use the part-to-whole information to determine percents. For example, because there is a 2 : 5 ratio of cars to total vehicles, $\frac{2}{5}$, or 40%, of the vehicles are cars. Likewise, $\frac{3}{5}$, or 60%, of the vehicles are trucks.

Do not use this logic to create percents using the part-to-part ratio. The result would be illogical. For example, there is a ratio of 2 : 3 cars to trucks, so $\frac{2}{3}$, or approximately 66.66%, of the cars are trucks? That doesn't make sense because this ratio represents two different parts, not one part of a whole.

Here are a few other common part-to-part ratios and the resulting percents of the whole.

Dogs : Cats	Dogs : Cats : Total	Dogs : Total	Cats : Total
1 : 1	1 : 1 : 2	1 : 2 50%	1 : 2 50%
1 : 2	1 : 2 : 3	1 : 3 33.33…%	2 : 3 66.66…%
1 : 3	1 : 3 : 4	1 : 4 25%	3 : 4 75%

Try this problem:

> A bouquet contains white roses and red roses. If the ratio of white to red roses is 5 : 3, what percent of all the roses are red?
>
> (A) 37.5%
> (B) 40%
> (C) 60%
> (D) 62.5%
> (E) 80%

First, glance at those answers. They're really far apart. Do you have a sense as to whether the answer should be at the lower or higher end of that range?

Since the red roses are the smaller number in the ratio, they should represent less than 50% of the roses. Only two answers are reasonable—and you haven't even done any work yet!

Set up a part-to-part-to-whole ratio using just the top row of the standard ratio table:

	White	Red	Total
R	5	3	8

The ratio of red to total is 3 to 8, or $\frac{3}{8}$. As a percent, $\frac{3}{8} = 37.5\%$. The correct answer is (A).

If you are given the percents of the two parts of the whole, you can set up a part-to-part-to-whole ratio with the percents. Here's an example:

If 20% of the animals in a certain zoo are skunks, what is the ratio of non-skunk animals to skunks in the zoo?

The Part + Part = Whole equation is Skunks + Non-skunks = Animals. Use 100% for the whole. If 20% of the animals are skunks, then 100% − 20% = 80% are non-skunks. Reduce those numbers as far as possible to get the ratio:

	S	N-S	Total
%	20%	80%	100%
R	1	4	5

The three numbers share a common factor of 20. Take out that common factor, and you have 1 : 4 : 5 as the ratio of skunks to non-skunks to total animals at the zoo.

Thus, the ratio of non-skunk animals to skunks is 4 to 1. When answering ratio questions, double-check the order in which you were asked to present the desired categories. A ratio of 1 to 4 would be incorrect for this question.

Check Your Skills

17. A flowerbed contains only roses and tulips. If 25% of the flowers are tulips and there are 27 roses, how many total flowers are in the flowerbed?

Answer can be found on page 205

Non-Integer Ratios

So far, the ratio problems have dealt with items that have to come in integer amounts, such as roses or animals or marbles. But some problems allow you to have non-integer amounts—for example, ingredients for recipes. You can still use the ratio box and unknown multiplier with non-integers. Just make sure that you're only using non-integer amounts if the details of the problem allow you to do so.

These kinds of ratio problems often come in the form of some kind of mixture. Consider this example:

A salad dressing recipe calls for 3 cups of olive oil, 1 cup of vinegar, $\frac{1}{2}$ cup of lemon juice, and $\frac{1}{2}$ cup of mustard. If you need 7.5 total cups of salad dressing, how many cups of olive oil will you need to use?

The standard recipe is like the "ratio" of the recipe. First, find the total volume for the recipe ratio:

	O	V	L	M	Total
R	3	1	0.5	0.5	**5**
×					
A	◯				7.5

Then, find the unknown multiplier and solve for the number of cups of olive oil. If needed, use the variable m to represent the unknown multiplier and create an equation to solve:

$$5m = 7.5$$
$$m = 1.5$$

	O	V	L	M	Total
R	3	1	0.5	0.5	5
×	× **1.5**				× **1.5**
A	(4.5)				7.5

It's completely fine to have a non-integer multiplier when you are working with something that can be split into fractional parts, such as the number of cups of salad dressing. You couldn't do this for a problem about people; you don't want to have half of a person!

If you...	Then you...	Like this:
Have a ratio mixture problem	Solve as usual, even if some parts of the problem aren't in integer form	Water : Oil : Total = 3 : 5 : 8 If you have 20 gallons of liquid total, then: Total: $8m = 20$ so $$m = \frac{20}{8} = 2.5$$

Check Your Skills

18. A cereal mixture consists of 2 cups oats to 3 cups almonds to 0.5 cups puffed rice. In order to make 27.5 cups of the mixture, how many cups of puffed rice are needed?

Answer can be found on page 205.

Check Your Skills Answer Key

1. **3.5%:** Make the number *seem* greater; move the decimal to the right two places. $0.035 = 3.5\%$.

2. $\frac{3}{8}$: If you notice that 375 and 1,000 are both divisible by 125, then you can perform the simplification in one step—but don't waste a lot of time trying to find the "perfect" number. Simplify by smaller numbers until you can't go any further. The solution first divides top and bottom by 5 and then divides top and bottom by 25.

$$0.375 = \frac{\overset{75}{\cancel{375}}}{\underset{200}{\cancel{1,000}}} = \frac{\overset{3}{\cancel{75}}}{\underset{8}{\cancel{200}}} = \frac{3}{8}$$

3. $\frac{6}{25}$: Place the number 24 over 100 and simplify.

$$24\% = \frac{\overset{6}{\cancel{24}}}{\underset{25}{\cancel{100}}} = \frac{6}{25}$$

4. **0.6:** This one is on the list to memorize. Alternatively, because the denominator is 5, you can multiply numerator and denominator by the same number to turn the denominator into a power of 10.

$$\frac{3}{5} \times \frac{2}{2} = \frac{6}{10} = 0.6$$

If you had multiple-choice answers, you might also be able to estimate. The fraction $\frac{3}{5}$ is a little more than 50%, since $\frac{2.5}{5}$ is equal to 50%.

5. **37.5%:** This one is on the list to memorize. You could also estimate. The fraction $\frac{2}{8}$ is equal to $\frac{1}{4}$ and the fraction $\frac{4}{8}$ is equal to $\frac{1}{2}$, so $\frac{3}{8}$ must be between 25% and 50%. In fact, it's exactly halfway between the other two, or 37.5%.

As a last resort, perform long division to find the decimal, then convert to a percent.

$$
\begin{array}{r}
0.375 \\
8\overline{)3.000} \\
\underline{24} \\
60 \\
\underline{56} \\
40 \\
\underline{40} \\
0
\end{array}
$$

6. **3,275.3:** Move the decimal two places to the right.

$$32.753 \times 10^2 = 3,275.3$$

7. **4.3681:** Move the decimal four places to the left.

$$43{,}681 \times 10^{-4} = 43{,}681 \div 10^4 = 4.3681$$

8. **3.173:** Make sure to carry over the 1 from the units digit column. $8 + 5 = 13$, so 3 becomes the units digit and the tens digit becomes $2 + 4 + 1 = 7$.

$$
\begin{array}{r}
3.128 \\
+ \quad 0.045 \\
\hline
3.173
\end{array}
$$

9. **1.5616:** Line up the decimals and subtract.

$$
\begin{array}{r}
1.8746 \\
- \quad 0.3130 \\
\hline
1.5616
\end{array}
$$

10. **0.66:**

$$0.6 \times 1.1 =$$
$$6 \times 11 = 66$$
$$0.6 \times 1.1 = 0.66$$

11. **0.000128:**

$$0.004 \times 0.032$$
$$4 \times 32 = 128$$
$$0.004 \times 0.032 = 0.000128$$

12. **208:** Move the decimal in 520,000 to the left four places, and move the decimal in 0.0004 to the right four places.

$$520{,}000 \times 0.0004 = 52 \times 4 = 208$$

13. **12:** Move each decimal to the right five places.

$$\frac{0.00084}{0.00007} = \frac{84}{7} = 12$$

14. **70:** The value 21 is 30% of some number, so expect the answer to be greater than 21. Translate the equation, then solve.

21	=	$\frac{30}{100}$	$\times$	y
21	is	30 percent	of	what number?

$$21 = \frac{30}{100}y$$

$$\frac{10\cancel{0}}{3\cancel{0}} \times 21 = y$$

$$\frac{10 \times \overset{7}{\cancel{21}}}{\underset{1}{\cancel{3}}} = y$$

$$70 = y$$

15. **20%:** To find the percent decrease, focus on the amount by which 90 was reduced. $90 - 72 = 18$, so:

$$\frac{18}{90} = \frac{\overset{2}{\cancel{18}}}{\underset{10}{\cancel{90}}} = \frac{2}{10} = 20\%$$

Alternatively, benchmark the answer. How do you go from 90 to 18?

$$90 = 100\%$$
$$9 = 10\%$$

Multiply 9 by 2 to get 18, so multiply 10% by 2 to get 20%.

16. **72:** Convert 80% and 75% to fractions. $80\% = \frac{4}{5}$ and $75\% = \frac{3}{4}$.

$$x = \frac{\cancel{4}}{5} \times \frac{3}{\cancel{4}} \times 120$$

$$x = \frac{\cancel{4}}{\cancel{5}} \times \frac{3}{\cancel{4}} \times \overset{24}{\cancel{120}}$$

$$x = 3 \times 24 = 72$$

17. **36:** This problem can be solved using ratios or percentages. Both solutions are shown.

    If 25% of the flowers are tulips, then 75% are roses. The ratio of T : R is 25% : 75% or 1 : 3. Use this information, coupled with the fact that there are 27 roses, to find the total number of flowers.

	T	R	Total
R	1	3	4
×		9	9
A		27	(36)

    Alternatively, use percentages to solve. The problem states that there are 27 roses. The problem also states that 25% of the flowers are tulips, so the 27 roses represent 75% of all flowers. You're missing the final 25% to get to 100%, or all of the flowers. How can you use the given information to find the value for 25% of the flowers?

    Benchmark! You know that 75% = 27. The value 25% is one-third of 75%, so the corresponding figure is one-third of 27, or 9. Since 75% + 25% = 100%, it's also the case that 27 + 9 = 36 represents the total number of flowers.

18. **2.5:** Find the unknown multiplier and solve for the amount of puffed rice needed.

	O	A	R	Total
R	2	3	0.5	5.5
×			× **5**	× **5**
A			(2.5)	27.5

Chapter Review: Drill Sets

Drill 1

1. Fill in the missing information in the chart below. See the third row for an example.

Fraction	Decimal	Percent
		1%
$\frac{1}{20}$		
$\frac{1}{10}$	0.1	10%
$\frac{1}{8}$		
	0.2	
		25%
	0.3	
		33.33…%
$\frac{3}{8}$		
		40%
$\frac{1}{2}$		
	0.6	
		66.66…%
		70%
	0.75	
$\frac{4}{5}$		
	0.875	
$\frac{9}{10}$		
		100%
	1.1	
$\frac{6}{5}$		
		125%
	1.5	

2. Fill in the missing information in the chart below. Use the first row as an example. Note that improper fractions cannot be converted into part-to-part ratios.

Ratio (Part : Part)	Fraction (Part-to-Whole)	Decimal	Percent
3 : 17	$\frac{3}{20}$	0.15	15%
9 : 11			45%
		0.20	
	$\frac{4}{5}$		
n/a		13.5	
6 : 14			30%
		0.33	
			0.02%
5 : 3			62.5%
n/a	$\frac{3}{2}$		

Drill 2

Simplify the following expressions. Answers in the explanations may appear in more than one form (e.g., fraction and decimal).

3. What is 30% of $3.50 ?

4. What is 0.7 times 110% ?

5. Simplify $\frac{2}{5} + 20\% + 0.7$.

6. Simplify $1.5 \div \left(\frac{5}{8} - 50\% \right)$.

7. Simplify $190\% - \left(1.2 \div \frac{4}{5} \right)$.

Drill 3

Simplify the following expressions.

8. 0.27×2

9. $0.48 + 0.02$

10. 20×0.35

11. $\dfrac{54.197}{10^2}$

12. $\dfrac{12.6}{0.3}$

Drill 4

Simplify the following expressions.

13. $\dfrac{6}{0.5}$

14. 2.1×0.04

15. $0.6(50) + 0.25(120)$

16. $\dfrac{0.49}{0.07}$

17. $100 \times 0.01 \times 0.01$

Drill 5

18. Which of the following equals 4.672×10^4 ?

    (A)   4,672

    (B)   46,720

    (C)   467,200

19. Which of the following equals 337×10^{-4} ?

    (A)   3,370,000

    (B)   0.0337

    (C)   0.0000337

20. Which of the following equals 8.25×10^5 ?

    (A) 825×10^7

    (B) 825×10^4

    (C) 825×10^3

21. Which of the following equals 0.003482 ?

    (A) 34.82×10^{-4}

    (B) 34.82×10^2

    (C) 34.82×10^4

22. Which of the following equals 12.12×10^{-3} ?

    (A) -1.21×10^3

    (B) 0.00001212×10^3

    (C) 0.01212×10^3

Drill 6

23. What is 15% of 40 ?

24. 12 is 5% of what number?

25. 4 is what percent of 32 ?

    (A)   7.5%

    (B)   10%

    (C)   12.5%

    (D)   20%

    (E)   25%

26. 7% of 9 is what percent of 7 ?

27. 25% of 30 is 75% of what number?

28. What percent of 6 is 37.5% of 160 ?

29. If 14 is added to 56, what is the percent increase?

30. What is the percent increase from 50 to 60 ?

31. What number is 40% more than 30 ?

32. What is 60% less than 60 ?

Drill 7

33. If m is reduced by 55%, the resulting number is 90. What is the value of m ?

    (A)   40.5

    (B)   163.63...

    (C)   200

34. If 75 reduced by x percent is 54, what is the value of x ?

    (A)   9

    (B)   28

    (C)   72

35. If x is 15% more than 20, what is 30% of x ?

    (A)   0.9

    (B)   4.9

    (C)   6.9

36. What is 50% of 12% of 50 ?

    (A)   3

    (B) $\dfrac{2,500}{3}$

    (C)   300

37. What is 120% of 30% of 400 ?

    (A)   24

    (B)   144

    (C)   364

38. If 45% of 80 is x percent more than 24, what is the value of x ?

39. 10% of 30% of what number is 200% of 6 ?

40. If $q \neq 0$, what percent of 25% of q is q percent of 20 ?

41. If $a \neq 0$, 200% of 4% of a is what percent of $\dfrac{a}{2}$?

42. If positive integer m is first increased by 20%, then decreased by 25%, and finally increased by 60%, the resulting number is what percent of m ?

Drill 8

43. A recipe calls for 1 cup of cheese and $\dfrac{1}{2}$ cup of sauce in order to make 1 pizza. If Bob used 15 cups of sauce to make pizzas, how many cups of cheese did he use?

44. On a safari, visitors saw only giraffes and lions. If they saw 7 giraffes for every 3 lions, and they saw 60 animals in total, how many lions did they see?

    (A)   18

    (B)   21

    (C)   42

45. The ratio of oranges to peaches to strawberries in a fruit jam recipe is 2 : 3 : 6. If Rav uses four and a half peaches in the recipe, how many oranges and how many strawberries should Rav add?

46. A certain automotive dealer sells only cars and trucks and currently has 51 trucks for sale. If the ratio of cars for sale to trucks for sale is 1 to 3, how many cars are for sale?

47. Mustafa has invented a new dance in which he moves 3 steps forward for every 4 steps he moves back. If the dance requires 49 steps in total, how many steps forward has Mustafa taken at the completion of the dance?

    (A)   7

    (B)   21

    (C)   28

48. A steel manufacturer combines 98 ounces of iron with 2 ounces of carbon to make 1 sheet of steel. How much iron, in ounces, is used in $\dfrac{1}{2}$ of a sheet of steel?

49. To make a 64-ounce smoothie, Malin must use 4 bananas, 2 apples, 6 cups of yogurt, and 8 teaspoons of protein powder. How many cups of yogurt will Malin need to make a 16-ounce smoothie?

    (A) $\dfrac{1}{2}$

    (B)   1

    (C) $1\dfrac{1}{2}$

Drill Sets Solutions

Drill 1

1.

Fraction	Decimal	Percent
$\frac{1}{100}$	0.01	1%
$\frac{1}{20}$	0.05	5%
$\frac{1}{10}$	0.1	10%
$\frac{1}{8}$	0.125	12.5%
$\frac{1}{5}$	0.2	20%
$\frac{1}{4}$	0.25	25%
$\frac{3}{10}$	0.3	30%
$\frac{1}{3}$	0.3333...	33.33...%
$\frac{3}{8}$	0.375	37.5%
$\frac{2}{5}$	0.40	40%
$\frac{1}{2}$	0.50	50%
$\frac{3}{5}$	0.6	60%
$\frac{2}{3}$	0.6666...	66.66...%
$\frac{7}{10}$	0.7	70%
$\frac{3}{4}$	0.75	75%
$\frac{4}{5}$	0.8	80%
$\frac{7}{8}$	0.875	87.5%

Fraction	Decimal	Percent
$\frac{9}{10}$	0.9	90%
1	1.0	100%
$\frac{11}{10}$	1.1	110%
$\frac{6}{5}$	1.2	120%
$\frac{5}{4}$	1.25	125%
$\frac{3}{2}$	1.5	150%

2.

Ratio (Part : Part)	Fraction (Part-to-Whole)	Decimal	Percent
3 : 17	$\frac{3}{20}$	0.15	15%
9 : 11	$\frac{9}{20}$	0.45	45%
1 : 4	$\frac{1}{5}$	0.20	20%
4 : 1	$\frac{4}{5}$	0.8	80%
n/a	$\frac{27}{2}$	13.5	1,350%
3 : 7	$\frac{3}{10}$	0.3	30%
33 : 67	$\frac{33}{100}$	0.33	33%
1 : 4,999	$\frac{1}{5,000}$	0.0002	0.02%
5 : 3	$\frac{5}{8}$	0.625	62.5%
n/a	$\frac{3}{2}$	1.5	150%

Drill 2

3. **$1.05:** Use benchmarks. How can you get to 30% using the "easy" percentages of 50%, 10%, 5%, or 1%?

    30% of $3.50

    | 100%: $3.50 | Take 10% of $3.50. |
    | 10%: $0.35 | |

    $(0.35)(3) = \$1.05$ Multiply by 3 to get 30%.

4. **0.77 or $\dfrac{77}{100}$ or 77%:** If you feel comfortable with percents, take 10% of 0.7 and add that to 100% of 0.7 to get 110% of 0.7.

    100% of 0.7 is 0.7

    10% of 0.7 is 0.07   Move the decimal one place to the left.

    $0.7 + 0.07 = 0.77$

    Alternatively, convert both values to fractions (because fractions are easier to simplify when multiplying), and solve from there.

    $$0.7 \times 110\% = \frac{7}{10} \times \frac{11\cancel{0}}{10\cancel{0}} = \frac{77}{100} = 77\% = 0.77$$

5. **130% or $\dfrac{13}{10}$ or 1.3:** All of the forms are part of your lists to memorize. It's easier to add integers (in the form of percents) than decimals or fractions, so convert all three terms to percents.

    $$\frac{2}{5} + 20\% + 0.7 =$$
    $$40\% + 20\% + 70\% = 130\%$$

6. **12 or 1,200%:**

$$1.5 \div \left(\frac{5}{8} - 50\%\right) =$$

Convert percent to fraction because 50% is easier to convert than $\frac{5}{8}$.

$$1.5 \div \left(\frac{5}{8} - \frac{1}{2}\right) =$$

$$1.5 \div \left(\frac{5}{8} - \frac{4}{8}\right) =$$

$$1.5 \div \left(\frac{1}{8}\right) =$$

Convert decimal to fraction because first, 1.5 is easier to convert than $\frac{1}{8}$; and second, it's easier to divide fractions, as factors often cancel.

$$\frac{3}{2} \div \frac{1}{8} =$$

$$\frac{3}{2} \times \frac{8}{1} =$$

$$\frac{3}{\cancel{2}} \times \frac{\cancel{8}^{4}}{1} = 12$$

7. **40% or 0.4 or $\frac{2}{5}$:**

$$190\% - \left(1.2 \div \frac{4}{5}\right) =$$

Convert decimal to fraction because fractions are easier to divide than decimals.

$$190\% - \left(\frac{6}{5} \div \frac{4}{5}\right) =$$

$$190\% - \left(\frac{6}{5} \times \frac{5}{4}\right) =$$

$$190\% - \left(\frac{3}{2}\right) =$$

$$190\% - 150\% = 40\%$$

Convert fraction to percent because $\frac{3}{2}$ is easier to convert than 190%, and percents are easier to subtract.

Drill 3

8. **0.54:** First, ignore the decimals and multiply the numbers. Then move the decimal two places to the left to account for the two decimals in the initial problem.

$$0.27 \times 2 =$$
$$27 \times 2 = 54$$
$$0.27 \times 2 = 0.54$$

9. **0.5:** Line up the decimal points and add.

$$\begin{array}{r} 0.48 \\ + \ 0.02 \\ \hline 0.50 \end{array}$$

10. **7:** Trade off a decimal place to make the math easier.

$$20 \times 0.35 =$$
$$2 \times 3.5 = \qquad \text{Take a decimal place from 20 and give it to 0.35.}$$
$$2 \times 3.5 = 7$$

It is possible to solve this problem by counting decimals, but the 0 in the units digit of 20 makes it easy to make a mistake. Here's what the math would look like.

$$20 \times 0.35 =$$
$$20 \times 35 = 700 \qquad \text{Ignore the decimals at first.}$$
$$20 \times 0.35 = 7.00$$

11. **0.54197:** Divide by 100, or move the decimal two places to the left.

$$\frac{54.197}{10^2} = \frac{54.197}{100} = 0.54197$$

12. **42:** Multiply the top and bottom by 10, or move each decimal one place to the right, in order to get integers.

$$\frac{12.6}{0.3} \times \frac{10}{10} = \frac{126}{3} = 42$$

Drill 4

13. **12:** Multiply the top and bottom by 10 (or move the decimal one place to the right on both the top and bottom).

$$\frac{6}{0.5} \times \frac{10}{10} = \frac{60}{5} = 12$$

14. **0.084:** Ignore the decimals and multiply the numbers. Then move the decimal three places to the left.

$$2.1 \times 0.04 =$$
$$21 \times 4 = 84$$
$$2.1 \times 0.04 = 0.084$$

15. **60:** Convert the decimals to fractions because fractions are easier to multiply, then simplify.

$$0.6(50) + 0.25(120) =$$
$$\left(\frac{3}{5} \times 50\right) + \left(\frac{1}{4} \times 120\right) =$$
$$\left(\frac{3}{\cancel{5}} \times \cancel{50}^{\,10}\right) + \left(\frac{1}{\cancel{4}} \times \cancel{120}^{\,30}\right) =$$
$$30 + 30 = 60$$

16. **7:** Multiply the top and bottom by 100 (or move the decimal point two places to the right on the top and bottom).

$$\frac{0.49}{0.07} \times \frac{100}{100} = \frac{49}{7} = 7$$

17. **0.01:** Ignore the decimals and multiply the numbers first, then insert the missing decimal by moving the decimal point four places to the left.

$$100 \times 0.01 \times 0.01 =$$
$$100 \times 1 \times 1 = 100$$
$$100 \times 0.01 \times 0.01 = 0.01$$

Drill 5

18. **(B) 46,720:** The answer choices don't use a power of 10, so multiply that 10^4 into the starting number. Since the exponent is a positive 4, move the decimal four places to the right:

$$4.6720$$

The number becomes 46,720.

19. **(B) 0.0337:** Multiplying by 10^{-4} will make the resultant number smaller, so eliminate answer choice (A). Because the exponent is -4, move the decimal four places to the left:

$$0337.$$

The number becomes 0.0337.

20. **(C) 825×10^3:** Glance at the answers. All are 825 multiplied by a power of 10. What math occurs to turn 8.25 into 825?

$$8.25$$

In other words, 8.25 increases by two decimal places to 825. In order to balance out that increase, reduce the exponent by 2, so 10^5 becomes 10^3. The new form of the number is 825×10^3.

21. **(A) 34.82×10^{-4}:** Glance at the answers. Two have positive powers of ten, and one has a negative power of ten. All start with the form 34.82, which is larger than 0.003482. To equate those numbers, 34.82 must be multiplied by a negative power of ten. Only answer choice (A) fits.

To compute the answer exactly, first figure out how many decimal places you need to move in order to get to 34.82.

$$0.003482$$

The number 0.003482 increased by four decimal places. In order to balance out that increase, multiply by a power of 10^{-4}, since the exponent is equivalent to the number of decimal places moved: 34.82×10^{-4}.

22. **(B) 0.00001212×10^3:** Glance at the answers. Each one uses the form 10^3. If you change the exponent from -3 to 3, how do you need to change the starting number?

To go from -3 to 3, add 6. If the exponent adds 6, then the starting number needs to lose six decimal places. Glance at the answers again. Answer (A) changes to a negative number, but moving the decimals won't change a positive number to a negative one. Answer (C) doesn't move the decimal enough places, so the answer must be (B).

$$000012.12$$

The number becomes 0.00001212×10^3.

Drill 6

23. **6:** Eyeball this. 15% is small, so the answer should be significantly less than 40 (but still positive).

To benchmark, find 10% and 5% of 40, then add to get 15%, which is what the problem asks for. $10\% = 4$ and $5\% = 2$, so $15\% = 4 + 2 = 6$.

Alternatively, use algebra.

What	is	15%	of	40 ?
x	$=$	$\dfrac{15}{100}$	$\times$	40

$$x = \frac{15}{100} \times 40$$
$$x = \frac{3}{20} \times 40$$
$$x = \frac{3}{20} \times \overset{2}{\cancel{40}}$$
$$x = 6$$

24. **240:** Because 12 is 5% of a number, expect the answer to be much greater than 12.

To benchmark, set 12 equal to 5% and find 100%.

$$5\% = 12$$
$$\times 20 \quad \times 20$$
$$100\% = 240$$

Alternatively, use algebra.

12	is	5%	of	what number?
12	=	$\frac{5}{100}$	$\times$	y

$$12 = \frac{5}{100}(y)$$
$$12 = \frac{1}{20}(y)$$
$$20(12) = y$$
$$y = 240$$

25. **(C) 12.5%:** The value 4 is considerably less than 32. In fact, it's less than 50% of 32, and you may even recognize (before doing the math) that it's less than 25%. To benchmark, find 10% and 20%, since those are in the answer choices.

$$100\% = 32$$
$$10\% = 3.2$$

10% is only 3.2, which is too small, so the answer can't be (A) or (B). And 20% would be twice that, or 6.4, which is too big, so the answer can't be (D) or (E). Only answer (C) makes sense. The GMAT rarely asks you to complete tough computations—the test writers know they haven't given you a calculator. Instead, messy calculations can often be estimated. Look for these opportunities!

26. **9%:** Translate and solve.

7%	of	9	is	what percent	of	7
$\frac{7}{100}$	$\times$	9	=	$\frac{x}{100}$	$\times$	7

$$\frac{7}{100} \times 9 = \left(\frac{x}{100}\right)(7) \qquad \text{Multiply both sides by 100.}$$
$$7 \times 9 = 7x \qquad \text{Divide both sides by 7.}$$
$$9 = x$$

27. **10:** Use your memorized equivalents for 25% and 75%.

25%	of	30	is	75%	of	what number?
$\frac{1}{4}$	$\times$	30	$=$	$\frac{3}{4}$	$\times$	x

$\frac{1}{4} \times 30 = \left(\frac{3}{4}\right)(x)$ Multiply both sides by 4.

$30 = 3x$ Divide both sides by 3.

$10 = x$

28. **1,000%:** 37.5% is on the list of common equivalents to memorize.

What percent	of	6	is	37.5%	of	160
$\frac{x}{100}$	$\times$	6	$=$	$\frac{3}{8}$	$\times$	160

$\frac{x}{100} \times 6 = \frac{3}{8} \times 160$ Simplify before you multiply.

$\frac{x}{100} \times 6 = \frac{3}{\cancel{8}} \times \cancel{160}^{20}$

$\frac{x}{100} \times 6 = 60$

$\frac{x}{100} = 10$

$x = 1,000$

29. **25%:** In this percent change problem, the change is 14 and the original number is 56. Use the percent change formula:

$$\% \text{ change} = \frac{\text{change}}{\text{original}}$$

$$= \frac{14}{56} = \frac{1}{4} = 25\%$$

30. **20%:** If the increase is from 50 to 60, then the change is $60 - 50 = 10$, and 50 is the original number. Use the percent change formula.

$$\% \text{ change} = \frac{\text{change}}{\text{original}}$$

$$\frac{10}{50} = \frac{1}{5} = 20\%$$

31. **42:** There are two ways to represent 40% more than 30. The first is a literal translation—30 plus an additional 40% of 30.

$$x = 30 + \left(\frac{40}{100} \times 30\right)$$

Find 40% of 30, then add to 30.

$$100\% = 30 \qquad \text{Find 10\% of 30.}$$
$$10\% = 3$$
$$(3)(4) = 12 \qquad \text{To find 40\%, multiply 10\% by 4.}$$
$$30 + 12 = 42 \qquad \text{Add } 30 + 40\% \text{ of 30.}$$

Alternatively, 40% more than a number is the same as 100% of that number plus an additional 40%, or 140%. In that case, 40% more than 30 can be represented this way.

$$x = \frac{140}{100} \times 30$$
$$x = \frac{14\cancel{0}}{10\cancel{0}} \times 3\cancel{0}$$
$$x = 42$$

32. **24:** You can use fractions to set this up.

$$x = 60 - \left(\frac{60}{100} \times 60\right)$$

60% of 60 is equivalent to 10% of 60 multiplied by 6. Do the math in two steps: 10% of 60 is 6, and $6 \times 6 = 36$. Therefore:

$$x = 60 - 36 = 24$$

Alternatively, 60% less than 60 is equivalent to 40% of 60.

$$x = \frac{40}{100} \times 60$$

40% of 60 is equivalent to 10% of 60 multiplied by 4. Do the math in two steps: 10% of 60 is 6, and $6 \times 4 = 24$.

5

Drill 7

33. **(C) 200:** Reducing a number by 55% will result in a smaller number, so the answer must be greater than 90. Answer choice (A) is impossible. To differentiate between answers (B) and (C), start calculating—but stop at any point that you realize only one answer makes sense. If m is *reduced* by 55%, then 45% of m remains. Use the 45% figure in the translation.

$$\frac{45}{100}m = 90 \qquad \text{Cross-multiply to isolate the } m.$$

$$m = 90 \times \frac{100}{45} \qquad \text{Simplify before you divide!}$$

$$m = {}^{2}\cancel{90} \times \frac{100}{\cancel{45}_{1}} \qquad \text{The 90 and 45 cancel. Don't forget about the two leftover zeros!}$$

$$m = 200$$

Glance at the answer in the context of the question stem. Does it make sense? When 200 is reduced by 55%, the answer is 90. In other words, 200 is reduced by a little more than half, so what remains (90) is a little less than half; the answer does make sense.

When you have lots of zeros or are shifting decimals, you may want to double-check the solution in this way to make sure that it makes sense. If you had gotten answer (A), 40.5, for example, you would have asked yourself: Does it make sense that 40.5 reduced by 55% equals 90? No! In this way, you might catch a careless mistake or even get yourself to the correct answer choice.

34. **(B) 28:** Glance at the answers; they're really far apart. 75 will get smaller by some percentage, and the new number will be 54. If 75 were reduced by 50% or greater, the result would be less than half of 75, which is less than about 40. Because the result is 54, the percent reduction should be less than 50%. Eliminate answer choice (C).

Next, if 75 were reduced by 10%, the result would be $75 - 7.5 = 67.5$, so 9% doesn't go far enough. The answer can't be (A) either; only answer (B) makes sense.

To solve directly, translate the problem. A reduction is translated as 75 minus x percent of 75. As always, simplify before you multiply.

$$75 - \left(\frac{x}{100} \times 75\right) = 54$$

$$75 - \frac{75x}{100} = 54$$

$$75 - 54 = \frac{75x}{100}$$

$$21 = \frac{3x}{4}$$

$$\overset{7}{\cancel{21}}\left(\frac{4}{\cancel{3}}\right) = x$$

$$28 = x$$

You could also use the percent change formula. If 75 has been reduced to 54, then the change is $75 - 54 = 21$. The original number is 75.

$$\% \text{ change} = \frac{\text{change}}{\text{original}}$$

$$\frac{21}{75} = \frac{7}{25}$$

Now what? Get the denominator to be 100 so that the number is expressed as a percent.

$$\frac{7}{25} = \frac{28}{100} = 28\%$$

35. **(C) 6.9:** The answers are quite far apart, so begin by estimating. The value of x is greater than 20, but not a lot greater, so let x equal 20 for now. What's 30% of 20? Benchmark: 10% is 2, so 30% is 6. The actual value of x is something greater than 20, so the actual answer must be greater than 6. Only answer (C) qualifies.

To solve algebraically, translate the given equation (x is 15% more than 20) and the question (what is 30% of x)? Also, the given equation specifies a variable (x), so translate the word *what* as a different variable, such as y.

$$x = \frac{115}{100} \times 20$$

$$y = \frac{3}{10} \times x$$

Solve the first equation for x.

$$x = \frac{115}{100} \times 20 = \frac{115}{\cancelto{5}{100}} \times \cancel{20} = \frac{115}{5} = 23$$

Plug $x = 23$ into the second equation to find y.

$$y = \frac{3}{10} \times 23$$

$$y = \frac{69}{10}$$

$$y = 6.9$$

36. **(A) 3:** Eyeball the problem. The answer has to be less than 50, so answer (C) won't work. Answer (B) also represents a value that is greater than 50; cross that one off, too. Only answer (A) makes sense.

To solve, benchmark. First, 50% of 50 is 25. Next, what's 12% of 25? $12\% = 10\% + 1\% + 1\% = 2.5 + 0.25 + 0.25 = 3$.

To solve algebraically, use a fractional representation because fractions are easier to multiply.

$$x = \frac{1}{2} \times \frac{12}{100} \times 50$$

$$x = \frac{1}{2} \times \frac{6}{\cancel{50}} \times \cancel{50}$$

$$x = 3$$

37. **(B) 144:** This one is harder to eyeball because one percentage increases the value and the other decreases it. The increase of 120% is really only adding 20% to the base figure of 400, while taking 30% will reduce that figure to less than half, so answer (C) is too large.

    To solve, benchmark. First, find 120% of 400. This is 20% more than 400, so find 20% and add: 10% = 40, so 20% = 80. The new value is 400 + 80 = 480. Next, find 30% of 480: 10% = 48, so 30% = answer (B), since answer (A) is less than 48. (In fact, 48 × 3 = 144.)

    Alternatively, solve algebraically. Note that there are four zeros on top and four on the bottom; cancel them all.

    $$x = \frac{120}{100} \times \frac{30}{100} \times 400$$
    $$x = \frac{12\cancel{0}}{1\cancel{00}} \times \frac{3\cancel{0}}{1\cancel{00}} \times 4\cancel{00}$$
    $$x = 12 \times 3 \times 4$$
    $$x = 144$$

38. **50:** Because the translation is decently long, it might be helpful to figure out the parts separately.

    For the left-hand side of the equation, 45% of 80, benchmark to find 45%: 10% + 10% + 10% + 10% + 5%, or (4)(10%) + 5%. First, find the individual percentages: 10% of 80 is 8 and 5% of 80 is half of 8, or 4. Then, find 45%.

    $$(4)(10\%) + 5\% = 45\%$$
    $$(4)(8) + 4 = 36$$

    Alternatively, you can make 45% by taking 50% − 5%, which is 40 − 4 = 36.

    Either way, the equation in the question stem is now "36 is x percent more than 24." This is a percent change problem! The original number is the "more than" number, 24. The change is the difference between the two numbers: 36 − 24 = 12. Use the percent change formula to solve.

    $$\text{Percent change} = \frac{12}{24} = \frac{1}{2} = 50\%$$

    Note that the answer is 50, not 50%. The question stem refers to x percent and asks for x by itself. If you plug 50 for x into the question stem, the equation will then say 50%.

39. **400:** If you've memorized the fact that 200% = 2, you can skip the first line of math and go straight to the second.

    $$\frac{1}{10} \times \frac{3}{10} \times y = \frac{200}{100} \times 6$$
    $$\frac{1}{10} \times \frac{3}{10} \times y = 2 \times 6$$
    $$y = \left(2 \times \cancel{6}^{2}\right)\left(\frac{100}{\cancel{3}}\right)$$
    $$y = 400$$

40. **80%:** The question already contains a variable (q), so use another variable, such as n, to represent the unknown *what percent*.

$$\frac{n}{100} \times \frac{1}{4} \times q = \frac{q}{100} \times 20$$

$$\frac{n}{\cancel{100}} \times \frac{1}{4} \times \cancel{q} = \frac{\cancel{q}}{\cancel{100}} \times 20 \qquad \text{Divide both sides by } q, \text{ and multiply both sides by 100.}$$

$$\frac{n}{4} = 20$$

$$n = 80$$

41. **16%:** If you've memorized the fact that $200\% = 2$, you can skip the first line of math and go straight to the second.

$$\frac{200}{100} \times \frac{4}{100} \times a = \frac{x}{100} \times \frac{a}{2}$$

$$2 \times \frac{4}{\cancel{100}} \times \cancel{a} = \frac{x}{\cancel{100}} \times \frac{\cancel{a}}{2} \qquad \text{Divide both sides by } a, \text{ and multiply both sides by 100.}$$

$$2 \times 4 = \frac{x}{2}$$

$$2 \times 2 \times 4 = x$$

$$16 = x$$

42. **144%:** Assign a new variable, such as y, for *what percent*. Also, both sides of the equation below are multiplied by the variable m. Cancel m immediately (since it is not zero), and solve for y.

$$\cancel{m} \times \frac{120}{100} \times \frac{75}{100} \times \frac{160}{100} = \frac{y}{100} \times \cancel{m}$$

$$\frac{6}{5} \times \frac{3}{4} \times \frac{8}{5} = \frac{y}{100}$$

$$\frac{6}{5} \times \frac{3}{\cancel{4}} \times \frac{\overset{2}{\cancel{8}}}{5} = \frac{y}{100}$$

$$\frac{36}{25} = \frac{y}{100}$$

$$\overset{4}{\cancel{100}} \times \frac{36}{\cancel{25}} = y$$

$$144 = y$$

Drill 8

43. **30:** The ratio of cheese to sauce is $1 : \frac{1}{2}$, and Bob uses 15 cups of sauce. Find the unknown multiplier and solve for the cheese.

	Cheese	Sauce	Total
R	1	$\frac{1}{2}$	
×	× 30	× 30	
A	(30)	15	

44. **(A) 18:** First, glance at those answers. Did the visitors see more giraffes or more lions? The ratio of giraffes to lions is 7 : 3, and they saw 60 animals total, so they saw more giraffes than lions. Fewer than half of the animals were lions, so the answer should be less than 30. Eliminate answer (C).

	G	L	Total
R	7	3	10
×		× 6	× 6
A		(18)	60

45. **3.5 oranges, 9 strawberries:** The ratio of oranges to peaches to strawberries is 2 : 3 : 6, and there are 4.5 peaches total. Set up a table to solve for the number of pieces of fruit.

	O	P	S	Total
R	2	3	6	11
×	× 1.5	× 1.5	× 1.5	
A	(3.5)	4.5	(9)	

46. **17:** The ratio of cars to trucks is 1 : 3, and there are 51 trucks for sale. Set up a table to solve for the number of cars.

	C	T	Total
R	1	3	
×	× 17	× 17	
A	(17)	51	

47.   **(B) 21:** The ratio of forward to back is 3 : 4, and the dance consists of 49 steps in total. Use a table to determine how many forward steps Mustafa takes.

	F	B	Total
R	3	4	7
×	× 7		× 7
A	(21)		49

48.   **49:** The ratio of iron to carbon to sheets of steel is 98 : 2 : 1. You can set up a table to find how much iron to use to make just $\frac{1}{2}$ of a sheet of steel. But reflect for a moment first.

You might notice that $\frac{1}{2}$ of a sheet of steel is exactly half of 1 full sheet of steel. As a result, you would need half as much iron, or $\frac{98}{2} = 49$ ounces. No table needed!

49.   **(C) $1\frac{1}{2}$:** The ratio of bananas to apples to yogurt to protein powder is 4 : 2 : 6 : 8 for a 64-ounce smoothie. Set up a table to determine the number of cups of yogurt needed to make a 16-ounce smoothie. Note that, in this case, the total is the 64 ounces of smoothie, not the total number of various ingredients, since the different ingredients are not measured using the same starting units.

	B	A	Y	P	Total Ounces in Smoothie
R	4	2	6	8	64
×			$\times \frac{1}{4}$		$\times \frac{1}{4}$
A			$1\frac{1}{2}$		16

Alternatively, logic it out. A 16-ounce smoothie is exactly one-quarter of a 64-ounce smoothie. As a result, Malin will need one-quarter as much yogurt, or $6 \times \frac{1}{4} = 1.5$, or $1\frac{1}{2}$ cups.

CHAPTER 6

Equations

In This Chapter

In this chapter, you will learn to manipulate expressions and equations in order to solve for unknowns, otherwise known as *variables*.

CHAPTER 6 Equations

Manipulating expressions and equations is at the core of algebra. The first step is to understand the distinction between expressions and equations.

Expressions Don't Have Equals Signs

An **expression** such as $3y + 8z$ ultimately represents a number. It has a value, although you may not know that value. The expression $3y + 8z$ contains numbers that are known (3 and 8) and unknown (y and z) and that are linked by arithmetic operations (in this case, $+$ and $\times$).

Here are more expressions:

$$12 \qquad x - y \qquad 2w^3 \qquad 4(n + 3)(n + 2) \qquad \frac{\sqrt{x}}{3b - 2}$$

These all have one thing in common: They do *not* have an equals sign. An expression never contains an equals sign.

When you **simplify** an expression, you reduce the number of separate terms. You might also pull out and cancel common factors. In other words, you make the expression simpler. However, *you never change the expression's value as you simplify*. Here are some examples:

Unsimplified		Simplified	How
$3x + 4x$	$\rightarrow$	$7x$	Combine like terms
$\frac{2y^2}{3} + \frac{3y^2}{5}$	$\rightarrow$	$\frac{19y^2}{15}$	Find a common denominator and then combine
$x + xy$	$\rightarrow$	$x(1 + y)$	Pull out a common factor
$\frac{3x^2}{6x}$	$\rightarrow$	$\frac{x}{2}$	Cancel common factors

At times you might go in reverse. For instance, you might **distribute** a common factor:

$$x(1 + y) \rightarrow x + xy$$

Or you might multiply the top and bottom of a fraction by the same number to change its look. The result may seem even *less* simplified, temporarily. But by finding a common denominator, you can add fractions and get a simpler final result. For example:

Unsimplified		Even Less Simple		Simplified
$\frac{w}{2} + \frac{w}{4}$	$\rightarrow$	$\frac{w \times 2}{2 \times 2} + \frac{w}{4}$	$\rightarrow$	$\frac{3w}{4}$

As you simplify an expression (or even complicate one), the value of the expression must never change.

When you *evaluate* an expression, you figure out its value—the actual number represented by that expression.

To evaluate an expression, *substitute numbers in for any variables, then simplify.* In other words, *swap out the variables,* replacing them with numbers. Then do the arithmetic. Some people say "plug and chug." You *plug* in the values of the variables, then you *chug* through the simplification. Some people also call this "subbing in."

Whatever you call this process, you have to know the values of the variables to evaluate the expression. Otherwise, you're stuck.

Finally, remember PEMDAS! Follow PEMDAS when you plug and chug. Consider this example:

Evaluate the expression $3\sqrt{2x}$ given that x has the value of 8.

First, substitute in 8 for x:

$3\sqrt{2x}$ becomes $3\sqrt{2(8)}$

Now simplify the expression:

$3\sqrt{2(8)} = 3\sqrt{16} = 3 \times 4 = 12$

You have now evaluated the expression $3\sqrt{2x}$ when $x = 8$. For that particular value of x, the value of the expression is 12.

Pay attention to negative signs, especially if the value you're subbing in is negative. Put in parentheses to obey PEMDAS. Consider this example:

If $y = -2$, what is the value of $3y^2 - 7y + 4$?

Substitute -2 for y. Put parentheses around the -2 to clarify that you're subbing in negative 2, not subtracting 2 somehow. Therefore:

$3y^2 - 7y + 4$ becomes $3(-2)^2 - 7(-2) + 4$

Now simplify:

$3(-2)^2 - 7(-2) + 4 = 3(4) - (-14) + 4 = 12 + 14 + 4 = 30$

Be sure to square the negative sign and to subtract -14 (in other words, add 14). The value of the expression is 30.

You might have to plug into expressions in answer choices. Try this problem:

If $y = 6$, then which of the following expressions has the value of 20 ?

(A) $y + 14$
(B) $y - 14$
(C) $20y$

When you substitute 6 in for y in the answer choices, only $y + 14$ results in 20. The answer is (A).

Other expressions involving y could also equal 20 when $y = 6$, such as $2y + 8$ or $y^2 - 16$. On the GMAT, you would not be forced to pick between these expressions, because they would all be correct answers to this question.

If you...	Then you...	Like this:
Simplify an expression	Combine like terms or perform other legal algebra moves	$3x + 4x$ becomes $7x$
Evaluate an expression	Substitute numbers in for unknowns, then simplify	When $x = 2$, $7x$ becomes $7(2)$ or 14

An Equation Says "Expression A $=$ Expression B"

An equation is a complete sentence that has a subject, a verb, and an object. The sentence always takes this form:

$$\text{Subject} \quad equals \quad \text{object.}$$
$$\text{One expression} \quad \text{equals} \quad \text{another expression.}$$

$$2x \quad - \quad z \quad = \quad y \quad + \quad 4$$
$$\text{Two } x \quad \text{minus} \quad z \quad \text{equals} \quad y \quad \text{plus} \quad \text{four}$$

An equation always sets one expression $(2x - z)$ equal to another expression $(y + 4)$.

Everything you know about simplifying or evaluating expressions applies in the world of equations, because equations are made up of expressions.

You can *simplify* an expression on just one side of an equation, because you are not changing the value of that expression. So the equation still holds true, even though you're ignoring the other side.

For instance, simplify the left side but leave the right side alone:

$$3x + 5x = y \qquad \text{becomes} \qquad 8x = y$$

You can also *evaluate* an expression on just one side. For instance, say you have $\frac{8x}{15} = y$ and you know that $x = 5$. Then you can plug and chug just on the left side:

$$\frac{8x}{15} = y \qquad \text{becomes} \qquad \frac{8(5)}{15} = y \qquad \text{and finally} \qquad \frac{8}{3} = y$$

Throughout all these changes on the left side, the right side has remained y.

If all you could do to equations was to simplify or evaluate expressions, then your toolset would be limited. However, you can do much more.

You can truly *change* both sides. You can actually alter the values of the two expressions on either side of an equation. You just have to follow the Golden Rule.

Golden Rule of Equations: Do the Same Thing to Both Sides

You can change the value of the *left* side any way you want…

…as long as you change the *right* side in *exactly the same way*.

Here's an example:

$$x + 5 = 8$$

If you subtract 5 from the left side, you must subtract 5 from the right side. You get a new equation with new expressions. As long as you do the same thing to both sides, then the second equation is true, too:

$$
\begin{aligned}
x + 5 &= 8 \\
-5 & -5 \\
\hline
x &= 3
\end{aligned}
$$

Here is a table of the major Golden Rule moves you can do to both sides of an equation:

1. **Add the same thing** to both sides. • That "thing" can be a number or a variable expression. • Show the addition underneath to minimize careless errors. (Later, if you feel comfortable with the steps, you can stop writing out the addition.)	$\begin{aligned} y - 6 &= 15 \\ +6 & +6 \\ \hline y &= 21 \end{aligned}$
2. **Subtract the same thing** from both sides.	$\begin{aligned} z + 4 &= k \\ -4 & -4 \\ \hline z &= k - 4 \end{aligned}$
3. **Multiply both sides by the same thing.** • Put parentheses in so that you multiply *entire* sides.	$n + m = \dfrac{3w}{4}$ $4 \times (n + m) = \left(\dfrac{3w}{4}\right) \times 4$ $4n + 4m = 3w$
4. **Divide both sides by the same thing** (except 0, of course). • Extend the fraction bar all the way so that you divide *entire* sides.	$a + b = 5d$ $\dfrac{a + b}{5} = \dfrac{5d}{5}$ $\dfrac{a + b}{5} = d$
5. **Square both sides,** cube both sides, etc. • Put parentheses in so you square or cube *entire* sides.	$x + \sqrt{2} = \sqrt{7w}$ $\left(x + \sqrt{2}\right)^2 = \left(\sqrt{7w}\right)^2$ $\left(x + \sqrt{2}\right)^2 = 7w$
6. **Take the square root of both sides,** the cube root of both sides, etc. • Extend the radical signs to cover the *entire* sides.	$z^3 = 64$ $\sqrt[3]{z^3} = \sqrt[3]{64}$ $z = 4$

One warning about square-rooting both sides of an equation: The equation usually splits into *two* separate equations. For example:

$$x^2 = 49 \qquad \text{Square-root both sides}$$
$$\sqrt{x^2} = \sqrt{49}$$

$$x = 7 \quad \textbf{OR} \quad x = -7$$

There are two numbers that, when squared, equal 49, since squaring a negative results in a positive. When you take a square root of a squared number, you'll need to find the negative solution as well. For example:

| When you square | $y = 6$ | you always get | $y^2 = 36.$ |
| But if you solve | $y^2 = 36$ | you get | $y = 6 \text{ OR } y = -6.$ |

Essentially, the even exponent hides the fact that the base could be positive or negative, leaving you with two possible solutions.

Perform the same action to an entire side of an equation. Pretend that the expression on each side of the equation is surrounded by parentheses—and actually write those parentheses in as necessary. For example:

$$x + 4 = \frac{x}{2}$$

To multiply both sides by 2, put parentheses around both expressions:

$$2(x + 4) = \left(\frac{x}{2}\right)2$$

Now simplify. Distribute the 2 to both the *x* and the 4 and cancel the 2's on the right:

$$2x + 8 = x$$

This is why multiplying to get rid of a denominator is sometimes called *cross-multiplication*. You can imagine that the 2 that *was* in the denominator on the right side moves to the left side, where it is multiplied by the $x + 4$. Cross-multiplication is even more useful when both sides have denominators to begin with.

Let's go back to the equation at hand. You can simplify further with more Golden Rule moves. Subtract *x* from both sides, and then subtract 8:

$$
\begin{array}{rcr}
2x + 8 = & & x \\
-x & & -x \\
\hline
x + 8 = & & 0 \\
-8 & & -8 \\
\hline
x = & & -8
\end{array}
$$

You now have *x* by itself on one side, so the equation says that the value of *x* is -8. What you did here was *isolate the variable* or *solve for the variable*.

To isolate *x, get x by itself on one side of the equation.* You want the equation to read "$x = \ldots$"

The expression on the right side is often a number, as in the case above ($x = -8$).

In a more complicated equation, the right side could be an expression that contains other variables. Either way, the important thing when you isolate x on the left is that the right side *cannot* have any terms containing x. Otherwise, you haven't truly *isolated* the x on the left side.

You can instead isolate the x on the *right* side if you want. If you do that, make sure that the *left* side has no terms containing x. When you get "$x = $ a number" or "a number $= x$," then you have *solved the equation*. The number you get is a *solution* to the equation.

When you plug a solution into an equation (i.e., into the *variable* in an equation), you make the equation true. For example:

4 is a solution to the equation $2x + 7 = 15$.

Since 4 is a solution to the equation, the equation is true when you plug 4 in for x. Try it: $2(4) + 7 = 15$, or $15 = 15$.

If you have more than one variable in an equation, you may still want to isolate just one variable for some reason. Consider this example:

If $3x + 5y = 12$, what is x in terms of y?

"What is x in terms of y" means "get x by itself and put y and everything else on the opposite side of the equals sign."

For example, "a in terms of b" is the right side of this equation:

$$a = 3 + b^2$$

There are only a's on the left side and there are no a's on the right side—but the variable b is on the right side. So this equation represents a in terms of b.

To get x in terms of y, isolate x on one side by applying Golden Rule moves to the given equation:

$$\begin{array}{rcl} 3x + 5y & = & 12 \\ -\,5y & & -5y \\ \hline 3x & = & 12 - 5y \end{array}$$

Finally, divide by 3 to get x completely by itself:

$$x = \frac{12 - 5y}{3}$$

So x in terms of y is this expression: $\frac{12 - 5y}{3}$. This contains no x's. If you're looking for "x in terms of y," the answer will contain y's, not x's.

If you want y in terms of x instead, then isolate y on one side:

$$\begin{array}{rcl} 3x + 5y & = & 12 \\ -3x & & -3x \\ \hline 5y & = & 12 - 3x \end{array}$$

Finally, divide by 5 to get y in terms of x:

$$y = \frac{12 - 3x}{5}$$

If you...	Then you...	Like this:
Want to change an expression on one side of an equation	Apply the Golden Rule: Change both sides in exactly the same way	$y - 3 = 9$ $\underline{+3 \quad +3}$ $y \quad\ = 12$
Want to isolate the variable x in an equation	Perform Golden Rule moves and simplify until the equation reads "$x = $ everything else"	$7x + 4 = 18$ $\underline{-4 \quad -4}$ $7x \quad\ = 14$ $\dfrac{7x}{7} = \dfrac{14}{7}$ $x = 2$
Need x in terms of y	Perform Golden Rule moves and simplify until the equation reads "$x = $ everything else"	$7x + 4 = \ y$ $\underline{-4 \quad\quad -4}$ $7x \quad\ = y - 4$ $\dfrac{7x}{7} = \dfrac{y-4}{7}$ $x = \dfrac{y-4}{7}$

Check Your Skills

1. If $\sqrt{x + 2} = 4$, what is x?

2. If $\dfrac{y - 3}{x} = 2$, what is y in terms of x?

$(x+2)^{1/2} = 4 \qquad x+2 = 16 \qquad \boxed{x=14}$

$2x+3=y$

Answers can be found on page 247.

Isolate a Variable: Do PEMDAS in Reverse

When an expression is complicated, it's easy to get confused about how to isolate the variable inside.

Consider the following equation:

$$5(x - 1)^3 - 30 = 10$$

If you already knew the value for x, you would follow PEMDAS to simplify the left side of the equation:

1) Start with x. x

2) Subtract 1. $x - 1$

3) Cube the result. $(x - 1)^3$

4) Multiply by 5. $5(x - 1)^3$

5) Subtract 30. $5(x - 1)^3 - 30$

The answer would get you to 10, because that's what the equation says:

$$5(x - 1)^3 - 30 = 10$$

If you don't know the value of x, though, you need to isolate it on the left, so *undo* the PEMDAS steps by working in the *reverse order*.

The last step of the recipe was to subtract 30. To undo that step first, add 30 to both sides:

$$5(x-1)^3 - \cancel{30} = \quad 10$$
$$\underline{ +30 \quad +30}$$
$$5(x-1)^3 \qquad = \quad 40$$

Now, undo the previous step of the original recipe, which was to multiply by 5. Divide both sides by 5:

$$\frac{\cancel{5}\,(x-1)^3}{\cancel{5}} = \frac{40}{5}$$
$$(x-1)^3 = 8$$

Next, undo the cubing. The opposite of exponents is roots, so take the cube root of both sides:

$$\sqrt[3]{(x-1)^3} = \sqrt[3]{8}$$
$$(x-1) = 2$$

The parentheses are no longer necessary, so drop them:

$$x - 1 = 2$$

Finally, undo the subtraction by adding 1 to both sides, and you get $x = 3$.

In short, to solve for the variable, do PEMDAS backwards. Try this problem:

If $4\sqrt{x-6} + 7 = 19$, what is the value of x?

Start with A/S: addition and subtraction. First, *subtract* 7 from both sides:

$$4\sqrt{x-6} + 7 = \quad 19$$
$$\underline{\phantom{4\sqrt{x-6}} -7 \quad -7}$$
$$4\sqrt{x-6} \quad = \quad 12$$

Next, do any M/D: multiplication and division. In this case, *divide* both sides by 4:

$$\frac{\cancel{4}\sqrt{x-6}}{\cancel{4}} = \frac{12}{4}$$
$$\sqrt{x-6} = 3$$

Exponents are up next. Undo the square root by *squaring* both sides:

$$\sqrt{x-6} = 3$$
$$\left(\sqrt{x-6}\right)^2 = 3^2$$
$$x - 6 = 9$$

Finally, deal with the parentheses. *Add* 6 to both sides, and you end up with $x = 15$.

The original equation did not contain explicit parentheses, but the square root symbol extended over two terms: $\sqrt{x-6}$. That's just like putting parentheses around $x - 6$ and raising that whole quantity to the $\frac{1}{2}$ power: $(x-6)^{\frac{1}{2}}$.

A square root sign acts like parentheses when it extends over more than one term. Likewise, a fraction bar acts like parentheses when it's stretched over multiple terms.

If you…	Then you…	Like this:
Want to isolate a variable inside an expression	Follow PEMDAS in reverse	$2y^3 - 3 = 51$ $\underline{\ +3\ \ +3}$ $2y^3\ = 54$ $\dfrac{2y^3}{2} = \dfrac{54}{2}$ $y^3 = 27$ $\sqrt[3]{y^3} = \sqrt[3]{27}$ $y = 3$

Check Your Skills

Solve for x.

3. $3(x+4)^3 - 5 = 19$

4. $\sqrt[3]{x+5} - 7 = -8$

Answers can be found on page 247.

Simplify an Equation: Combine Like Terms and Eliminate Denominators

What happens when x shows up in multiple places in an equation? Consider this example:

If $\dfrac{5x - 3(4 - x)}{2x} = 10$, what is x?

To isolate x, you have to get all the x's together on one side. How? *Combine like terms.*

If there is a denominator, get rid of that right away, especially if the denominator contains the variable you are trying to isolate. *Always get variables out of denominators.*

Get rid of the fraction by multiplying both sides by the entire denominator, which is $2x$:

$$2x\left(\frac{5x - 3(4-x)}{2x}\right) = (10)\,2x$$

$$5x - 3(4-x) = 20x$$

Next, get x out of the parentheses. Distribute the -3 on the left side (be careful with the negative sign):

$$5x - 3(4-x) = 20x$$

$$5x - 12 + 3x = 20x$$

Now, combine like terms. First, add $5x$ and $3x$ on the left side:

$$5x - 12 + 3x = 20x$$
$$8x - 12 = 20x$$

Next, subtract $8x$ from both sides:

$$\begin{array}{rcr} 8x - 12 = & 20x \\ -8x \qquad & -8x \\ \hline -12 = & 12x \end{array}$$

Finally, divide by 12 to isolate x on the right:

$$\frac{-12}{12} = \frac{12x}{12}$$
$$-1 = x$$

You now have the answer to the question: $x = -1$.

If you...	Then you...	Like this:
Have a variable in multiple places in an equation	Combine like terms, which might be on different sides of the equation	$\begin{array}{rcr} 9y + 30 = & 12y \\ -9y \qquad & -9y \\ \hline 30 = & 3y \end{array}$
Have a variable in a denominator	Multiply to get the variable out of the denominator	$\dfrac{2z - 3}{z} = 4$ $z\left(\dfrac{2z - 3}{z}\right) = (4)z$ $2z - 3 = 4z$

Check Your Skills

Solve for x.

5. $\dfrac{2x + 6(9 - 2x)}{x - 4} = -3$

Answer can be found on page 247.

Variables in the Exponent: Make the Bases Equal

If a variable is in the exponent, the typical PEMDAS moves aren't going to help much. Consider this example:

If $3^x = 27^4$, what is x?

The key is to rewrite the terms so they have the *same base*. Usually, the best way to do this is to *factor bases into primes*.

On the left side, 3 is already a prime, so leave it alone. On the right side, 27 is not prime. Since $27 = 3^3$, replace 27 with 3^3. Put in parentheses to keep the exponents straight:

$$3^x = \left(3^3\right)^4$$

Simplify the right side by applying the "power to a power" rule: $\left(3^3\right)^4 = 3^{3\times4} = 3^{12}$. Therefore:

$$3^x = 3^{12}$$

In other words, 3 raised to the power of x is equal to 3 raised to the power of 12. This is only true if x itself is equal to 12:

$$3^x = 3^{12} \quad \rightarrow \quad x = 12$$

Once the bases are the same, the exponents must be the same.

This rule has exactly three exceptions: a base of 1, a base of 0, and a base of -1. The exceptions occur because more than one exponent of these particular bases results in the same number. For example:

$$1^2 = 1^3 = \ldots = 1 \qquad 0^2 = 0^3 = \ldots = 0 \qquad (-1)^2 = (-1)^4 = 1, \text{ while } (-1)^1 = (-1)^3 = -1$$

However, for every other base, the rule works. Try this problem:

If $4^y = 8^{y+1}$, what is the value of 2^y ?

(A) -8

(B) $\dfrac{1}{8}$

(C) $\dfrac{1}{4}$

(D)   1

(E)   8

Before you solve anything, glance at the answer choices. Since none of them contain a y, that's your clue that the equation can be manipulated to find the value of y. You just need to figure out how!

Look at the given equation. The variable y is in two exponents:

$$4^y = 8^{y+1}$$

To get the y out of the exponents, make the bases the same. Rewrite both 4 and 8 as powers of 2:

$$4 = 2^2 \qquad 8 = 2^3$$
$$4^y = 8^{y+1} \quad \rightarrow \quad \left(2^2\right)^y = \left(2^3\right)^{y+1}$$

Next, apply the "power to a power" rule on both sides:

$$\left(2^2\right)^y = \left(2^3\right)^{y+1}$$
$$2^{2y} = 2^{3(y+1)}$$

Now that the bases are the same (and the common base is not 1, 0, or -1), you can write a brand-new equation that sets the exponents equal to each other and then solve:

$$2y = 3(y+1)$$
$$2y = 3y + 3$$
$$-3 = y$$

Now that you've solved for *y*, find the value of 2^y, which is what the question asked. To do this, replace *y* with -3:

$$2^y = 2^{-3} = \frac{1}{8}$$

The correct answer is (B).

If you...	Then you...	Like this:
Have a variable in an exponent or exponents	Make the bases equal, usually by breaking the given bases down to primes, then set the exponents equal to each other	$3^x = 27^4$ $3^x = \left(3^3\right)^4$ $3^x = 3^{12}$ $x = 12$

Check Your Skills

6. If $4^6 = 64^{2x}$, what is the value of 16^x ?

Answer can be found on page 248.

Systems of Equations

Many GMAT problems will require you to deal with two variables. In most of those cases, you'll also have two equations. For example:

(a) $2x - 3y = 16$ (b) $y - x = -7$

A group of more than one equation is often called a *system of equations*. Solving a system of two equations with two variables, for example, *x* and *y*, means finding values for *x* and *y* that make both equations true *at the same time*.

The systems of equations discussed in this section have only one solution. That is, only one set of values of *x* and *y* makes the system work.

To solve a system of two equations and two unknowns, you can use either of two good strategies: (1) Isolate, then substitute or (2) Add or subtract to eliminate variables.

These strategies are similar at a high level. In both, here's what you do:

1. Kill off one equation and one unknown.

2. Solve the remaining equation for the remaining unknown.

3. Plug back into one of the original equations to solve for the other unknown.

However, the two strategies take very different approaches to step 1: how to kill off one equation and one unknown. Let's examine these approaches in turn.

Solve for an Unknown: Isolate, Then Substitute

This strategy is also known as *substitution*. Consider this system again:

(a) $2x - 3y = 16$ (b) $y - x = -7$

To follow the substitution strategy, first isolate one variable in one of the equations. Next, substitute into the other equation.

Which variable should you isolate? The one you *don't* ultimately want. If the problem asks for x, first isolate y.

Why y? Because the variable you isolate in your first step is the one that will then be killed off. You're left with one equation containing the other variable—the variable you want. This way, you save yourself time and effort.

Conversely, if the problem asks for y, first isolate x, so you can kill it off.

Let's say that the question asks for the value of x. In this case, isolate y in one of the given equations. In which equation should you isolate your variable? Whichever one you think is easier to deal with.

In the example above, it looks easier to isolate y in equation (b). All you have to do is add x to both sides:

(a) $2x - 3y = 16$

$$
\begin{array}{l}
\text{(b)} \quad y - x = -7 \\
\underline{ + x \qquad + x} \\
y \qquad = -7 + x \\
y \qquad = x - 7
\end{array}
$$

By the way, when you use this method, it's good practice to write the two equations in the system side by side. That way, you can do algebra down your page to isolate one variable without running into the other equation.

Now you have expressed y in terms of x. Since $y = x - 7$, you can replace y with $(x - 7)$ anywhere you see y. This will remove any references to y in the first equation:

(a) $2x - 3y = 16$

$$
\begin{array}{l}
\text{(b)} \quad y - x = -7 \\
\underline{ + x \qquad + x} \\
y \qquad = -7 + x \\
y \qquad = x - 7
\end{array}
$$

$2x - 3(x - 7) = 16$

Note: When you sub in an expression such as $x - 7$, place parentheses around that expression in order to avoid PEMDAS errors.

Now that you have killed off one variable (y) and one equation (the second one), you have just one variable left (x) in one equation. Solve for that variable:

$$
\begin{array}{ll}
2x - 3(x - 7) = 16 & \\
2x - 3x + 21 = 16 & \text{Distribute the } -3. \\
-x + 21 = 16 & \text{Combine } 2x \text{ and } -3x \text{ (like terms).} \\
-x = -5 & \text{Subtract 21 from both sides.} \\
x = 5 & \text{Multiply both sides by } -1.
\end{array}
$$

At this point, you're done if the question asks only for x—and the GMAT commonly asks for just one of the variables. If the question asks for $x + y$ or some other expression involving both x and y, then you can solve for y by plugging your value of x into either of the original equations.

Which equation should you plug back into? The one in which you isolated y. In fact, you should plug into the revised form of that equation—the one that looks like "$y = $ something." This is the easiest way to solve for y, so this is the equation you'll want to use:

$$y = x - 7$$

To start, swap out x and replace it with 5, since you found that $x = 5$:

$$y = (5) - 7$$
$$y = -2$$

Now you have the complete solution: $x = 5$ and $y = -2$. These are the *solutions* of the system; in other words, this is the only pair of values that makes *both* of the original equations true at the same time:

(a) $2x - 3y = 16$ (b) $y - x = -7$

         $2(5) - 3(-2) = 16$ $(-2) - (5) = -7$

            True   True

If you...	Then you...	Like this:
Have two equations and two unknowns	Isolate one unknown, then substitute into the other equation	$2x - 3y = 16$ and $y - x = -7$ $y = -7 + x = x - 7$ $2x - 3(x - 7) = 16$ ... $x = 5$ $y = -2$

Check Your Skills

Solve for x and y by substitution.

    7. $6y + 15 = 3x$

        $x + y = 14$

Answer can be found on page 248.

Solve for an Unknown: Add or Subtract Equations

Substitution will always work. But some GMAT problems can be solved more easily with another method, known as *elimination*.

Here's how elimination works. You are always allowed to add two equations together. Just add the left sides up and put the result on the left, then add the right sides up and put that result on the right:

$$x = 4$$
$$\underline{+\; y = 7}$$
$$x + y = 4 + 7$$

Why is this allowed? You are actually adding the same thing to both sides of an equation. Since y equals 7, you can legally add y on the left side of $x = 4$ and add 7 on the right side. You are making the same change on each side of the first equation.

In this example, the resulting equation $(x + y = 11)$ is more complicated than the individual starting equations. However, adding equations can sometimes actually eliminate a variable—*and that's why you do it.*

Here's an example:

$$a + b = 11 \qquad a - b = 5$$

What happens when you add these equations together?

$$\begin{aligned} a + b &= 11 \\ + \quad a - b &= 5 \\ \hline 2a &= 16 \end{aligned}$$

The b's cancel out of the resulting equation completely, allowing you to solve for a:

$$2a = 16 \quad \rightarrow \quad a = 8$$

Finally, if you need b, plug a back into one of the original equations:

$$a + b = 11 \quad \rightarrow \quad (8) + b = 11 \quad \rightarrow \quad b = 3$$

So the complete solution to the original system of equations is $a = 8$ and $b = 3$.

If you solve this particular system by substitution, you will need to take a few more steps to isolate one variable. In this case, then, elimination is the faster approach.

You can also subtract equations to eliminate a variable. Take a look at this system of equations:

$$5n + m = 17 \qquad 2n + m = 11$$

Since "$+ m$" shows up in both equations, kill m by subtracting the second equation from the first:

$$\begin{aligned} 5n + m &= 17 \\ - \quad 2n + m &= 11 \\ \hline 3n &= 6 \end{aligned}$$

Subtract the whole left side, as well as the right side. To make sure that you follow PEMDAS, put a set of parentheses around the whole equation:

$$\begin{aligned} 5n + m &= 17 \\ - \quad (2n + m &= 11) \\ \hline 3n &= 6 \end{aligned}$$

Now you can solve for n, then plug back in to get m:

$$3n = 6 \quad \rightarrow \quad n = 2$$
$$2n + m = 11 \quad \rightarrow \quad 2(2) + m = 11 \quad \rightarrow \quad m = 7$$

To set up a good elimination, you can even multiply a whole equation by a number. That's the same thing as multiplying the left side and the right side by the same number. This is a Golden Rule move.

Consider this system of equations from earlier:

(a) $2x - 3y = 16$ (b) $y - x = -7$

To take the elimination approach, first rewrite the equations vertically and line up the variables. Here, you *want* to write one equation below the other (not *next to* each other as you do with substitution). Space the terms to line up x with x and y with y:

$$2x - 3y = 16$$
$$-x + y = -7$$

If you add the equations now, neither variable will drop out; you're left with $x - 2y = 9$. That's not helpful.

However, look what happens when you multiply the second equation by 2 on both sides:

$$2x - 3y = 16 \quad \rightarrow \quad 2x - 3y = 16$$
$$2(-x + y) = (-7)2 \quad \rightarrow \quad -2x + 2y = -14$$

Now add. The $2x$ term will cancel with the $-2x$ term:

$$2x - 3y = 16$$
$$+(-2x + 2y = -14)$$
$$\overline{-y = 2}$$
$$y = -2$$

Finally, use this value of y in one of the original equations to solve for x:

$$y - x = -7 \quad \rightarrow \quad (-2) - x = -7 \quad \rightarrow \quad -x = -5 \quad \rightarrow \quad x = 5$$

This is the same solution as before. In this case, the amount of work was about the same for each method. However, some problems are *much* easier to solve by elimination. Take a look at this example:

For this system of equations, what is the value of $x + y$?

$$\frac{1}{2}x + \frac{1}{3}y = 3$$

$$2x + y = 11$$

If you try direct substitution, you will need to make a lot of messy calculations involving fractions.

Instead, try elimination. First, multiply the top equation through by 6 to eliminate all fractions:

$$6\left(\frac{1}{2}x + \frac{1}{3}y\right) = (3)6$$
$$3x + 2y = 18$$

Next line up the two given equations, which now look much better:

$$3x + 2y = 18$$
$$2x + y = 11$$

Before going further, consider: What does the question specifically ask for? It does not ask for x or y separately. Rather, it asks for $x + y$.

You can certainly solve for one of the variables, then find the other. But there's a shortcut.

If you look carefully, you might notice that you can solve for the combination $x + y$ directly by *subtracting* the equations:

$$
\begin{array}{r}
3x + 2y = 18 \\
-\,(2x + y = 11) \\
\hline
x + y = 7
\end{array}
$$

Ta-da! The answer to the question is 7.

Elimination isn't always the best way. Know how to use both substitution and elimination.

If you...	Then you...	Like this:
Have two equations and two unknowns	Add or subtract equations to eliminate a variable	$2x - 3y = 16$ and $y - x = -7$ Multiply 2nd equation by 2: $2(y - x) = (-7)2$ Add equations: $\begin{array}{r} -2x + 2y = -14 \\ +\,(2x - 3y = 16) \\ \hline -y = 2 \\ y = -2 \end{array}$

Check Your Skills

Solve for x and y by elimination.

8. $x + 4y = 10$

$y - x = -5$

Answer can be found on page 248.

Three or More Variables: Isolate the Expression You Want

A few problems involve even more than two variables. Fortunately, all the procedures you've learned so far still work.

Focus on exactly what the question asks for and isolate that on one side of the equation. Here's an example:

If $\sqrt{\dfrac{a}{b}} = c$ and $abc \neq 0$, what is the value of b in terms of a and c?

Since the question asks for b, isolate b. The answer will contain a and c, because the question asks for an expression "in terms of" a and c.

$$\frac{a}{c^2} =$$

Take the given equation and do Golden Rule moves to isolate b:

$$\sqrt{\frac{a}{b}} = c$$

$$\left(\sqrt{\frac{a}{b}}\right)^2 = c^2 \qquad \text{Square both sides.}$$

$$\frac{a}{b} = c^2$$

$$b\left(\frac{a}{b}\right) = \left(c^2\right)b \qquad \text{Multiply both sides by } b \text{ to get it out of the denominator.}$$

$$a = c^2 b$$

$$\frac{a}{c^2} = \frac{c^2 b}{c^2} \qquad \text{Divide both sides by } c^2.$$

$$\frac{a}{c^2} = b$$

The answer to the question is $\dfrac{a}{c^2}$.

By the way, the "non-equation" $abc \neq 0$ was only there to prevent division by zero. When the test tells you that something does *not* equal zero ($\neq 0$), it's usually just trying to avoid a situation where something could be divided by zero.

Try this tough problem:

$$\frac{w}{x-y} = 3 \qquad\qquad y - x = 4$$

In the system of equations above, what is the value of w ?

(A) -12

(B) $-\dfrac{3}{4}$

(C) $\dfrac{3}{4}$

(D) $\dfrac{4}{3}$

(E) 12

You are asked for the value of w, so manipulate the equations to get "$w = \ldots$"

Glance at the answers; they are all numbers. The other variables, x and y, must disappear along the way.

One approach to this problem is to isolate x or y in the second equation, then substitute into the first equation. This way, you at least get rid of one variable. It's a start—let's see where it goes.

Isolate y by adding x to both sides of the second equation:

$$
\begin{array}{rl}
y - x = & 4 \\
\underline{+\,x \qquad +\,x} & \\
y \quad = & x + 4
\end{array}
$$

Now substitute $x + 4$ into the first equation in place of y. Be sure to put parentheses around $x + 4$:

$$\frac{w}{x - y} = 3$$

$$\frac{w}{x - (x + 4)} = 3$$

$$\frac{w}{x - x - 4} = 3$$

$$\frac{w}{-4} = 3$$

Look what happened—the variable x disappeared as well. This is a good sign. If x didn't cancel out, then you'd be in trouble. Solve for w now:

$$\frac{w}{-4} = 3$$

$$w = -12$$

The answer is (A).

Another approach to the problem is to recognize that $x - y$ is very similar to $y - x$. In fact, one is the negative version of the other:

$$x - y = -(y - x)$$

If you recognize this, then multiply the second equation by -1 so that it has $x - y$ on one side:

$$y - x = 4$$

$$(-1)(y - x) = (4)(-1)$$

$$x - y = -4$$

Now you can *substitute for the whole expression*. That is, you can swap out $x - y$ in the first equation and replace that entire thing with -4:

$$\frac{w}{x - y} = 3 \qquad \overset{\frown}{\underset{\smile}{x - y = -4}}$$

$$\frac{w}{(-4)} = 3$$

This gets you to the same point as the first method, and you end up with $w = -12$.

However many variables and equations you have, pay close attention to what the question is asking for:

- If the question asks for the variable x, isolate x, so you have "$x = \ldots$"
- If the question asks for x in terms of y, after you isolate x on one side of the equation, the other side should contain y.
- If the question asks for $x + y$, isolate that expression, so you have "$x + y = \ldots$"
- If a variable does not appear in the answer choices, help it vanish. Isolate it and substitute for it in another equation.
- If an expression such as $x - y$ shows up in two different equations, feel free to substitute for it so that the whole thing disappears.

If you...	Then you...	Like this:
Have three or more unknowns	Isolate whatever the question asks for, and use substitution to eliminate unwanted variables	If $a + b - c = 12$ and $c - b = 8$, what is the value of a? Isolate a: $a = 12 + c - b$ Substitute for $c - b$: $a = 12 + (8)$ $a = 20$

Check Your Skills

9. If $\dfrac{a}{c} + \dfrac{b}{3c} = 1$, what is c in terms of a and b?

Answer can be found on page 249.

$$\frac{3a + b}{3c} = 1$$

$$3a + b = 3c$$

$$a + \frac{b}{3} = c$$

6

Check Your Skills Answer Key

1. **14:** In order to find the value of x, isolate x on one side of the equation.

$$\sqrt{x+2} = 4$$
$$\left(\sqrt{x+2}\right)^2 = 4^2$$
$$x + 2 = 16$$
$$x = 14$$

2. **$2x + 3$:** In order to find y in terms of x, isolate y on one side of the equation.

$$\frac{y-3}{x} = 2$$
$$\cancel{x}\left(\frac{y-3}{\cancel{x}}\right) = (2)x$$
$$y - 3 = 2x$$
$$y = 2x + 3$$

3. **-2:**

$$3(x+4)^3 - 5 = 19$$
$$3(x+4)^3 = 24 \qquad \text{Add 5 to both sides.}$$
$$(x+4)^3 = 8 \qquad \text{Divide both sides by 3.}$$
$$(x+4) = 2 \qquad \text{Take the cube root of both sides.}$$
$$x = -2 \qquad \text{Remove the parentheses and subtract 4 from both sides.}$$

4. **-6:**

$$\sqrt[3]{(x+5)} - 7 = -8$$
$$\sqrt[3]{(x+5)} = -1 \qquad \text{Add 7 to both sides.}$$
$$x + 5 = -1 \qquad \text{Cube both sides and remove parentheses.}$$
$$x = -6 \qquad \text{Subtract 5 from both sides.}$$

5. **6:**

$$\frac{2x + 6(9 - 2x)}{x - 4} = -3$$
$$2x + 6(9 - 2x) = -3(x - 4) \qquad \text{Multiply by the denominator } (x - 4).$$
$$2x + 54 - 12x = -3x + 12 \qquad \text{Simplify grouped terms by distributing (be careful of the signs!).}$$
$$-10x + 54 = -3x + 12 \qquad \text{Combine like terms } (2x \text{ and } -12x).$$
$$54 = 7x + 12 \qquad \text{Add } 10x \text{ to both sides.}$$
$$42 = 7x \qquad \text{Subtract 12 from both sides.}$$
$$6 = x \qquad \text{Divide both sides by 7.}$$

6. **16:** You could answer this question by breaking each base down to a power of 2, but you'll save some time if you notice that all three bases in the question (4, 64, and 16) are powers of 4. Solve for x by rewriting 64 as 4^3.

$$4^6 = 64^{2x}$$
$$4^6 = \left(4^3\right)^{2x}$$
$$4^6 = 4^{6x}$$
$$6 = 6x$$
$$1 = x$$

If $x = 1$, then $16^x = 16^1 = 16$.

7. **$x = 11, y = 3$:**

(a) $6y + 15 = 3x$

$2y + 5 = x$ Divide the first equation by 3 to isolate x.

(b) $x + y = 14$

$(2y + 5) + y = 14$ Substitute $(2y + 5)$ for x in the second equation and solve for y.

$$3y + 5 = 14$$
$$3y = 9$$
$$y = 3$$

$2(3) + 5 = x$ Substitute $y = 3$ in the isolated form of the first equation to solve for x.

$$11 = x$$

8. **$x = 6, y = 1$:** The first equation has a positive x and the second equation has a negative x. Rearrange the second equation to put the variables in the same order and line it up under the first equation.

$$x + 4y = 10$$
$$+(-x + y = -5)$$
$$\overline{5y = 5}$$
$$y = 1$$

Now that you know $y = 1$, plug it back into either equation to solve for x.

$$x + 4y = 10$$
$$x + 4(1) = 10$$
$$x = 6$$

9. $\dfrac{3a + b}{3}$ **or** $a + \dfrac{b}{3}$**:** The wording "c in terms of ..." means that you have to isolate c on one side of the equation. To do so, first combine the fractions on the left side of the equation to combine like terms.

$$\frac{a}{c} + \frac{b}{3c} = 1$$

$$\frac{3a}{3c} + \frac{b}{3c} = 1$$

$$\frac{3a + b}{3c} = 1$$

$$\frac{3a + b}{3} = c$$

If you had answers, you would check them now for a match. If you didn't see a match, there's one more step you could take: Split the numerator.

$$\frac{\cancel{3}a}{\cancel{3}} + \frac{b}{3} = c$$

$$a + \frac{b}{3} = c$$

Chapter Review: Drill Sets

Drill 1

1. If $x = 2$, what is the value of $x^2 - 4x + 3$?

2. If $x = 3$, what is the value of $x + \sqrt{48x}$?

3. If $x = -4$, what is the value of $\dfrac{5 - x}{3} - x^2$?

4. If $c = 100$ and $p = 30c^2 - c$, what is the value of p ?

5. If $y - 3 = \dfrac{xy}{2}$ and $x = 3$, what is the value of y ?

Drill 2

Solve for the variable in the following equations.

6. $3(7 - x) = 4(1.5)$

7. $7x + 13 = 2x - 7$

8. $3t^3 - 7 = 74$

9. $1{,}200x + 6{,}000 = 13{,}200$

10. $\sqrt{x} = 3 \times 5 - 20 \div 4$

11. $-(y)^3 = -27$

12. $2 - \dfrac{\sqrt{2x + 2}}{2} = -1$

13. $5\sqrt{x} + 6 = 51$

Drill 3

Solve for x in the following equations.

14. $4(-3x - 8) = 8(-x + 9)$

15. $3x + 7 - 4x + 8 = 2(-2x - 6)$

16. $2x(4 - 6) = -2x + 12$

17. $\dfrac{3(6 - x)}{2x} = -6$

18. $\dfrac{13}{x + 13} = 1$

19. $\dfrac{10(-3x + 4)}{10 - 5x} = 2$

20. $\dfrac{15 - 3(5x - 3)}{3 - 2x} = 6$

21. $\dfrac{50(10 + 3x)}{50 + 7x} = 50$

Drill 4

Solve for the values of both variables in each system of equations using substitution. The explanations will use substitution to solve.

22. $7x - 3y = 5$
    $y = 10$

23. $y = 4x + 10$
    $y = 7x - 5$

24. $2h - 4k = 0$
    $k = h - 3$

25. $5x - 3y = 17$
    $2x - y = 8$

26. $12b = 2g$
    $4g - 3b = 63$

Drill 5

Solve for the values of both variables in each system of equations using elimination. The explanations will use elimination to solve.

27. $x - y = 4$
    $2x + y = 5$

28. $x + 2y = 5$
    $x - 4y = -7$

29. $a + b = 8$
    $2a + b = 13$

30. $7m - 2n = 1$
    $3m + n = 6$

31. $y - 2x - 1 = 0$
    $x - 3y - 1 = 0$

32. $\dfrac{1}{3}r - \dfrac{1}{6}s = 0$
    $2r + \dfrac{1}{2}s - 3 = 0$

Drill 6

Solve for the values of both variables in each system of equations. Decide whether to use substitution or elimination. The explanation will use one of the two methods and explain why that solution method was chosen.

33. $5x + 2y = 12$

$y = \dfrac{1}{2}x + 3$

34. $y - 1 = x + 2$

$2y = x + 1$

35. $8a - 2b - 5 = 0$

$2b + 4a - 4 = 0$

36. $3x = 6 - y$

$6x - y = 3$

37. $x = 2y - \dfrac{1}{2}$

$y - x = -\dfrac{3}{2}$

Drill 7

Solve for the indicated value in each system of equations.

38. $4x + y + 3z = 34$

$4x + 3z = 21$

What is y ?

39. $\dfrac{x - 3}{y - 2} = 4z$

If $y \neq 2$, what is the value of x in terms of y and z ?

(A) $4yz - 8z + 3$

(B) $4yz - 5$

(C) $y - 8z - 3$

40. $3x + 5y + 2z = 20$

$6x + 4z = 10$

What is y ?

41. $\dfrac{a - b}{4} = c + 1$

$c = b + 2$

What is b in terms of a ?

Drill 8

Solve for the indicated value in each system of equations.

42. $\dfrac{a + b}{c + d} = 10$

$3d = 15 - 3c$

If $c \neq d$, what is $a + b$?

43. $x = \dfrac{y}{5}$

$2z - 1 = \dfrac{x + y}{2}$

What is z in terms of x ?

44. $2^{x+y} = \sqrt{z - 2}$

$x = 2 - y$

What is z ?

45. $\dfrac{3x}{z + 4} = 4$

$4z = 3y$

If $z \neq -4$, what is $x - y$?

(A) $4z + 3$

(B) $\dfrac{16}{3}$

(C) $\dfrac{4}{3}$

6

Drill Sets Solutions

Drill 1

1. **−1:** To evaluate the expression, replace x with 2.

    $$x^2 - 4x + 3 =$$
    $$(2)^2 - 4(2) + 3 =$$
    $$4 - 8 + 3 = -1$$

2. **15:** To evaluate the expression, replace x with 3. Rather than multiply 48 by 3, pull a 3 out of 48 in order to create a perfect square that you can then remove from under the square-root symbol.

    $$x + \quad \sqrt{48x} =$$
    $$3 + \quad \sqrt{48(3)} =$$
    $$3 + \sqrt{(16)(3)(3)} =$$
    $$3 + \quad (4)(3) =$$
    $$3 + \quad\quad 12 = 15$$

3. **−13:** To evaluate the expression, replace x with −4 everywhere in the equation. Be extra careful with the negative signs; use parentheses to help keep track.

    $$\frac{5 - x}{3} - x^2 =$$
    $$\frac{5 - (-4)}{3} - (-4)^2 =$$
    $$\frac{5 + 4}{3} - (16) =$$
    $$\frac{9}{3} - 16 =$$
    $$3 - 16 = -13$$

4. **299,900:** To find the value of p, first replace c with 100.

    $$30c^2 - c =$$
    $$30(100)^2 - 100 =$$
    $$30(10,000) - 100 =$$
    $$300,000 - 100 = 299,900$$

5. **−6:** First, replace x with 3 in the equation.

    $$y - 3 = \frac{xy}{2}$$
    $$y - 3 = \frac{(3)\,y}{2}$$

Next, to find the value of y, isolate y on one side of the equation.

$$y - 3 = \frac{3y}{2}$$
$$2(y - 3) = 3y$$
$$2y - 6 = 3y$$
$$-6 = y$$

Drill 2

6. **5:** Apply PEMDAS in reverse.

    $$3(7 - x) = 4(1.5)$$

$21 - 3x = 6$	Distribute.
$-3x = -15$	Subtract 21.
$x = 5$	Divide by -3.

7. **−4:** Apply PEMDAS in reverse.

    $$7x + 13 = 2x - 7$$

$5x + 13 = -7$	Subtract $2x$.
$5x = -20$	Subtract 13.
$x = -4$	Divide by 5.

8. **3:** Apply PEMDAS in reverse.

    $$3t^3 - 7 = 74$$

$3t^3 = 81$	Add 7.
$t^3 = 27$	Divide by 3.
$t = 3$	Take the cube root.

9. **6:** Apply PEMDAS in reverse.

    $$1{,}200x + 6{,}000 = 13{,}200$$

$1{,}200x = 7{,}200$	Subtract 6,000.
$x = 6$	Divide by 1,200.

10. **100:** The right side contains no variables. First, simplify that side using standard PEMDAS rules.

$\sqrt{x} = 3 \times 5 - 20 \div 4$	
$\sqrt{x} = (3 \times 5) - (20 \div 4)$	
$\sqrt{x} = 15 - 5$	
$\sqrt{x} = 10$	
$x = 100$	Square both sides.

11. **3:** Be careful with the negatives!

$$-(y)^3 = -27$$

$$(y)^3 = 27 \qquad \text{Divide by } -1.$$

$$y = 3 \qquad \text{Take the cube root.}$$

12. **17:**

$$2 - \frac{\sqrt{2x+2}}{2} = -1$$

$$-\frac{\sqrt{2x+2}}{2} = -3 \qquad \text{Subtract 2 from both sides.}$$

$$\sqrt{2x+2} = 6 \qquad \text{Multiply by } -2.$$

$$2x + 2 = 36 \qquad \text{Square.}$$

$$2x = 34 \qquad \text{Subtract 2.}$$

$$x = 17 \qquad \text{Divide by 2.}$$

13. **81:**

$$5\sqrt{x} + 6 = 51$$

$$5\sqrt{x} = 45 \qquad \text{Subtract 6.}$$

$$\sqrt{x} = 9 \qquad \text{Divide by 5.}$$

$$x = 81 \qquad \text{Square.}$$

Drill 3

14. **−26:**

$$4(-3x - 8) = 8(-x + 9)$$

$$-12x - 32 = -8x + 72$$

$$-32 = 4x + 72$$

$$-104 = 4x$$

$$-26 = x$$

15. **−9:**

$$3x + 7 - 4x + 8 = 2(-2x - 6)$$

$$-x + 15 = -4x - 12$$

$$3x + 15 = -12$$

$$3x = -27$$

$$x = -9$$

16. **−6:**

$$2x(4-6) = -2x + 12$$
$$2x(-2) = -2x + 12$$
$$-4x = -2x + 12$$
$$-2x = 12$$
$$x = -6$$

17. **−2:**

$$\frac{3(6-x)}{2x} = -6$$
$$3(6-x) = -6(2x)$$
$$18 - 3x = -12x$$
$$18 = -9x$$
$$-2 = x$$

18. **0:**

$$\frac{13}{x+13} = 1$$
$$13 = 1(x+13)$$
$$13 = x + 13$$
$$0 = x$$

19. **1:**

$$\frac{10(-3x+4)}{10-5x} = 2$$
$$10(-3x+4) = 2(10-5x)$$
$$-30x + 40 = 20 - 10x$$
$$40 = 20 + 20x$$
$$20 = 20x$$
$$1 = x$$

20. **2:**

$$\frac{15 - 3(5x-3)}{3-2x} = 6$$
$$15 - 3(5x-3) = 6(3-2x)$$
$$15 - 15x + 9 = 18 - 12x$$
$$24 - 15x = 18 - 12x$$
$$24 = 18 + 3x$$
$$6 = 3x$$
$$2 = x$$

21. **−10:**

$$\frac{50(10 + 3x)}{50 + 7x} = 50$$

$$\cancel{50}(10 + 3x) = \cancel{50}(50 + 7x)$$

$$10 = 50 + 4x$$

$$-40 = 4x$$

$$-10 = x$$

Drill 4

22. $x = 5, y = 10$: Equation 1 is $7x - 3y = 5$ and equation 2 is $y = 10$.

Eq. (1):	Eq. (2):
$7x - 3y = 5$	$y = 10$

$$7x - 3(10) = 5 \qquad \text{Substitute 10 for } y \text{ in Eq. (1), and solve for } x.$$

$$7x - 30 = 5$$

$$7x = 35$$

$$x = 5$$

23. $x = 5, y = 30$:

Eq. (1):	Eq. (2):
$y = 4x + 10$	$y = 7x - 5$

$$(4x + 10) = 7x - 5 \qquad \text{Substitute } (4x + 10) \text{ for } y \text{ in Eq. (2), and solve for } x.$$

$$10 = 3x - 5$$

$$15 = 3x$$

$$5 = x$$

$$y = 4(5) + 10 \qquad \text{Substitute 5 for } x \text{ in Eq. (1), and solve for } y.$$

$$y = 30$$

24. $h = 6, k = 3$:

Eq. (1):	Eq. (2):
$2h - 4k = 0$	$k = h - 3$

$$2h - 4(h - 3) = 0 \qquad \text{Substitute } (h - 3) \text{ for } k \text{ in Eq. (1), and solve for } h.$$

$$2h - 4h + 12 = 0$$

$$-2h = -12$$

$$h = 6$$

$$k = (6) - 3 \qquad \text{Substitute 6 for } h \text{ in Eq. (2), and solve for } k.$$

$$k = 3$$

25. $x = 7, y = 6$:

	Eq. (1):	Eq. (2):
	$5x - 3y = 17$	$2x - y = 8$

$-y = -2x + 8$ Isolate y in Eq. (2).

$y = 2x - 8$

$5x - 3(2x - 8) = 17$ Substitute $(2x - 8)$ for y in Eq. (1), and solve for x.

$5x - 6x + 24 = 17$

$-x + 24 = 17$

$7 = x$

$y = 2(7) - 8$ Substitute 7 for x in the rephrased Eq. (2) to solve for y.

$y = 14 - 8$

$y = 6$

26. $b = 3, g = 18$:

	Eq. (1):	Eq. (2):
	$12b = 2g$	$4g - 3b = 63$

$6b = g$ Isolate g in Eq. (1).

$4(6b) - 3b = 63$ Substitute $(6b)$ for g in Eq. (2), and solve for b.

$24b - 3b = 63$

$21b = 63$

$b = 3$

$6(3) = g$ Substitute 3 for b in the rephrased Eq. (1) to solve for g.

$18 = g$

Drill 5

27. $x = 3, y = -1$: The first equation has the term $-y$ while the second equation has the term $+y$. If you add the equations together, these two terms will cancel.

$$x - y = 4$$
$$\underline{+(2x + y = 5)}$$
$$3x \qquad = 9$$

Therefore, $x = 3$. Plug this value into the first equation.

$$x - y = 4$$
$$(3) - y = 4$$
$$-y = 1$$
$$y = -1$$

28. $x = 1, y = 2$: Both equations have the term $+x$, so eliminate the variable x by subtracting one equation from the other.

$$\begin{array}{r} x + 2y = 5 \\ - \ (x - 4y = -7) \\ \hline 6y = 12 \end{array}$$

The equation simplifies to $y = 2$. Plug this value for y into the first equation and solve for x.

$$\begin{aligned} x + 2y &= 5 \\ x + 2(2) &= 5 \\ x + 4 &= 5 \\ x &= 1 \end{aligned}$$

When doing the initial subtraction, be very careful to change the sign of each term in the second equation when subtracting; for example, $-4y$ becomes $-(-4y) = +4y$, and -7 becomes $-(-7) = +7$. Alternatively, you could multiply the entire second equation $(x - 4y = -7)$ by -1 to get $-x + 4y = 7$, and then add this equation to the first.

$$\begin{array}{r} x + 2y = 5 \\ + \ (-x + 4y = 7) \\ \hline 6y = 12 \end{array}$$

This yields the same solution: $y = 2$ and $x = 1$.

29. $a = 5, b = 3$: Both equations have the term $+b$, so eliminate the variable b by subtracting one equation from the other.

$$\begin{array}{r} a + b = 8 \\ -(2a + b = 13) \\ \hline - a \quad\quad = -5 \end{array}$$

Hence, $a = 5$. Plug this value for a into the first equation, and solve for b.

$$\begin{aligned} a + b &= 8 \\ (5) + b &= 8 \\ b &= 3 \end{aligned}$$

30. $m = 1, n = 3$: None of the variables have the same coefficient (the number in front of a variable), so no variables will cancel as the equations are currently written. Multiply one of the equations by a constant so that one of the variables will then cancel when you add or subtract the equations.

The second equation has an n, and the first equation has a $-2n$. Multiply the second equation by 2 to get $6m + 2n = 12$, then add the two equations.

$$\begin{array}{r} 7m - 2n = 1 \\ + (6m + 2n = 12) \\ \hline 13m \quad\quad = 13 \end{array}$$

Therefore, $m = 1$. Plug this value into the second equation, and solve for n.

$$3m + n = 6$$
$$3(1) + n = 6$$
$$n = 3$$

31. $x = -\dfrac{4}{5}$, $y = -\dfrac{3}{5}$: First, manipulate the two equations so that the variables are nicely aligned on the left-hand side and the constant terms are all on the right.

$$-2x + y = 1$$
$$x - 3y = 1$$

Multiply one of the equations by a constant that will allow you to cancel one of the variables when you add or subtract. There are multiple correct ways to do this; one way is to multiply the second equation by 2, thereby replacing it with the equation $2x - 6y = 2$, then add the original first equation to the new second equation.

$$-2x + y = 1$$
$$\underline{+(2x - 6y = 2)}$$
$$-5y = 3$$

Therefore, $y = -\dfrac{3}{5}$. Glance at the two original equations. Which one is easiest to use to solve for x? In either case, you're still going to have to deal with the fractional value for y, but the second equation is a little easier because the x variable has a coefficient of 1. In other words, you won't have to divide at the end in order to solve.

$$x - 3\left(-\frac{3}{5}\right) = 1$$
$$x + \frac{9}{5} = 1$$
$$x = \frac{5}{5} - \frac{9}{5}$$
$$x = -\frac{4}{5}$$

32. $r = 1$, $s = 2$: The coefficients in this problem are messy—three fractions. Take a little time to think about the best way to proceed. Get rid of the messy part: Eliminate the fractions in both equations before proceeding.

If you multiply the first equation by 6, you will cancel out all denominators. Do the same for the second equation by multiplying everything by 2.

$$2r - s = 0$$
$$4r + s - 6 = 0$$

The coefficients on s are both 1 but with different signs, so when you add the equations, they will cancel. Next, rearrange the second equation so that the constant term is on the right side, then add the two equations.

$$2r - s = 0$$
$$\underline{+(4r + s = 6)}$$
$$6r \quad\;\; = 6$$

Finally, plug $r = 1$ into the first equation to solve for s.

$$2(1) - s = 0$$
$$2 = s$$

Drill 6

33. **$x = 1, y = 3.5$:** When one of the two equations is already arranged to isolate one of the variables, substitution is usually the better method. In this particular problem, the second equation isolates y, so take the right-hand side of the second equation and substitute it for y in the first equation.

$$5x + 2\left(\frac{1}{2}x + 3\right) = 12$$
$$5x + x + 6 = 12$$
$$6x = 6$$
$$x = 1$$

Plug this value for x into either of the original equations to solve for y; in this case, it will be easier to use the second equation (since it is already solved for y).

$$y = \frac{1}{2}x + 3$$
$$y = \frac{1}{2}(1) + 3$$
$$y = 3.5 \text{ or } 3\frac{1}{2}$$

34. **$x = -5, y = -2$:** For this system of equations, either method would be appropriate. Both equations would require some manipulation before you could stack-and-add, and neither equation already isolates one of its variables. When neither method seems to have an advantage, pick whichever you like best.

If you use substitution, it might be easier to solve the first equation for y: $y = x + 3$. Then substitute this into the second equation.

$$2(x + 3) = x + 1$$
$$2x + 6 = x + 1$$
$$x = -5$$

Plug this into the equation in which y has been isolated.

$$y = x + 3$$
$$y = (-5) + 3$$
$$y = -2$$

If you use elimination, manipulate the first equation to combine the constant terms $(y = x + 3)$. Now the two equations have the same order, and the coefficient on x is the same, so if you subtract the second from the first, the variable x will disappear.

$$\begin{aligned} y &= x + 3 \\ -(2y &= x + 1) \\ \hline -y &= 2 \end{aligned}$$

Therefore, $y = -2$; plug this value for y into either equation to solve for x: $x = -5$.

35. $a = \dfrac{3}{4}$, $b = \dfrac{1}{2}$: Even though some manipulations will be required to line up the variables nicely in this system, elimination is the optimal method because the $-2b$ in the first equation will cancel the $+2b$ in the second. Start by rearranging the two equations.

$$8a - 2b = 5$$
$$4a + 2b = 4$$

Then add them together.

$$\begin{aligned} 8a - 2b &= 5 \\ +(4a + 2b &= 4) \\ \hline 12a &= 9 \end{aligned}$$

Therefore, $a = \dfrac{9}{12} = \dfrac{3}{4}$. Plug this value into one of the two original equations, and solve for b. In this case, the second equation has a $4a$ term, which will cancel nicely with the fraction $\dfrac{3}{4}$.

$$2b + 4\left(\frac{3}{4}\right) - 4 = 0$$
$$2b + 3 - 4 = 0$$
$$2b = 1$$
$$b = \frac{1}{2}$$

36. $x = 1$, $y = 3$: For this system of equations, either method would be appropriate. Both equations would require some manipulation before you could stack-and-add, and neither equation is already solved for one of its variables. When neither method seems to have an advantage, pick whichever you like best.

To solve via substitution, isolate y in the first equation to get $y = 6 - 3x$, and then substitute this for y in the second equation.

$$6x - (6 - 3x) = 3$$
$$6x - 6 + 3x = 3$$
$$9x = 9$$
$$x = 1$$

Next, plug $x = 1$ into the same equation in which y has been isolated.

$$y = 6 - 3x$$
$$y = 6 - 3(1)$$
$$y = 3$$

To solve using the elimination method, rearrange the first equation to get $3x + y = 6$, and then add the equations together to eliminate y.

$$3x + y = 6$$
$$\underline{+(6x - y = 3)}$$
$$9x \quad\;\; = 9$$

Plug $x = 1$ into either of the equations to solve for y.

$$3x + y = 6$$
$$3(1) + y = 6$$
$$y = 3$$

37. $x = 3.5, y = 2$: The first equation is solved for x, which points to substitution. At the same time, the first equation has $+x$ on the left side while the second equation has $-x$. Either method will work; choose the one you like best.

To solve using the elimination method, rearrange the equations to line up all the variables. Next, add the two equations in order to eliminate the x term.

$$x - 2y = -\frac{1}{2}$$
$$\underline{+\left(-x + \;\; y = -\frac{3}{2}\right)}$$
$$-y = -2$$

Therefore, $y = 2$. Plug this into the first equation, since that equation already isolates x.

$$x = 2y - \frac{1}{2}$$
$$x = 2(2) - \frac{1}{2}$$
$$x = 3.5$$

To solve by substitution, take the first equation, which is already solved for x, and substitute into the second equation.

$$y - \left(2y - \frac{1}{2}\right) = -\frac{3}{2}$$
$$y - 2y + \frac{1}{2} = -\frac{3}{2}$$
$$-y = -\frac{3}{2} - \frac{1}{2}$$
$$-y = -2$$

Therefore, $y = 2$. As before, plug this value into the first equation to get $x = 3.5$.

Drill 7

38. **13:** This question contains only two equations, but three variables. To isolate y, you need to get rid of both x and z. Try to eliminate both variables at the same time. The coefficients of x and z are the same in both equations, so subtract the second equation from the first to eliminate both.

$$\begin{array}{r} 4x + y + 3z = 34 \\ -(4x \quad\quad + 3z = 21) \\ \hline y = 13 \end{array}$$

39. **(A) $4yz - 8z + 3$:** To find the value of x in terms of the other variables, isolate x on one side of the equation, and put everything else on the other side. Begin by getting rid of the fraction.

$$\frac{x - 3}{y - 2} = 4z$$
$$x - 3 = 4z(y - 2)$$
$$x = 4z(y - 2) + 3$$

The answers don't contain parentheses, so distribute to get rid of them.

$$x = 4yz - 8z + 3$$

40. **3:** In order to isolate y, you need to eliminate both x and z. Find a way to eliminate both variables at the same time.

Notice that the coefficients for x and z in the second equation (6 and 4, respectively) are exactly double their coefficients in the first equation (3 and 2, respectively). If you divide the second equation by 2, the coefficients will be the same.

$$\begin{array}{rcl} 3x + 5y + 2z = 20 & \rightarrow & 3x + 5y + 2z = 20 \\ 6x + 4z = 10 & \rightarrow & 3x + 2z = 5 \end{array}$$

Now, subtract the second equation from the first.

$$\begin{array}{r} 3x + 5y + 2z = 20 \\ -(3x \quad\quad + 2z = 5) \\ \hline 5y = 15 \\ y = 3 \end{array}$$

41. **$b = \dfrac{a - 12}{5}$:** To solve for b in terms of a, isolate b on one side of the equation, and put everything else on the other side. Furthermore, you will need to eliminate the variable c, because the question does not mention c.

The second equation is already solved for c, so start by substituting $b + 2$ for c in the first equation.

$$\frac{a - b}{4} = c + 1$$
$$\frac{a - b}{4} = (b + 2) + 1$$

Great—c is now gone. Now isolate b on one side of the equation.

$$\frac{a-b}{4} = b + 3$$
$$a - b = 4(b + 3)$$
$$a - b = 4b + 12$$
$$a - 12 = 5b$$
$$\frac{a-12}{5} = b$$

Drill 8

42. **50:** In order to find the value of $a + b$, you need to eliminate c and d from the first equation, but there are only two equations total. Is there a way to eliminate $c + d$ at once?

Try manipulating the second equation to give you $c + d$.

$$3d = 15 - 3c$$
$$3c + 3d = 15$$
$$c + d = 5$$

Next, substitute $c + d = 5$ into the first equation and isolate $a + b$.

$$\frac{a+b}{c+d} = 10$$
$$\frac{a+b}{(5)} = 10$$
$$a + b = 50$$

43. $z = \dfrac{3x+1}{2}$: In order to solve for z in terms of x, isolate z on one side of an equation that has only one other variable, x. In other words, get rid of the y somehow. Therefore, kill off y by isolating it in the first equation.

$$x = \frac{y}{5}$$
$$5x = y$$

Now substitute. In the second equation, replace y with $5x$.

$$2z - 1 = \frac{x+y}{2}$$
$$2z - 1 = \frac{x+(5x)}{2}$$

Finally, isolate z.

$$2z - 1 = \frac{6x}{2}$$
$$2z - 1 = 3x$$
$$2z = 3x + 1$$
$$z = \frac{3x+1}{2}$$

44. **18:** In order to isolate z, you have to eliminate x and y. The combination $x + y$ is in the exponent of the first equation, so try to isolate that same combination (or combo) in the second equation.

$$x = 2 - y$$
$$x + y = 2$$

In the first equation, substitute 2 for the combo $x + y$, then solve for z.

$$2^{x+y} = \sqrt{z - 2}$$
$$2^{(2)} = \sqrt{z - 2}$$
$$4 = \sqrt{z - 2}$$
$$(4)^2 = \left(\sqrt{z - 2}\right)^2$$
$$16 = z - 2$$
$$18 = z$$

45. **(B)** $\dfrac{16}{3}$: Both equations contain a z, but the question stem asks only for $x - y$. How can you get rid of that z?

First, get rid of the fraction in the first equation.

$$\frac{3x}{z + 4} = 4$$
$$3x = 4(z + 4)$$
$$3x = 4z + 16$$

You could solve the second equation for z and then substitute. However, both equations now have a $4z$ term. You can directly substitute $3y$ for $4z$ into the equation to get rid of the z variable, then solve for $x - y$.

$$3x = 4z + 16$$
$$3x = (3y) + 16$$
$$3x - 3y = 16$$
$$x - y = \frac{16}{3}$$

Quadratic Equations

In This Chapter

In this chapter, you will learn to manipulate and solve quadratic equations in order to solve for unknowns, otherwise known as variables. You'll also learn how to recognize three special quadratics and what to do when you see one on the test.

CHAPTER 7 Quadratic Equations

In high school algebra, you learned a number of skills for dealing with quadratic equations. You will need those skills again on the GMAT—but, as always, don't solve for everything. See what the questions ask and what form the answers are in, and react accordingly.

Mechanics of Quadratic Equations

A **quadratic expression** contains a squared variable, such as x^2, and no higher power. The word *quadratic* comes from the Latin word for *square*. Here are a few quadratic expressions:

$$z^2 \qquad\qquad y^2 + y - 6 \qquad\qquad x^2 + 8x + 16 \qquad\qquad w^2 - 9$$

A quadratic expression can also be disguised. You might not see an actual exponent in the form given to you. Here are some disguised quadratic expressions:

$$z \times z \qquad\qquad (y + 3)(y - 2) \qquad\qquad (x + 4)^2 \qquad\qquad (w - 3)(w + 3)$$

If you multiply these expressions out—that is, if you distribute them—then you will have exponents on the variables. Note that each expression in the second list equals the corresponding expression in the first list (once you multiply out).

A **quadratic equation** contains a quadratic expression and an equals sign:

Quadratic expression = something else

A quadratic equation usually has two solutions. That is, in most cases, *two* different values of the variable each make the equation true. Solving a quadratic equation means finding those values.

Before you can solve quadratic equations, you have to be able to distribute and factor quadratic expressions.

Distribute $(a + b)(x + y)$: Use FOIL

Recall that distributing means applying multiplication across a sum. For example:

$$5 \qquad \times \qquad (3 + 4) \qquad = \qquad 5 \times 3 \qquad + \qquad 5 \times 4$$

Five | times | the quantity three plus four | equals | five times three | plus | five times four.

Here are more examples:

$$3(x + 2) = 3x + 6 \qquad\qquad (z - 12)y = zy - 12y \qquad\qquad w(a + b) = wa + wb$$

What if you have to distribute the product of *two* sums? Try this example:

$$(a + b)(x + y) =$$

Multiply every term in the first sum by every term in the second sum, then add up all the products. This is just distribution on steroids.

To make the products, use the acronym **FOIL**: **F**irst, **O**uter, **I**nner, **L**ast (or **F**irst, **O**utside, **I**nside, **L**ast). For example:

$(\boldsymbol{a} + b)(\boldsymbol{x} + y)$	F – multiply the First term in each of the parentheses:	$(a)(x) = ax$
$(\boldsymbol{a} + b)(x + \boldsymbol{y})$	O – multiply the Outer term in each:	$(a)(y) = ay$
$(a + \boldsymbol{b})(\boldsymbol{x} + y)$	I – multiply the Inner term in each:	$(b)(x) = bx$
$(a + \boldsymbol{b})(x + \boldsymbol{y})$	L – multiply the Last term in each:	$(b)(y) = by$

Now add up the products:

$$(a + b)(x + y) = ax + ay + bx + by$$

By the way, you can also FOIL numbers. For example:

What is 102×301 ?

If you express 102 as $100 + 2$ and 301 as $300 + 1$, you can rewrite the question as a product of two sums:

What is $(100 + 2)(300 + 1)$?

Now FOIL it out:

$$(100 + 2)(300 + 1) = (100 \times 300) + (100 \times 1) + (2 \times 300) + (2 \times 1)$$
$$= 30,000 + 100 + 600 + 2$$
$$= 30,702$$

You get the same answer if you multiply these numbers in longhand—but FOILing allows you to work with easier numbers.

Now try to FOIL this disguised quadratic expression: $(x + 2)(x + 3)$:

$(\boldsymbol{x} + 2)(\boldsymbol{x} + 3)$	F – multiply the First term in each of the parentheses:	$(x)(x) = x^2$
$(\boldsymbol{x} + 2)(x + \boldsymbol{3})$	O – multiply the Outer term in each:	$(x)(3) = 3x$
$(x + \boldsymbol{2})(\boldsymbol{x} + 3)$	I – multiply the Inner term in each:	$(2)(x) = 2x$
$(x + \boldsymbol{2})(x + \boldsymbol{3})$	L – multiply the Last term in each:	$(2)(3) = 6$

Add up the products:

$$(x + 2)(x + 3) = x^2 + 3x + 2x + 6$$

You can combine the two like terms in the middle ($3x$ and $2x$) into one term ($5x$):

$$(x + 2)(x + 3) = x^2 + \underline{3x + 2x} + 6 = x^2 + 5x + 6$$

Now compare the expression you started with and the expression you ended up with:

$$(x + 2)(x + 3) \qquad\qquad x^2 + 5x + 6$$

Examine how the numbers on the left relate to the numbers on the right:

The 2 and the 3 *multiply* to give you the 6.

The 2 and the 3 *add* to give you the 5 in $5x$.

What if you have subtraction? Do it the same way, but account for the fact that you now have some negative numbers in the mix. Try this problem:

$$(y - 5)(y - 2) =$$

First, FOIL. Keep track of minus signs. Put them in the products:

$(\boldsymbol{y} - 5)(\boldsymbol{y} - 2)$	F – multiply First terms:	$(y)(y) = y^2$
$(\boldsymbol{y} - 5)(y - \boldsymbol{2})$	O – multiply Outer terms:	$(y)(-2) = -2y$
$(y - \boldsymbol{5})(\boldsymbol{y} - 2)$	I – multiply Inner terms:	$(-5)(y) = -5y$
$(y - \boldsymbol{5})(y - \boldsymbol{2})$	L – multiply Last terms:	$(-5)(-2) = 10$

Finally, add the products and combine like terms:

$$(y - 5)(y - 2) = y^2 - 2y - 5y + 10 = y^2 - 7y + 10$$

Again, examine how the numbers on the left relate to the numbers on the right:

The -5 and the -2 *multiply* to give you the positive 10.

The -5 and the -2 *add* to give you the -7 in $-7y$.

Here's one last wrinkle. In the course of doing these problems, you might encounter a sum written as $4 + z$ rather than as $z + 4$. You can FOIL it as is, or you can flip the sum around so that the variable is first. Either way works fine.

If you...	Then you...	Like this:
Want to distribute $(x + 5)(x - 4)$	FOIL it out and combine like terms	$= (x + 5)(x - 4)$ $= x^2 - 4x + 5x - 20$ $= x^2 + x - 20$

Check Your Skills

FOIL the following expressions.

1. $(x + 4)(x + 9)$
2. $(y + 3)(y - 6)$

Answers can be found on page 290.

Factor $x^2 + 5x + 6$: Find the Original Numbers in $(x + \ldots)(x + \ldots)$

FOILing is a form of distribution. So going in reverse is a form of *factoring*. To factor a quadratic expression such as $x^2 + 5x + 6$ means to *rewrite the expression as a product of two sums*. For example:

$$x^2 + 5x + 6 = (x + \ldots)(x + \ldots)$$

The form on the right is called the **factored form**. (You can call $x^2 + 5x + 6$ the **distributed form**.)

You already know the answer, because earlier you turned $(x + 2)(x + 3)$ into $x^2 + 5x + 6$:

$$(x + 2)(x + 3) = x^2 + 3x + 2x + 6 = x^2 + \boldsymbol{5x + 6}$$

Consider the relationship between the numbers one more time:

$2 + 3 = 5$, the coefficient of the x term $2 \times 3 = 6$, the constant term on the end

This is true in general. The two numbers in the factored form *add* to the x coefficient, and they *multiply* to the constant.

Now think about how to work backwards:

$x^2 + \mathbf{5}x + \mathbf{6} = (x + \ldots)(x + \ldots)$

You need two numbers that multiply together to 6 and sum to 5.

Look first for factor pairs of the constant—in this case, two numbers that multiply to 6. Then check the sum:

2 and 3 are a factor pair of 6, because $2 \times 3 = 6$.

2 and 3 also sum to 5, so this is the correct pair.

$x^2 + \mathbf{5}x + \mathbf{6} = (x + \mathbf{2})(x + \mathbf{3})$

Try this problem:

$y^2 + \mathbf{7}y + \mathbf{6} = (y + \ldots)(y + \ldots)$

The constant is the same: 6. So you need a factor pair of 6. But now the pair has to sum to 7.

Therefore, 2 and 3 no longer work. But 1 and 6 are also a factor pair of 6. So factor $y^2 + 7y + 6$ like this:

$y^2 + \mathbf{7}y + \mathbf{6} = (y + 1)(y + 6)$

Now try to factor this quadratic:

$z^2 + 7z + 12 = (z + \ldots)(z + \ldots)$

Again, start with the constant. Look for a factor pair of 12 that sums to 7.

It might help to list the factor pairs of 12:

1×12

2×6

3×4

The only factor pair of 12 that sums to 7 is 3 and 4. Therefore:

$3 \times 4 = 12$ $3 + 4 = 7$

$z^2 + 7z + \mathbf{12} = (z + \mathbf{3})(z + \mathbf{4})$

What if you have subtraction? The same principles hold. Just think of the minus signs as part of the numbers themselves. Try this example:

$x^2 - \mathbf{9}x + \mathbf{18} = (x + \ldots)(x + \ldots)$

You need two numbers that multiply to 18, but now they have to add up to -9. Think about rules of negatives and positives. If the sum is negative, then at least one of the terms must be negative. And for the product to be positive, the terms must have the same sign. Therefore, both numbers must be negative.

Again, consider listing out the factor pairs of 18; include the fact that both terms are negative:

$$-1 \times -18$$
$$-2 \times -9$$
$$-3 \times -6$$

The answer is -3 and -6. $-3 \times -6 = 18$ $-3 + (-6) = -9$

Now, write the factored form of the quadratic expression:

$$x^2 - \mathbf{9}x + \mathbf{18} = (x - \mathbf{3})(x - \mathbf{6})$$

If the constant (in the case above, 18) is positive, then the two numbers in the factored form must *both be positive* or *both be negative*. How do you know which it is? Check the sign of the x term.

If the sign of the constant (in the case below, 12) and the x term ($7z$) are both positive, then the two numbers in factored form are positive:

$$z^2 + \mathbf{7}z + \mathbf{12} = (z + \mathbf{3})(z + \mathbf{4})$$

If, on the other hand, the constant (18) is positive but the x term ($9x$) is negative, then both numbers in factored form are negative:

$$x^2 - \mathbf{9}x + \mathbf{18} = (x - \mathbf{3})(x - \mathbf{6})$$

What if the constant is negative? Again, *think of the minus sign as part of the number.* Try this problem:

$$w^2 + \mathbf{3}w - \mathbf{10} = (w + \ldots)(w + \ldots)$$

You need two numbers that multiply to negative 10 and that sum to 3.

For the product to be -10, one number must be positive and the other one must be negative. That's the only way to get a negative product of two numbers:

$$\text{Pos} \times \text{Neg} = \text{Neg}$$

If the constant is negative, then in the factored form, *one number is positive and the other one is negative.*

This means that you are adding a positive and a negative to get 3:

$$\text{Pos} + \text{Neg} = 3$$

If you think about the negative as subtracting a positive, then you are looking for a difference of 3. Here's how to find the two numbers you want. First, pretend that the constant is positive. Think of the normal, positive factor pairs of 10. Which pair *differs* by 3?

$$1 \times 10$$
$$2 \times 5$$

The answer is 5 and 2.

But now you must decide which is positive and which is negative. Because the sum is positive, it must be the case that the greater factor is the positive one:

The answer is 5 and −2. $5 \times (-2) = -10$ $5 + (-2) = 3$

Notice that −5 and 2 would give you the correct product (-10) but the incorrect sum $(-5 + 2 = -3)$.

Now you know where to place the signs in the factored form. Place the minus sign with the 2:

$$w^2 + 3w - 10 = (w + 5)(w - 2)$$

When the *constant is negative*, start by testing the positive factor pairs and asking which pair *differs* by the correct amount.

Once you find a good factor pair for the constant (say 5 and 2), then determine which term is negative and which is positive by asking whether the sum is negative or positive. If the sum is positive, the greater term must be positive. If the sum is negative, the greater term must be negative. If you're not sure, try both possibilities! For example:

$$y^2 - 4y - 21 = (y + \ldots)(y + \ldots)$$

You need two numbers that multiply to *negative* 21 and that sum to *negative* 4.

One number must be positive, while the other is negative.

Start by looking for the positive factor pairs of 21. Which pair *differs* by 4?

Only 7 and 3 work. $7 \times 3 = 21$ $7 - 3 = 4$

Next, make one number negative so that the product is −21 and the *sum* is −4. Which number should be negative?

The sum is negative, so make the greater term negative. In this case, 7 needs to be the negative term:

$$(-7) \times 3 = -21 \qquad (-7) + 3 = -4$$

Write the factored form:

$$y^2 - 4y - 21 = (y - 7)(y + 3)$$

Finally, the GMAT can make factoring a quadratic expression harder in a couple of ways.

In the first way, every term in the expression is multiplied through by a common numerical factor, including the x^2 term:

$$3x^2 + 21x + 36 = \ldots$$

In this case, *pull out the common factor first*. Put parentheses around what's left. Then factor the quadratic expression as usual:

$$3x^2 + 21x + 36$$
$$3(x^2 + 7x + 12)$$
$$3(x + 3)(x + 4)$$

If the x^2 term is negative, factor out a common factor of -1 *first*. That will flip the sign on every term:

$$-x^2 + 9x - 18 = \ldots$$

Pull out the -1, which becomes a minus sign outside a set of parentheses. Don't forget to flip the sign of every term. Then factor the quadratic expression as usual:

$$-x^2 + 9x - 18 = -(x^2 - 9x + 18) = -(x - 3)(x - 6)$$

The second way quadratics can be made harder is this: The x^2 term has a coefficient, but you can't divide it out without creating fractions for the other terms. *Avoid fractional coefficients at all costs.* If you cannot pull out a common factor from the x^2 term without turning coefficients into fractions, then keep a coefficient on one or even both x's in your factored form. At this point, *experiment with factor pairs of the constant* until you get a match. For example:

$$2z^2 - z - 15 = \ldots$$

Don't factor a 2 out of all the terms, since that will give you fractions for the second and third terms. Rather, set up the parentheses on the right. Put a $2z$ in one set and a z in the other set:

$$2z^2 - z - 15 = (2z + \ldots)(z + \ldots)$$

At least you've got the F of FOIL covered.

What you already know how to do is still useful. Since the constant is negative (-15), one of the numbers must be negative, while the other must be positive.

Pretend for a minute that the constant is positive 15. You still need a factor pair of 15. There are only two factor pairs of 15:

$$1 \times 15 = 15 \qquad\qquad 3 \times 5 = 15$$

But which pair do you want? Which number becomes negative? And where does that one go—with the $2z$ or the z?

The middle term is your guide. The coefficient on the z term is only -1, so it's very unlikely that the "1 and 15" factor pair will work. The numbers are probably 3 and 5, with a minus sign on exactly one of them. This covers the L of FOIL.

Finally, experiment. Try the numbers in different configurations. Examine only the OI of FOIL (Outer and Inner) to see whether you get the right middle term:

$$2z^2 - z - 15 = (2z + \ldots)(z + \ldots) = (2z + 3)(z - 5)? \quad \textbf{FOIL} \to (2z)(-5) + 3z \text{ does not equal } -z$$

Think about what happened. Since $-10z$ and $3z$ are too far apart, swap the numbers:

$$2z^2 - z - 15 = (2z + \ldots)(z + \ldots) = (2z - 5)(z + 3)? \quad \textbf{FOIL} \to (2z)(3) - 5z \text{ equals } z, \text{ not } -z$$

Closer! The coefficient, 1, is correct this time, but not the sign. So switch the signs:

$$2z^2 - z - 15 = (2z + \ldots)(z + \ldots) = (2z + 5)(z - 3)? \quad \textbf{FOIL} \to (2z)(-3) + 5z \text{ equals } -z \quad \text{YES}$$

So $2z^2 - z - 15$ factors into $(2z + 5)(z - 3)$. You can always check your work by FOILing the result:

$$(2z + 5)(z - 3) = 2z^2 - 6z + 5z - 15 = 2z^2 - z - 15$$

If you're ever in doubt, FOIL it back out! Luckily, you won't often see a FOIL this complex on the GMAT. (You might even decide that a problem this complex isn't worth your time.)

If you...	Then you...	Like this:
Want to factor $x^2 + 11x + 18$	Find a factor pair of 18 that sums to 11	$x^2 + 11x + 18$ $= (x + 9)(x + 2)$ $9 + 2 = 11$ $9 \times 2 = 18$
Want to factor $x^2 - 8x + 12$	Find a factor pair of 12 that sums to 8, then make both numbers negative	$x^2 - 8x + 12$ $= (x - 6)(x - 2)$ $(-6) + (-2) = -8$ $(-6) \times (-2) = 12$
Want to factor $x^2 + 6x - 16$	Find a factor pair of 16 that *differs* by 6, then make the bigger number *positive* so that the sum is 6 and the product is -16	$x^2 + 6x - 16$ $= (x + 8)(x - 2)$ $8 + (-2) = 6$ $8 \times (-2) = -16$
Want to factor $x^2 - 5x - 14$	Find a factor pair of 14 that *differs* by 5, then make the bigger number *negative* so that the sum is -5 and the product is -14	$x^2 - 5x - 14$ $= (x - 7)(x + 2)$ $(-7) + 2 = -5$ $(-7) \times 2 = -14$
Want to factor $-2x^2 + 16x - 24$	Factor out -2 from all terms first, then factor the quadratic expression normally	$-2x^2 + 16x - 24$ $= -2(x^2 - 8x + 12)$ $= -2(x - 6)(x - 2)$

Check Your Skills

Factor the following expressions.

3. $x^2 + 14x + 33$

4. $x^2 - 14x + 45$

5. $x^2 + 3x - 18$

6. $x^2 - 5x - 66$

Answers can be found on page 290.

Solve a Quadratic: Set Equal to 0

So far, you've dealt with quadratic *expressions*—distributing them and factoring them.

Now how do you solve quadratic *equations*? Try this example:

If $x^2 + x = 6$, what are the possible values of x?

You are asked for *possible* values of *x*. Usually, two different values of *x* will make a quadratic equation true. In other words, expect the equation to have two solutions.

The best way to solve most quadratic equations involves a particular property of the number 0. For example:

If $ab = 0$, then either $a = 0$ or $b = 0$ (or both, potentially).

In words, if the product of two numbers is 0, then you know that at least one of the numbers is 0. (This is known as the Zero Product rule.)

This is true no matter how complicated the factors. For example:

If $(a + 27)(b - 12) = 0$, then either $a + 27 = 0$ or $b - 12 = 0$ (or both).

In words, if the quantity $a + 27$ times the quantity $b - 12$ equals 0, then at least one of those quantities must equal 0.

This gives you a pathway to solve quadratic equations:

1. Rearrange the equation to make one side equal 0. The other side will contain a quadratic expression.

2. Factor the quadratic expression. The equation will look like this:
   (Something)(Something else) = 0

3. Set each factor equal to 0.
   Something = 0   or   Something else = 0

These two equations will be easier to solve. Each one will give you one possible solution for the original equation.

Try this with the problem above. First, rearrange the equation to make one side equal to 0:

$$\begin{array}{rr} x^2 + x = & 6 \\ -6 & -6 \\ \hline x^2 + x - 6 = & 0 \end{array}$$

Next, factor the quadratic expression on the left side:

$$x^2 + x - 6 = 0$$
$$(x + 3)(x - 2) = 0$$

Finally, set each factor (the quantities in parentheses) equal to 0, and solve for *x* in each case:

$$\begin{array}{ccc} x + 3 = 0 & & x - 2 = 0 \\ x = -3 & \text{or} & x = 2 \end{array}$$

Now you have the two possible values of *x*, which are the two solutions to the original equation:

$$x^2 + x = 6 \qquad x \text{ could be } -3 \qquad (-3)^2 + (-3) = 6$$
$$\text{or } x \text{ could be } 2 \qquad 2^2 + 2 = 6$$

By the way, the two equations you get at the end can't both be true at the same time. What $x^2 + x = 6$ tells you is that x must equal either -3 or 2. The value of x is one or the other; it's not both simultaneously. The variable has multiple *possible* values.

The solutions of a quadratic equation are also called its **roots**. This is not the same thing as a square root. When you hear someone say "take the root," they usually mean take the square root. (The real test will specify to take the square root in this circumstance.) But when you hear someone say "find the root," they're usually talking about solving a quadratic.

If an additional condition is placed on the variable, you can often narrow down to one solution. Try this example:

If $y < 0$ and $y^2 = y + 30$, what is the value of y?

First, solve the quadratic equation. Rearrange it so that one side equals 0:

$$\begin{array}{rl} y^2 & = \quad y + 30 \\ \underline{-y - 30} & \underline{-y - 30} \\ y^2 - y - 30 & = 0 \end{array}$$

Next, factor the quadratic expression:

$$y^2 - y - 30 = 0$$
$$(y - 6)(y + 5) = 0$$

Set each factor equal to 0:

$$\begin{array}{lll} y - 6 = 0 & & y + 5 = 0 \\ y = 6 & \text{or} & y = -5 \end{array}$$

At this point, y is either 6 or -5. Go back to the question stem, which gives the additional condition that $y < 0$. Since y is negative, y cannot be 6. Thus, the answer to the question is -5. The quadratic equation gave you two possibilities, but only one of them fits the constraint that $y < 0$.

Occasionally, a quadratic equation has only one solution on its own. Try this example:

If $w^2 - 8w + 16 = 0$, what is the value of w?

The quadratic equation already has one side equal to 0, so go ahead and factor the quadratic expression:

$$w^2 - 8w + 16 = 0$$
$$(w - 4)(w - 4) = 0$$

The two factors in parentheses happen to be identical. In this special case, you don't get two separate equations and two separate roots—you get just one:

$$w - 4 = 0 \quad \text{so} \quad w = 4$$

The only solution is 4.

Lastly, *never factor before you set one side equal to 0*. Try this example:

If z is positive and $z^2 + z - 8 = 4$, what is the value of z?

You might be tempted to factor the left side right away. But the Zero Product rule only works when the product equals 0. Why? Imagine the easier example that $xy = 4$. Must it be true that $x = 4$ or $y = 4$? No! They could both equal 2, or you could have a combination of non-integer values. You can only solve these quadratics when you first set them equal to 0.

Avoid the temptation to factor right away, and instead rearrange to make the right side 0 first:

$$\begin{array}{rcl} z^2 + z - 8 &=& 4 \\ -4 & & -4 \\ \hline z^2 + z - 12 &=& 0 \end{array}$$

Now you can factor the left side:

$$z^2 + z - 12 = 0$$
$$(z + 4)(z - 3) = 0$$

Finally, set each factor equal to 0 and solve for z:

$$z + 4 = 0 \quad \text{or} \quad z - 3 = 0$$

Thus, z equals either -4 or 3. Since you are told that z is positive, z must be 3.

It's technically legal to factor a quadratic expression whenever you want to. But if you factor the expression *before* setting one side equal to zero, your factors don't tell you anything useful. You can't set them individually to 0.

When you solve a quadratic equation, always set one side equal to 0 *before* you factor.

If you...	Then you...	Like this:
Want to solve $x^2 + 11x = -18$	Rearrange to make one side 0, factor the quadratic side, then set the factors equal to 0	$x^2 + 11x = -18$ $x^2 + 11x + 18 = 0$ $(x + 9)(x + 2) = 0$ $(x + 9) = 0$ or $(x + 2) = 0$ $x = -9$ or $x = -2$

Check Your Skills

Solve the following quadratic equations.

7. $x^2 + 2x - 35 = 0$

8. $x^2 - 15x = -26$

Answers can be found on page 291.

Solve a Quadratic with No *x* Term: Take Positive and Negative Square Roots

Occasionally, you encounter a quadratic equation with an x^2 term but no x term. For example:

If x is negative and $x^2 = 9$, what is x?

Here's the fast way to solve: *Take positive and negative square roots:*

$$x^2 = 9$$
$$\sqrt{x^2} = \sqrt{9}$$
$$x = 3 \text{ or } -3$$

Since you are told that x is negative, the answer to the question is -3.

You can also solve this problem using the method of the previous section. Although the method is longer in this case, it's worth seeing how it works.

First, rearrange the equation to make one side 0:

$$x^2 = 9$$
$$x^2 - 9 = 0$$

Next, factor the quadratic expression. The strange thing is that there is no x term, but imagine that it has a coefficient of 0:

$$x^2 - 9 = 0$$
$$x^2 + 0x - 9 = 0$$

Because the constant (-9) is negative, you need a factor pair of 9 that *differs* by 0. In other words, you need 3 and 3. Make one of these numbers negative to fit the equation as given:

$$x^2 + 0x - 9 = 0$$
$$(x + 3)(x - 3) = 0$$

FOIL the result back out to see how the x terms cancel in the middle:

$$(x + 3)(x - 3) = x^2 - 3x + 3x - 9 = x^2 - 9$$

Finally, set each of the factors in parentheses equal to 0 and solve for x:

$$x + 3 = 0 \quad \text{or} \quad x - 3 = 0$$

Thus, x equals either -3 or 3. Since the question tells you that x is negative, -3 is the answer.

Obviously, the second method is overkill in this case. It's important to understand, though, so that you know why $x^2 - 9$ factors to $(x + 3)(x - 3)$. You'll come back to that point later in this chapter.

If you...	Then you...	Like this:
Want to solve $x^2 = 25$	Take the positive and negative square roots of both sides	$x^2 = 25$ $\sqrt{x^2} = \sqrt{25}$ $(x + 5)(x - 5) = 0$ $x = -5 \, \text{or} \, x = 5$

Check Your Skills

9. If $x^2 - 3 = 1$, what are all the possible values of x?

Answer can be found on page 291.

Solve a Quadratic with Squared Parentheses: Take Positive and Negative Square Roots

Sometimes, you'll be given a problem that looks like you should FOIL, but don't jump in yet. There may be an easier way to solve. For example:

If $(y + 1)^2 = 16$, what are the possible values of y?

Notice that the variable y only shows up once in the expression, not as both a y and a y^2 term. This means you can do this problem in either of two ways.

One way is to treat $y + 1$ as if it were a new variable, z. In other words, $z = y + 1$.

Solve $z^2 = 16$ by taking positive and negative square roots:

$$(y + 1)^2 = 16$$
$$z^2 = 16$$
$$z = 4 \, \text{or} \, z = -4$$

Then plug $z = y + 1$ into that last step:

$$z = 4 \, \text{or} \, z = -4$$
$$y + 1 = 4 \, \text{or} \, y + 1 = -4$$

If you prefer, you can take the positive and negative square roots right away:

$$(y + 1)^2 = 16$$
$$y + 1 = 4 \, \text{or} \, y + 1 = -4$$

Finally, solve the simpler equations for y: $y = 3$ or $y = -5$.

You can also expand $(y + 1)^2$ into $(y + 1)(y + 1)$. Next, FOIL this product out:

$$(y + 1)(y + 1) = y^2 + y + y + 1 = y^2 + 2y + 1$$

7

Now solve the quadratic equation normally. Set one side equal to 0 and factor:

$$y^2 + 2y + 1 = 16$$
$$y^2 + 2y - 15 = 0$$
$$(y + 5)(y - 3) = 0$$
$$y = -5 \text{ or } 3$$

Which way is faster depends on the numbers involved. With big numbers, the first way is usually easier.

If you...	Then you...	Like this:
Want to solve $(z - 7)^2 = 225$	Take the positive and negative square roots of both sides	$(z - 7)^2 = 225$ $\sqrt{(z - 7)^2} = \sqrt{225}$ $z - 7 = 15 \text{ or } z - 7 = -15$ $z = 22 \text{ or } z = -8$

Check Your Skills

10. If $(z + 2)^2 = 144$, what are the possible values of z?

Answer can be found on page 291.

Higher Powers: Solve Like a Normal Quadratic

If you have a higher power of x in the equation, look for solutions as if the equation were a typical quadratic: Set one side equal to 0, factor as much as you can, and then set the factors equal to 0. Try this problem:

What are all of the roots of the equation $x^3 = 3x^2 - 2x$?

(A) $-3, 1, 2$

(B) $-2, 1, 3$

(C) $0, 1, 2$

(D) $0, 1$

(E) $1, 2$

Recall that a *root* of an equation is a solution—a value for the variable that makes the equation true.

First, set one side of the equation equal to 0:

$$x^3 = 3x^2 - 2x$$
$$x^3 - \left(3x^2 - 2x\right) = 0$$
$$x^3 - 3x^2 + 2x = 0$$

You may notice that every term on the left contains an x. In other words, x is a common factor. You might be tempted to divide both sides by x to eliminate that factor.

Resist that temptation. *Never divide an equation by* x *unless you know for sure that* x *is not 0* (x ≠ 0). You could be dividing by 0 without realizing it.

The problem doesn't tell you that $x \neq 0$. So, rather than divide away the x, pull it out to the left and keep it around:

$$x^3 - 3x^2 + 2x = 0$$
$$x\left(x^2 - 3x + 2\right) = 0$$

Now factor the quadratic expression in the parentheses normally and rewrite the equation:

$$x\left(x^2 - 3x + 2\right) = 0$$
$$x(x - 2)(x - 1) = 0$$

There are *three* factors on the left side: x, $(x - 2)$, and $(x - 1)$. Set each one of them equal to 0 to get *three* solutions to the original equation:

$$x = 0 \qquad\qquad x - 2 = 0 \qquad\qquad x - 1 = 0$$
$$x = 2 \qquad\qquad\qquad x = 1$$

If you had divided away the x earlier, you would have missed the $x = 0$ solution. The question asks for all of the roots, so the answer is 0, 1, and 2, or (C).

By the way, the presence of a cubed term, x^3, is the signal that there could be three solutions. The number of the largest exponent generally signals the number of solutions (unless two of the solutions turn out to be the same number, as in the example you saw earlier in this chapter).

If you...	Then you...	Like this:
Want to solve $x^3 = x$	Solve like a normal quadratic: set the equation equal to 0, factor, and set factors equal to 0	$x^3 = x$ $x^3 - x = 0$ $x\left(x^2 - 1\right) = 0$ $x(x + 1)(x - 1) = 0$ $x = 0$ or $x + 1 = 0$ or $x - 1 = 0$ $x = 0, -1,$ or 1

Check Your Skills

11. What are all of the possible solutions to the equation $x^3 - 2x^2 = 3x$?

Answer can be found on page 292.

$$x^3 - 2x^2 - 3x = 0$$
$$x\left(x^2 - 2x - 3\right) = 0$$
$$x(x - 3)(x + 1) = 0$$
$$0 \qquad 3 \qquad -1$$

Other Instances of Quadratics

You will come across quadratic expressions in various circumstances other than the ones already given. Fortunately, the skills of FOILing and factoring are still relevant as you simplify the problem.

See a Quadratic Expression in a Fraction: Factor and Cancel

Try this problem:

If $x \neq -1$, then $\dfrac{x^2 - 2x - 3}{x + 1}$ is equivalent to which of the following?

(A) $x + 1$

(B) $x + 3$

(C) $x - 3$

(handwritten: $(x - 3)(x + 1)$)

This question doesn't involve a typical quadratic expression. However, the numerator of the fraction is a quadratic. To simplify the fraction, *factor* the quadratic expression:

$$x^2 - 2x - 3 = (x - 3)(x + 1)$$

Now substitute the factored form back into the fraction:

$$\frac{x^2 - 2x - 3}{x + 1} = \frac{(x - 3)(x + 1)}{x + 1}$$

Finally, cancel any common factors from the top and bottom of the fraction. The common factor is the entire quantity $x + 1$. Since that is the denominator, the whole thing cancels out, and the expression is no longer a fraction:

$$\frac{x^2 - 2x - 3}{x + 1} = \frac{(x - 3)(x + 1)}{(x + 1)} = \frac{(x - 3)\,\cancel{(x + 1)}}{\cancel{(x + 1)}} = x - 3$$

The correct answer is (C).

The constraint that $x \neq -1$ is mentioned only to prevent division by 0 in the fraction. You don't have to use this fact directly.

If you see a quadratic expression in a numerator or denominator, try factoring the expression. Then cancel common factors.

Common factors can be disguised, of course, even when you don't have quadratics. Try this problem:

If $x \neq y$, then $\dfrac{y - x}{x - y}$ is equivalent to which of the following?

(A) -1

(B) $x^2 - y^2$

(C) $y^2 - x^2$

(handwritten: $\dfrac{(y - x)}{-1(y - x)}$)

The numerator $y - x$ may look different from the denominator $x - y$. However, these two expressions are actually identical except for a sign change throughout:

$$y - x = -(x - y) \qquad \text{because} \qquad -(x - y) = -x + y = y - x$$

In other words, these expressions differ only by a factor of -1. The GMAT loves this little disguise. Expressions that differ only by a sign change are different by a factor of -1.

Rewrite the numerator. Pull a -1 out of the $(x - y)$ term:

$$\frac{y - x}{x - y} = \frac{-(x - y)}{x - y}$$

Now you can cancel $x - y$ from both the top and bottom. You are left with -1 on top, so the whole fraction is equal to -1:

$$\frac{y - x}{x - y} = \frac{-(x - y)}{x - y} = \frac{-\cancel{(x - y)}}{\cancel{(x - y)}} = -1$$

The correct answer is (A).

Try this last problem:

> If $y \ne -8$, then $\dfrac{(y + 7)^2 + y + 7}{y + 8}$ is equivalent to which of the following?
>
> (A) $y + 7$
>
> (B) $y + 8$
>
> (C) $2y + 14$

[handwritten: $y^2 + 14y + 49 + y + 7 \rightarrow \dfrac{y^2 + 15y + 56}{(y + 8)} \rightarrow \dfrac{\cancel{(y+8)}(y+7)}{\cancel{(y+8)}}$]

The long way to solve this is to expand $(y + 7)^2$, then add $y + 7$, then factor and cancel. This approach will work. Fortunately, there's a faster way.

Put parentheses around the last $y + 7$ on top of the fraction:

$$\frac{(y + 7)^2 + y + 7}{y + 8} = \frac{(y + 7)^2 + (y + 7)}{y + 8}$$

This subtle change can help you see that you can factor the numerator. You can *pull out a common factor—* namely, $(y + 7)$—from both the $(y + 7)^2$ and from the $(y + 7)$.

When you pull out $(y + 7)$ from the $(y + 7)^2$ term, you are left with $(y + 7)$. And when you pull out $(y + 7)$ from the $(y + 7)$ term, you are left with 1:

$$\frac{(y + 7)^2 + (y + 7)}{y + 8} = \frac{(y + 7)\left[(y + 7) + 1\right]}{y + 8}$$

Since $y + 7 + 1 = y + 8$, you can simplify the second factor on top:

$$\frac{(y + 7)^2 + (y + 7)}{y + 8} = \frac{(y + 7)\left[(y + 7) + 1\right]}{y + 8} = \frac{(y + 7)(y + 8)}{y + 8}$$

Finally, cancel the $y + 8$ quantity from the top and bottom as a common factor of both:

$$\frac{(y + 7)^2 + (y + 7)}{y + 8} = \frac{(y + 7)\left[(y + 7) + 1\right]}{y + 8} = \frac{(y + 7)(y + 8)}{y + 8} = \frac{(y + 7)\cancel{(y + 8)}}{\cancel{(y + 8)}} = y + 7$$

The correct answer is (A).

The recurring principle is this: Look for ways to pull out common factors from complicated fractions and cancel them.

If you...	Then you...	Like this:
See a quadratic expression in a fraction	Factor the quadratic and cancel common factors	If $z \neq -3$, $$\frac{z^2 + 5z + 6}{z + 3} = \frac{(z + 2)(z + 3)}{(z + 3)}$$ $$= z + 2$$

Check Your Skills

Simplify the following fraction by factoring the quadratic expression.

12. If $x \neq -3$, $\dfrac{x^2 + 7x + 12}{x + 3} =$ $\dfrac{(x+4)(x+3)}{x+3} = x+4$

Answer can be found on page 292.

See a Special Product: Convert to the Other Form

Three quadratic expressions are so important on the GMAT that they are called **special products**. Here they are:

$$(x + y)^2 = x^2 + 2xy + y^2 \qquad (x - y)^2 = x^2 - 2xy + y^2 \qquad (x + y)(x - y) = x^2 - y^2$$

Square of a sum   **Square of a difference**   **Difference of squares**

First, memorize these forms. Second, whenever you see one of these forms, *write down both forms*. Ask yourself which form is the better one to use for this particular problem; often, the better one is the one that the problem did *not* give you outright.

The GMAT often disguises these forms using different variables, numbers, roots, and so on. For example:

$(x+4)(x+4)$ $(a-2b)(a-2b^2)$

$x^2 + 8x + 16 \qquad\qquad a^2 - 4ab + 4b^2 \qquad\qquad \left(1 + \sqrt{2}\right)\left(1 - \sqrt{2}\right) \quad 1^2 - 2 = -1$

Square of a sum   Square of a difference   Difference of squares

The first example can be factored normally:

$$x^2 + 8x + 16 = (x + 4)(x + 4) = (x + 4)^2$$

The test likes "square of a sum" and "square of a difference" because the two forms can be used to create quadratic equations that have only one solution. For instance:

$$x^2 + 8x + 16 = 0$$

$$(x + 4)^2 = 0$$

There is only one solution: $x + 4 = 0$, or $x = -4$. The only number that makes $x^2 + 8x + 16 = 0$ true is -4.

The second example above, $a^2 - 4ab + 4b^2$, is tougher to factor. First, recognize that the first and last term are both perfect squares:

$$a^2 = \text{the square of } a \qquad 4b^2 = \text{the square of } 2b$$

This can provide a hint as to how to factor. Set up $(a - 2b)^2$ and FOIL it to check that it matches:

$$(a - 2b)^2 = (a - 2b)(a - 2b)$$
$$= a^2 - 2ab - 2ab + 4b^2$$
$$= a^2 - 4ab + 4b^2$$

It does, so $(a - 2b)^2$ is indeed the other form.

The third example above, $\left(1 + \sqrt{2}\right)\left(1 - \sqrt{2}\right)$, matches the factored form of the "difference of squares," the most important of the three special products.

In distributed or expanded form, the difference of squares has no middle term. In the process of FOILing, the Outer and the Inner terms cancel:

$$\left(1 + \sqrt{2}\right)\left(1 - \sqrt{2}\right) = 1^2 + \left(-\sqrt{2}\right) + \sqrt{2} + \sqrt{2}\left(-\sqrt{2}\right)$$
$$= 1^2 - \left(\sqrt{2}\right)^2$$
$$= 1 - 2$$
$$= -1$$

It's not a coincidence that the middle terms cancel. If you multiply $x + y$ by $x - y$, the middle terms, or *cross-terms*, will be $-xy$ and $+xy$, which sum to 0 and drop out every time.

As a result, don't FOIL every time you multiply a sum of two things by the difference of those same two things. You'll waste time.

Rather, match up to the "difference of squares" template. Square the first thing, then subtract the square of the second thing. Try this problem:

$$\left(3 + 2\sqrt{3}\right)\left(3 - 2\sqrt{3}\right) = \qquad$$

The two terms are 3 and $2\sqrt{3}$. Always make sure that the other expression in parentheses is the *difference* of the same exact terms. Since that's true, you can square the 3, square the $2\sqrt{3}$, and subtract the second from the first:

$$\left(3 + 2\sqrt{3}\right)\left(3 - 2\sqrt{3}\right) = 3^2 - \left(2\sqrt{3}\right)^2$$
$$= 9 - (4)(3)$$
$$= 9 - 12$$
$$= -3$$

Try this same type of special product, but going in the other direction:

$$16x^4 - 9y^2 = \qquad$$

To treat this as a difference of squares, figure out what each term is the square of:

$$16x^4 = (4x^2)^2 \qquad 9y^2 = (3y)^2$$

So the first term of the difference ($16x^4$) is the square of $4x^2$, and the second term ($9y^2$) is the square of $3y$.

Take those square roots ($4x^2$ and $3y$) and place them in two sets of parentheses:

$$16x^4 - 9y^2 = (4x^2 \quad 3y)(4x^2 \quad 3y)$$

Put a $+$ sign in one set of parentheses and a $-$ sign in the other:

$$16x^4 - 9y^2 = (4x^2 + 3y)(4x^2 - 3y)$$

Now you have factored a difference of squares.

Earlier in this chapter, you saw $x^2 - 9 = 0$. The left side of that equation is the difference of squares:

$$x^2 - 9 = (x + 3)(x - 3)$$

Now you can factor difference-of-squares equations much more quickly; you don't have to do the full reverse FOIL process.

To match a special products template, you might have to rearrange an equation. For example:

If $x^2 + y^2 = -2xy$, what is the sum of x and y ?

You need to find $x + y$. Look at the equation. It doesn't exactly match a special product, but it does have some squares. Try rearranging the equation:

$$
\begin{aligned}
x^2 \qquad\quad + y^2 &= -2xy \\
\underline{+\ 2xy \qquad\quad + 2xy} & \\
x^2 + 2xy + y^2 &= 0
\end{aligned}
$$

$x^2 + 2xy + y^2$

$(x+y)(x+y) = \boxed{0}$

The left side now matches the square of a sum:

$$x^2 + 2xy + y^2 = 0$$
$$(x + y)^2 = 0$$

The right side is also now equal to 0—a double benefit of the first move you made.

Since the square of $x + y$ equals 0, you know that $x + y$ itself must equal 0. That is the answer to the question.

If you...	Then you...	Like this:
See a special product	Write down both forms, then decide which is better to use	$4w^2 - 25z^4 = $ ~~4w 2w~~ $5z^2$ $(2w)^2 - (5z^2)^2 = $ $(2w + 5z^2)(2w - 5z^2)$
See something close to a special product	Rearrange the equation to try to fit the special product template	$m^2 + n^2 = 2mn$ $m^2 - 2mn + n^2 = 0$ $(m - n)(m - n) = 0$ $m - n = 0$ $m = n$

Check Your Skills

Factor the following quadratic expressions.

13. $25a^4b^6 - 4c^2d^2$

14. $4x^2 + 8xy + 4y^2$

Answers can be found on page 292.

Check Your Skills Answer Key

1. $x^2 + 13x + 36$:

$$(x + 4)(x + 9)$$

$(\boldsymbol{x} + 4)(\boldsymbol{x} + 9)$	F – multiply First terms:	$(x)(x) = x^2$
$(x + 4)(x + \boldsymbol{9})$	O – multiply Outer terms:	$(x)(9) = 9x$
$(x + \boldsymbol{4})(\boldsymbol{x} + 9)$	I – multiply Inner terms:	$(4)(x) = 4x$
$(x + \boldsymbol{4})(x + \boldsymbol{9})$	L – multiply Last terms:	$(4)(9) = 36$

$$x^2 + 9x + 4x + 36 \longrightarrow x^2 + 13x + 36$$

2. $y^2 - 3y - 18$:

$$(y + 3)(y - 6)$$

$(\boldsymbol{y} + 3)(\boldsymbol{y} - 6)$	F – multiply First terms:	$(y)(y) = y^2$
$(y + 3)(y - \boldsymbol{6})$	O – multiply Outer terms:	$(y)(-6) = -6y$
$(y + \boldsymbol{3})(\boldsymbol{y} - 6)$	I – multiply Inner terms:	$(3)(y) = 3y$
$(y + \boldsymbol{3})(y - \boldsymbol{6})$	L – multiply Last terms:	$(3)(-6) = -18$

$$y^2 - 6y + 3y - 18 \longrightarrow y^2 - 3y - 18$$

3. $(\boldsymbol{x} + \boldsymbol{3})(\boldsymbol{x} + \boldsymbol{11})$: Find a pair of numbers that multiplies to $+33$ and sums to $+14$. The factor pairs 1×33 and 3×11 multiply to $+33$, but only 3 and 11 also sum to $+14$.

4. $(\boldsymbol{x} - \boldsymbol{5})(\boldsymbol{x} - \boldsymbol{9})$: Ultimately, both numbers will need to be negative because the product (45) is positive, but ignore the signs for now. Find a pair that multiplies to $+45$ and sums to 14. The factor pairs of 45 are 1×45, 3×15, and 5×9. Only 5 and 9 sum to 14. Turn both negative: $-5 + -9 = -14$.

5. $(\boldsymbol{x} + \boldsymbol{6})(\boldsymbol{x} - \boldsymbol{3})$: The final term is negative, so you need one positive and one negative number. The middle term is positive, so the greater of the two numbers has to be the positive one. Finally, you need something that multiplies to -18 but whose *difference* is $+3$. The factor pairs 1×18, 2×9, and 3×6 multiply to 18. The difference of 3 and 6 is 3, so this is the correct pair. Make the 6 positive and the 3 negative in order for the sum to be $+3$.

6. $(\boldsymbol{x} + \boldsymbol{6})(\boldsymbol{x} - \boldsymbol{11})$: The final term is negative, so there will be one positive and one negative number. The middle term is also negative, so the greater of the two numbers has to be negative. You need something that multiplies to -66 and whose difference is -5. The pairs 1×66, 2×33, 3×22, and 6×11 multiply to 66. The difference of 6 and 11 is 5, so this is the correct pair. Make the 11 negative and the 6 positive in order for the sum to be -5.

7. **5 or −7:** Find a factor pair that multiplies to 35 and has a difference of 2. The pair 5 times 7 multiplies to 35, and the difference between the factors is 2. The $2x$ term is positive, so the greater of the two numbers (7) is the positive one in the parentheses.

$$x^2 + 2x - 35 = 0$$
$$(x - 5)(x + 7) = 0$$

$$x - 5 = 0 \qquad \qquad x + 7 = 0$$
$$x = 5 \qquad \text{or} \qquad x = -7$$

Note that the two factors switch signs once you actually set each expression equal to 0 and solve for the variable: The factor −5 becomes the solution 5, and the factor 7 becomes the solution −7.

8. **2 or 13:** Set the equation equal to 0, then solve.

$$x^2 - 15x = -26$$
$$x^2 - 15x + 26 = 0$$

The factors −2 and −13 multiply to 26 and sum to −15.

$$x^2 - 15x + 26 = 0$$
$$(x - 2)(x - 13) = 0$$

$$x - 2 = 0 \qquad \qquad x - 13 = 0$$
$$x = 2 \qquad \text{or} \qquad x = 13$$

9. **2 or −2:** Since there is no x term, isolate x^2 on one side of the equation. Then take the square root of both sides.

$$x^2 - 3 = 1$$
$$x^2 = 4$$
$$\sqrt{x^2} = \sqrt{4}$$
$$x = \pm 2$$

10. **10 or −14:** Begin by taking the square root of both sides. Remember to include the negative solution.

$$(z + 2)^2 = 144$$
$$\sqrt{(z + 2)^2} = \sqrt{144}$$

$$z + 2 = 12 \qquad \qquad z + 2 = -12$$
$$z = 10 \qquad \text{or} \qquad z = -14$$

11. **−1, 0, or 3:** Begin by setting the equation equal to 0.

$$x^3 - 2x^2 = 3x$$
$$x^3 - 2x^2 - 3x = 0$$

All the terms contain x, so factor x out of the left side of the equation. Then factor the quadratic and solve for x.

$$x^3 - 2x^2 - 3x = 0$$
$$x\left(x^2 - 2x - 3\right) = 0$$
$$x(x - 3)(x + 1) = 0$$

$$x = 0 \qquad x - 3 = 0 \qquad x + 1 = 0$$
$$x = 3 \qquad\qquad x = -1$$

12. **$x + 4$:** Factor the numerator, then cancel.

$$\frac{x^2 + 7x + 12}{x + 3} = \frac{\cancel{(x + 3)}\,(x + 4)}{\cancel{(x + 3)}} = x + 4$$

13. **$(5a^2b^3 + 2cd)(5a^2b^3 - 2cd)$:** Any expression that contains one term subtracted from another can be expressed as a difference of squares. Take the square root of each term. The square root of $25a^4b^6$ is $5a^2b^3$, and the square root of $4c^2d^2$ is $2cd$. In one set of parentheses, add the square roots; in the other, subtract: $(5a^2b^3 + 2cd)(5a^2b^3 - 2cd)$.

14. **$(2x + 2y)^2$ or $4(x + y)^2$:** This is a more complicated version of the form $(x + y)^2 = x^2 + 2xy + y^2$. Take the square root of the first term to get $(2x)$ and the square root of the last term to get $(2y)$: $4x^2 + 8xy + 4y^2 = (2x + 2y)(2x + 2y)$. You could also factor a 4 out first and solve from there: $4x^2 + 8xy + 4y^2 = 4(x^2 + 2xy + y^2) = 4(x + y)(x + y) = 4(x + y)^2$.

7

Chapter Review: Drill Sets

Drill 1

Distribute the following expressions.

1. $(x + 2)(x - 3)$

2. $(2s + 1)(s + 5)$

3. $(5 + a)(3 + a)$

4. $(3 - z)(z + 4)$

5. $(3p + 2q)(p - 2q)$

Drill 2

Solve the following equations. List all possible solutions.

6. $x^2 - 2x = 0$

7. $z^2 = -5z$

8. $y^2 + 4y + 3 = 0$

9. $r^2 - 10r = 24$

10. $y^2 + 3y = 0$

11. $y^2 + 12y + 36 = 0$

12. $a^2 - a - 12 = 0$

13. $x^2 + 9x - 90 = 0$

14. $2a^2 + 6a + 4 = 0$

15. $2b^3 + 6b^2 - 36b = 0$

Drill 3

Simplify the following expressions.

16. If $a \neq b$, then $\dfrac{a^2 - b^2}{a - b}$ is equivalent to which of the following?

    (A) $a - b$

    (B) $a + b$

    (C) $a^2 + b^2$

17. If $|r| \neq |s|$, then $\dfrac{r^2 + 2rs + s^2}{r^2 - s^2}$ is equivalent to which of the following?

    (A) $\dfrac{r + s}{r - s}$

    (B) $\dfrac{r - s}{r + s}$

    (C) $2rs$

18. If $x \neq 1$, then $\dfrac{5x^3}{x - 1} - \dfrac{5x^2}{x - 1}$ is equivalent to which of the following?

    (A) $\dfrac{x}{x - 1}$

    (B) $5x^2$

    (C) 0

19. If $y \neq -5$, then $\dfrac{y + 5 - (y + 5)^2}{y + 5}$ is equivalent to which of the following?

    (A) $y + 5$

    (B) $2y$

    (C) $-y - 4$

20. If $m \neq -7$, then $\dfrac{m^2 + 2m}{m + 7} + \dfrac{49 + 12m}{m + 7}$ is equivalent to which of the following?

    (A) 1

    (B) $(m + 7)^2$

    (C) $m + 7$

Drill 4

Simplify the following expressions.

21. If $z \neq -1$, then $\dfrac{4z^2 - 12z - 16}{2z + 2}$ is equivalent to which of the following?

    (A) $z - 4$

    (B) $2z - 8$

    (C) $4z - 16$

7

22. If $c \neq -5$, then $\dfrac{5ab + abc}{abc^2 + 10abc + 25ab}$ is equivalent to which of the following?

    (A) $\dfrac{1}{(c + 5)^2}$

    (B) $c + 5$

    (C) $\dfrac{1}{c + 5}$

23. If $x \neq 0$ or 1, then $\dfrac{\left(x^5 - x^3\right)}{\left(x^3 - x^2\right)} \times \dfrac{x}{5}$ is equivalent to which of the following?

    (A) $\dfrac{x^2}{5}$

    (B) $\dfrac{x^3 - x^2}{5}$

    (C) $\dfrac{x^3 + x^2}{5}$

24. If $x \neq 2$, then $\left(x^2 - 7x + 10\right) \times \dfrac{x + 5}{x - 2}$ is equivalent to which of the following?

    (A) $x^2 - 25$

    (B) $x^2 + 10x + 25$

    (C) $x + 5$

25. If $x \neq 3$, then $\dfrac{x^2 - 6x + 9}{3 - x}$ is equivalent to which of the following?

    (A) $x - 3$

    (B) $3 - x$

    (C) $(x - 3)^2$

7

Drill Sets Solutions

Drill 1

1. $x^2 - x - 6$:

$$(x + 2)(x - 3)$$
$$x^2 - 3x + 2x - 6$$
$$x^2 - x - 6$$

2. $2s^2 + 11s + 5$:

$$(2s + 1)(s + 5)$$
$$2s^2 + 10s + s + 5$$
$$2s^2 + 11s + 5$$

3. $15 + 8a + a^2$:

$$(5 + a)(3 + a)$$
$$15 + 5a + 3a + a^2$$
$$15 + 8a + a^2$$

4. $-z^2 - z + 12$:

$$(3 - z)(z + 4)$$
$$3z + 12 - z^2 - 4z$$
$$-z^2 - z + 12$$

5. $3p^2 - 4pq - 4q^2$:

$$(3p + 2q)(p - 2q)$$
$$3p^2 - 6pq + 2pq - 4q^2$$
$$3p^2 - 4pq - 4q^2$$

Drill 2

6. $x = 0$ or 2:

$$x^2 - 2x = 0$$
$$x(x - 2) = 0$$
$$x = 0 \qquad x - 2 = 0$$
$$x = 2$$

7. **$z = 0$ or -5:**

$$z^2 = -5z$$
$$z^2 + 5z = 0$$
$$z(z + 5) = 0$$

$$z = 0 \qquad\qquad (z + 5) = 0$$
$$z = -5$$

8. **$y = -1$ or -3:**

$$y^2 + 4y + 3 = 0$$
$$(y + 1)(y + 3) = 0$$

$$(y + 1) = 0 \qquad\qquad (y + 3) = 0$$
$$y = -1 \qquad\qquad\qquad y = -3$$

9. **$r = -2$ or 12:**

$$r^2 - 10r = 24$$
$$r^2 - 10r - 24 = 0$$
$$(r + 2)(r - 12) = 0$$

$$(r + 2) = 0 \qquad\qquad (r - 12) = 0$$
$$r = -2 \qquad\qquad\qquad r = 12$$

10. **$y = 0$ or -3:**

$$y^2 + 3y = 0$$
$$y(y + 3) = 0$$

$$y = 0 \qquad\qquad (y + 3) = 0$$
$$y = -3$$

11. **$y = -6$:** When the two factors are identical (both are $y + 6$), then you don't have to solve each one. The answer will be the same.

$$y^2 + 12y + 36 = 0$$
$$(y + 6)(y + 6) = 0$$

$$(y + 6) = 0$$
$$y = -6$$

12. $a = 4$ **or** -3**:**

$$a^2 - a - 12 = 0$$
$$(a - 4)(a + 3) = 0$$

$$(a - 4) = 0 \qquad (a + 3) = 0$$
$$a = 4 \qquad\qquad a = -3$$

13. $x = -15$ **or** 6**:**

$$x^2 + 9x - 90 = 0$$
$$(x + 15)(x - 6) = 0$$

$$(x + 15) = 0 \qquad (x - 6) = 0$$
$$x = -15 \qquad\qquad x = 6$$

14. $a = -2$ **or** -1**:** Ignore the 2 term that you pull out front because it doesn't contain a variable.

$$2a^2 + 6a + 4 = 0$$
$$2(a^2 + 3a + 2) = 0$$
$$2(a + 2)(a + 1) = 0$$

$$(a + 2) = 0 \qquad (a + 1) = 0$$
$$a = -2 \qquad\qquad a = -1$$

15. $b = 0, -6,$ **or** 3**:** The $2b$ term that you pull out front does have to be set equal to 0, because it contains a variable.

$$2b^3 + 6b^2 - 36b = 0$$
$$2b(b^2 + 3b - 18) = 0$$
$$2b(b + 6)(b - 3) = 0$$

$$2b = 0 \qquad (b + 6) = 0 \qquad (b - 3) = 0$$
$$b = 0 \qquad\quad b = -6 \qquad\quad b = 3$$

Drill 3

16. **(B)** $a + b$**:** The key to simplifying this expression is to recognize the special product. The numerator is in the form of the difference of squares.

$$a^2 - b^2 = (a + b)(a - b)$$

After replacing the original numerator with $(a + b)(a - b)$, cancel the $(a - b)$ in the numerator with the $(a - b)$ in the denominator.

$$\frac{a^2 - b^2}{a - b} = \frac{(a + b)\,\cancel{(a - b)}}{\cancel{(a - b)}} = a + b$$

17. **(A)** $\dfrac{r+s}{r-s}$: Both the numerator and the denominator contain special products. Factor each one to find common terms to cancel.

$$\frac{r^2 + 2rs + s^2}{r^2 - s^2} = \frac{\cancel{(r+s)}(r+s)}{\cancel{(r+s)}(r-s)} = \frac{r+s}{r-s}$$

18. **(B)** $5x^2$: First perform the subtraction to combine the two terms.

$$\frac{5x^3}{x-1} - \frac{5x^2}{x-1} = \frac{5x^3 - 5x^2}{x-1}$$

Next, pull out the common term $(5x^2)$ in the numerator.

$$\frac{5x^3 - 5x^2}{x-1} = \frac{5x^2 \cancel{(x-1)}}{\cancel{(x-1)}} = 5x^2$$

19. **(C)** $-y - 4$: It is tempting to expand the quadratic term in the numerator, but examine this first. None of the answer choices are fractions, so the denominator must cancel out somehow. And notice all those y's and 5's elsewhere? Look to cancel first.

Group the $y + 5$ terms.

$$\frac{(y+5) - (y+5)^2}{(y+5)}$$

You can cancel a $(y+5)$ from each term (remember that the numerator contains two separate terms, separated by the subtraction sign).

$$\frac{\cancel{(y+5)} - (y+5)^{\cancel{2}}}{\cancel{(y+5)}} = \frac{1 - (y+5)}{1} = -y - 4$$

If you think you might make a mistake doing it that way, you can first pull the common $y + 5$ term out of the numerator.

$$\frac{(y+5) - (y+5)^2}{(y+5)} = \frac{\cancel{(y+5)}[1 - (y+5)]}{\cancel{(y+5)}} = 1 - y - 5 = -y - 4$$

20. **(C)** $m + 7$: The denominators are the same, so add the fractions.

$$\frac{m^2 + 2m}{m+7} + \frac{49 + 12m}{m+7} = \frac{m^2 + 14m + 49}{m+7}$$

None of the answer choices are fractions, so find a way to eliminate the denominator. Start by factoring the numerator.

$$\frac{m^2 + 14m + 49}{m+7} = \frac{\cancel{(m+7)}(m+7)}{\cancel{(m+7)}} = m + 7$$

Drill 4

21. **(B) $2z - 8$:** Simplify this problem by factoring a 4 out of the numerator and a 2 out of the denominator.

$$\frac{4z^2 - 12z - 16}{2z + 2} = \frac{\overset{2}{\cancel{4}}\left(z^2 - 3z - 4\right)}{\cancel{2}\left(z + 1\right)} = \frac{2\left(z^2 - 3z - 4\right)}{\left(z + 1\right)}$$

The answers don't contain fractions, so the denominator must cancel somehow. Factor the numerator.

$$\frac{2\left(z^2 - 3z - 4\right)}{\left(z + 1\right)} = \frac{2\cancel{(z + 1)}\left(z - 4\right)}{\cancel{(z + 1)}} = 2(z - 4) = 2z - 8$$

22. **(C) $\dfrac{1}{c + 5}$:** This might seem nearly impossible to factor. Glance at the answers. Notice anything?

The variables a and b have disappeared. These variables must cancel out as you solve! Notice that every term has an ab piece; factor it out and cancel.

$$\frac{5ab + abc}{abc^2 + 10abc + 25ab} = \frac{\cancel{ab}\left(5 + c\right)}{\cancel{ab}\left(c^2 + 10c + 25\right)} = \frac{5 + c}{c^2 + 10c + 25}$$

That's more manageable. There is no match in the answers yet, so factor the denominator to see whether you can cancel any further.

$$\frac{5 + c}{c^2 + 10c + 25} = \frac{5 + c}{(c + 5)(c + 5)} = \frac{\cancel{(c + 5)}}{\cancel{(c + 5)}(c + 5)} = \frac{1}{c + 5}$$

23. **(C) $\dfrac{x^3 + x^2}{5}$:** There's no obvious way to proceed through this question. The best bet is to try to simplify before multiplying. Notice that x^3 can be factored out of the numerator and x^2 can be factored out of the denominator.

$$\frac{\left(x^5 - x^3\right)}{\left(x^3 - x^2\right)} \times \frac{x}{5} = \frac{x^3\left(x^2 - 1\right)}{x^2\left(x - 1\right)} \times \frac{x}{5}$$

Cancel x^2 from the top and bottom. Also, the numerator now contains $\left(x^2 - 1\right)$. Factor this.

$$\frac{x^3\left(x^2 - 1\right)}{x^2\left(x - 1\right)} \times \frac{x}{5} =$$

$$\frac{x^{\cancel{3}^1}\left(x + 1\right)\cancel{\left(x - 1\right)}}{\cancel{x^2}\cancel{\left(x - 1\right)}} \times \frac{x}{5} =$$

$$\frac{x\left(x + 1\right) \times x}{5}$$

Glance at the answer choices. None of the numerators in the answer choices have parenthetical expressions, so multiply the numerator out.

$$\frac{x(x+1)(x)}{5} = \frac{x^2(x+1)}{5} = \frac{x^3 + x^2}{5}$$

24. **(A) $x^2 - 25$:** Simplify before you multiply. Are there any common factors to cancel?

$$\left(x^2 - 7x + 10\right) \times \frac{x+5}{x-2} =$$

$$\frac{(x-5)\,\cancel{(x-2)}}{1} \times \frac{x+5}{\cancel{x-2}} =$$

$$(x-5)(x+5) =$$

$$x^2 - 25$$

25. **(B) $3 - x$:** Glance at the answers: no fractions. Factor the numerator to try to cancel the denominator.

$$\frac{x^2 - 6x + 9}{3 - x} = \frac{(x-3)(x-3)}{3-x}$$

Neither one of the expressions in the numerator matches the denominator. However, $(x-3) = -(3-x)$. Factor out a (-1) from the denominator, then cancel.

$$\frac{(x-3)(x-3)}{3-x} = \frac{\cancel{(x-3)}(x-3)}{-\cancel{(x-3)}} = \frac{x-3}{-1} = -x + 3 \text{ or } 3 - x$$

7

Inequalities and Absolute Value

In This Chapter

- An Inequality with a Variable: A Range on the Number Line

- Many Values "Solve" an Inequality

- Solve Inequalities: Isolate Variable by Transforming Each Side

- Multiply or Divide an Inequality by a Negative: Flip $>$ to $<$ or Vice Versa

- Absolute Value: The Distance from Zero

- Replace $|x|$ with x in One Equation and with $-x$ in Another

- Inequalities and Absolute Values: Set Up Two Inequalities

In this chapter, you will learn to manipulate and solve inequalities and you'll learn how to work with absolute value.

CHAPTER 8 Inequalities and Absolute Value

Inequalities are almost the same as equations—with a couple of key differences. **Absolute value** is a way to think about how far a number is from zero, without regard to whether that number is positive or negative.

An Inequality with a Variable: A Range on the Number Line

Inequalities use $<$, $>$, $\leq$, or $\geq$ to describe the relationship between two expressions. For example:

$$5 > 4 \qquad y \leq 7 \qquad x < 5 \qquad 2x + 3 \geq 0$$

Like equations, *inequalities are full sentences.* Always read from left to right:

$x < y$	x is less than y.
$x > y$	x is greater than y.
$x \leq y$	x is less than or equal to y. x is at most y.
$x \geq y$	x is greater than or equal to y. x is at least y.

You can also have two inequalities in one statement. Make a compound sentence:

$9 < g < 200$	9 is less than g, and g is less than 200.
$-3 < y \leq 5$	-3 is less than y, and y is less than or equal to 5.
$7 \geq x > 2$	7 is greater than or equal to x, and x is greater than 2.

To visualize an inequality that involves a variable, *draw the inequality on a number line.* On a number line, *greater than* means *to the right of* and *less than* means *to the left of*:

$$y > 5 \qquad\qquad\qquad\qquad b \leq 2$$

y is to the right of 5, which is *not* included in the number line (as shown by the empty circle around 5) because y is greater than 5, but not equal to 5.

b is to the left of 2 (or on top of 2). Here, 2 is included in the solution, because b can equal 2. A solid black circle indicates that you include the point itself.

Any number covered by the black arrow (or a filled-in circle) will make the inequality true and so is a possible solution to the inequality. Any number not covered by the black arrow (or covered with an empty circle) is not a solution.

If you...	Then you...	Like this:
Want to visualize an inequality	Put it all on a number line, where $<$ means *to the left of* and $>$ means *to the right of*	

Check Your Skills

Draw a number line on your notepaper, then draw the given equation on the number line.

1. $x > 3$

2. $b \geq -2$

3. $y = 4$

Translate the following into inequality statements.

4. z is greater than v.

$z > v$

5. The total amount is greater than \$2,000.

$x > 2000$

Answers can be found on page 312.

Many Values "Solve" an Inequality

What does it mean to *solve* an inequality?

It means the same thing as to solve an equation: Find the value or values of x that make the inequality true. When you plug a solution back into the original equation or inequality, you get a *true statement*.

Here's what's different: Equations have only one (or just a few) values as a solutions. In contrast, *inequalities give a whole range of values as solutions*—often way too many to list individually. For example:

Equation: $x + 3 = 8$	**Inequality:** $x + 3 < 8$
The solution to $x + 3 = 8$ is $x = 5$, which is the *only* number that will make the equation true.	The solution to $x + 3 < 8$ is $x < 5$. The number 5 itself is not a solution because $5 + 3 < 8$ is not a true statement. But 4 is a solution because $4 + 3 < 8$ is true. For that matter, 4.99, 3, 2, 2.87, -5, and -100 are also solutions. The list goes on.
Plug back in to check: $5 + 3 = 8$. True.	For all of the correct answers: (any number less than 5) $+ 3 < 8$. True.

Check Your Skills

6. If $x < 10$, what is a possible value of x?

(A) -3

(B) $-\dfrac{3}{2}$

(C) 2.5

(D) 9.999

(E) All of the above

Answer can be found on page 312.

Solve Inequalities: Isolate Variable by Transforming Each Side

As with equations, your objective is to isolate a variable on one side of the inequality. When the variable is by itself, you can see what the solution (or range of solutions) really is.

For example, $2x + 6 < 12$ and $x < 3$ provide the same information. But it's easier to understand the full range of solutions when you see the second inequality, which literally says that x is less than 3.

Many manipulations are the same for inequalities as for equations. First of all, you are *always* allowed to simplify an expression on just one side of an inequality. Just don't change the expression's value:

$$2x + 3x < 45 \qquad \text{is the same as} \qquad 5x < 45$$

The inequality sign isn't involved in this simplification.

Next, some Golden Rule moves work the same way for inequalities as for equations. For instance, you can *add* anything you want to both sides of an inequality. Just make sure you do the same thing to both sides. You can also *subtract* the same thing from both sides of an inequality:

$$
\begin{array}{ll}
a - 4 > 6 & y + 7 < 3 \\
\underline{+4 \quad +4} & \underline{-7 \quad -7} \\
a \quad > 10 & y \quad < -4
\end{array}
$$

You can also add or subtract variables from both sides of an inequality. It doesn't matter what the signs of the variables might be.

If you...	Then you...	Like this:
Want to add or subtract the same quantity on both sides of an inequality	Go ahead and do so	$x + y > -4$ $\underline{-y \quad -y}$ $x > -4 - y$

Check Your Skills

Isolate the variable in the following inequalities.

7. $x - 6 < 13$

8. $y + 11 \geq -13$

9. $x + 7 > 7$

Answers can be found on page 312.

Multiply or Divide an Inequality by a Negative: Flip $>$ to $<$ or Vice Versa

If you multiply both sides of an inequality by a *positive* number, leave the inequality sign alone. The same is true for division. For example:

$$\frac{x}{3} < 7 \qquad\qquad 4y > 12$$

$$3\left(\frac{x}{3}\right) < (7)3 \qquad\qquad \frac{4y}{4} > \frac{12}{4}$$

$$x < 21 \qquad\qquad\qquad y > 3$$

However, if you multiply or divide both sides of an inequality by a *negative* number, *flip the inequality sign*. The *greater than* symbol turns into the *less than* symbol and vice versa:

$$-2x > 10 \qquad\qquad\qquad -b \geq 10$$

$$\left(\frac{-2x}{-2}\right) < \left(\frac{10}{-2}\right) \qquad (-1)(-b) \leq (10)(-1)$$

$$x < -5 \qquad\qquad\qquad\qquad b \leq -10$$

If you didn't switch the sign, then inequalities such as $5 < 7$ would become false when you multiply them by, say, -1. You must flip the sign:

$$5 < 7 \qquad\text{but}\qquad -5 > -7$$

5 is less than 7 $\qquad\qquad\qquad$ -5 is greater than -7

What about multiplying or dividing an inequality by a variable? If you aren't given the sign of the variable, then avoid taking this step! If you don't know the sign of the "hidden number" that the variable represents, then you don't know whether to switch the sign.

If the problem tells you the variable is positive, or if the variable has to be positive (e.g., it counts people or measures a length), then you can go ahead and multiply or divide. If you're told the variable is negative, flip the sign when you multiply or divide by that variable. If you're not told, don't multiply or divide by that variable.

If you...	Then you...	Like this:
Multiply or divide both sides of an inequality by a *negative* number	Flip the inequality sign	$45 < -5w$ $$\left(\frac{45}{-5}\right) > \left(\frac{-5w}{-5}\right)$$ $-9 > w$

Check Your Skills

Isolate the variable in the following inequalities.

> 10. $x + 3 \geq -2$
>
> 11. $-2y < -8$
>
> 12. $a + 4 \geq 2a$

Answers can be found on pages 312–313.

Absolute Value: The Distance from Zero

The **absolute value** of a number describes how far that number is from 0. It is the distance between that number and 0 on a number line.

The symbol for absolute value is |number|. For instance, write the absolute value of -5 as $|-5|$.

The absolute value of 5 is 5. This is how it would look on a number line:

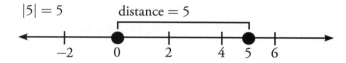

The absolute value of -5 is also 5:

$$|-5| = 5 \qquad \text{distance} = 5$$

In either case, the number is 5 units away from 0 on the number line.

When you face an expression like $|4 - 7|$, treat the absolute value symbol like parentheses. Solve the math *inside* first, and then find the absolute value of the answer:

$$|4 - 7| = ?$$
$$4 - 7 = -3$$
$$|-3| = 3$$

Almost every absolute value is positive. There is one exception:

$$|0| = 0$$

Except for 0, every absolute value is positive.

8

Check Your Skills

Mark the following expressions as TRUE or FALSE.

13. $|3| = 3$
14. $|-3| = -3$
15. $|3| = -3$
16. $|-3| = 3$
17. $|3 - 6| = 3$
18. $|6 - 3| = -3$

Answers can be found on page 313.

Replace $|x|$ with x in One Equation and with $-x$ in Another

You may see a variable inside the absolute value sign:

$$|y| = 3$$

This equation has *two solutions*. There are two numbers that are 3 units away from 0, namely 3 and −3. Both of these numbers could be the value of *y*, but not simultaneously, so *y* is *either* 3 *or* −3.

When you see a variable inside an absolute value, look for the variable to have two possible values. Here is a step-by-step process for finding both solutions:

$|y| = 3$	Step 1: Isolate the absolute value expression on one side of the equation. Here, the expression is already isolated.
$+(y) = 3$ or $-(y) = 3$	Step 2: Drop the absolute value signs and *set up two equations*. The first equation has the positive value of what's inside the absolute value. The second equation has the negative value.
$y = 3$ or $-y = 3$	Step 3: Solve both equations.
$y = 3$ or $y = -3$	There are two possible values for *y*.

For equations only (*not* inequalities), you can take a shortcut and go right to "*y* equals plus or minus 3." This shortcut works as long as the absolute value expression is by itself on one side of the equation. As always, only use a shortcut if you know it well enough to avoid careless mistakes.

Here's what to do if the absolute value expression is *not* by itself on one side of the equation:

$$6 \times |2x + 4| = 30$$

To solve this problem, use the same approach:

$6 \times \|2x + 4\| = 30$			
$\|2x + 4\| = 5$			Step 1: Isolate the absolute value expression on one side of the equation or inequality.
$+(2x + 4) = 5$	or	$-(2x + 4) = 5$	Step 2: Set up two equations—one positive and one negative.
		$-2x - 4 = 5$	
$2x = 1$	or	$-2x = 9$	Step 3: Solve both equations/inequalities.
$x = \frac{1}{2}$	or	$x = -\frac{9}{2}$	There are two possible values for x.

If you...	Then you...	Like this:
Have a variable inside absolute value signs	Drop the absolute value and set up two equations, one positive and one negative	$\|z\| = 4$ $+(z) = 4$ or $-(z) = 4$ $z = 4$ or $z = -4$

Check Your Skills

Solve the following equations with absolute values in them.

19. $|a| = 6$ 6 ; -6
20. $|x + 2| = 5$ 3 ; -7
21. $|3y - 4| = 17$ $-(3y-4)$ $-3y+4=17$ $-3y=13$ $y=-\frac{13}{3}$ $y=7$
22. $4\left|x + \frac{1}{2}\right| = 18$ $\frac{9}{2} = -x + \frac{1}{2}$ $x=4$ or -5

Answers can be found on pages 313–314.

Inequalities and Absolute Values: Set Up Two Inequalities

Some tough problems include both inequalities and absolute values. To solve these problems, combine what you have learned about inequalities with what you have learned about absolute values. For example:

$$|x| \geq 4$$

8

The basic process for dealing with absolute values is the same for inequalities as it is for equations. The absolute value is already isolated on one side, so now drop the absolute value signs and *set up two inequalities*. The first inequality has the positive value of what was inside the absolute value signs, while the second inequality has the negative value:

$$+(x) \geq 4 \qquad \text{or} \qquad -(x) \geq 4$$

Next, isolate the variable in each inequality, as necessary:

$$+(x) \geq 4 \quad \text{or} \quad -(x) \geq 4$$
$$x \geq 4 \qquad\qquad -x \geq 4 \qquad \text{In the negative example, divide by } -1.$$
$$x \leq -4 \qquad \text{Flip the sign when dividing by a negative.}$$

The two solutions to the original equation are $x \geq 4$ or $x \leq -4$. Draw those two inequalities on a number line:

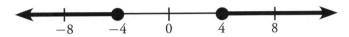

As before, any number that is covered by the black arrow will make the inequality true. Because of the absolute value, there are now two arrows instead of one, but nothing else has changed. Both -4 and any number to the left of -4 will make the inequality true, as will both 4 and any number to the right of 4.

Looking back at the inequality $|x| \geq 4$, you can also interpret it in terms of distance. For example, $|x| \geq 4$ means "x is at least 4 units away from 0, in either direction." The black arrows indicate all numbers for which that statement is true.

Here is a harder example, with a twist:
$$|y + 3| < 5$$

(handwritten) $-y-3<5$ $y<8$ $\qquad y<2$ or $y<8$

Once again, the absolute value is already isolated on one side, so set up the two inequalities:

$$+(y + 3) < 5 \quad \text{or} \quad -(y + 3) < 5$$

Next, isolate the variable:

$$y + 3 < 5 \qquad \text{or} \qquad -y - 3 < 5$$
$$y < 2 \qquad\qquad\qquad -y < 8$$
$$y > -8$$

The two inequalities are $y < 2$ and $y > -8$. If you plot those results, something curious happens:

It seems as if every number should be a solution to the equation. But try plugging $y = 5$ into $|y + 3| < 5$. What happens?

It doesn't work: $|5 + 3|$ is not less than 5. In fact, the only numbers that make the inequality true are those that are true for *both* inequalities. The number line should look like this:

In other words, $-8 < y < 2$.

When the two solutions overlap, as in this example, only the numbers that fall in the range of *both* arrows will be solutions to the inequality, so combine the solutions into one big inequality: $-8 < y < 2$.

If your two arrows do *not* overlap, as in the first example, any number that falls in the range of *either* arrow will be a solution to the inequality, so leave the solutions as two separate inequalities: $x \geq 4$ or $x \leq -4$.

You can also interpret $|y + 3| < 5$ in terms of distance: "$(y + 3)$ is less than 5 units away from from 0, in either direction." The shaded segment indicates all numbers y for which this is true. As inequalities get more complicated, don't worry about interpreting their meaning—just solve them algebraically!

If you...	Then you...	Like this:
Have an inequality with a variable inside absolute value signs	Drop the absolute value and set up two inequalities, one positive and one negative. If the two solutions don't overlap, write *or* between them and leave them as two inequalities.	$|z| > 4$ $+(z) > 4 \quad -(z) > 4$ no overlap $z > 4$ or $z < -4$
	If the two solutions do overlap, combine them into one big inequality.	$|a| < 4$ $+(a) < 4 \quad -(a) < 4$ $a < 4 \quad a > -4$ overlap! $-4 < a < 4$

Check Your Skills

Solve the following inequalities with absolute values in them.

23. $|x + 1| > 2$ $-x-1>2$ $x>1$ or $x<-3$
     $x<-3$
24. $|-x - 4| \geq 8$ $(-x-4) \times \leq 12$ or $x \geq 4$ $4 \leq x \leq 12$
25. $|x - 7| < 9$

$-2<x<16$
$-x+7<9$
$x>-2$

Answers can be found on page 314.

Check Your Skills Answer Key

1. [number line: open circle at 3, arrow pointing right; marks at 2, 3, 4]

2. [number line: closed circle at −2, arrow pointing right; marks at −3, −2, −1]

3. [number line: closed circle at 4; marks at 3, 4, 5]

4. $z > v$: *z is greater than v is translated as* $z > v$.

5. $t > 2{,}000$: Let $t =$ total amount. *The total amount is greater than $2,000* is translated as $t > 2{,}000$.

6. **(E) All of the above:** All of the numbers in the answer choices are to the left of 10 on the number line, so all of them are possible values for x.

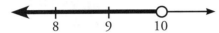

[number line: open circle at 10, arrow pointing left; marks at 8, 9, 10]

7. $x < 19$:

$$x - 6 < 13$$
$$\underline{+6 \quad +6}$$
$$x < 19$$

8. $y \geq -24$:

$$y + 11 \geq -13$$
$$\underline{-11 \quad\;\; -11}$$
$$y \geq -24$$

9. $x > 0$:

$$x + 7 > 7$$
$$\underline{-7 \; -7}$$
$$x > 0$$

10. $x \geq -5$:

$$x + 3 \geq -2$$
$$\underline{-3 \quad\; -3}$$
$$x \geq -5$$

11. $y > 4$:

$$-2y < -8$$
$$\left(\frac{-2y}{-2}\right) > \left(\frac{-8}{-2}\right)$$
$$y > 4$$

12. $4 \geq a$ or $a \leq 4$:

$$a + 4 \geq 2a$$
$$\underline{-a \qquad -a}$$
$$4 \geq a$$

13. **True:** The absolute value of 3 is 3.

14. **False:** The absolute value of -3 is *not* -3; rather, the absolute value of -3 is 3. An absolute value is *never* negative.

15. **False:** The absolute value of 3 is 3. An absolute value is never negative.

16. **True:** The absolute value of -3 is 3.

17. **True:** $|3 - 6| = |-3| = 3$

18. **False:** $|6 - 3| = |3| = 3$. An absolute value is never negative.

19. $a = 6$ or -6:

$$|a| = 6$$

$$+(a) = 6 \qquad \text{or} \qquad -(a) = 6$$
$$a = 6 \qquad\qquad\qquad a = -6$$

20. $x = 3$ or -7:

$$|x + 2| = 5$$

$$+(x + 2) = 5 \qquad \text{or} \qquad -(x + 2) = 5$$
$$x + 2 = 5 \qquad\qquad\qquad -x - 2 = 5$$
$$x = 3 \qquad\qquad\qquad -x = 7$$
$$x = -7$$

21. $y = 7$ or $-\dfrac{13}{3}$:

$$|3y - 4| = 17$$

$$+(3y - 4) = 17 \qquad \text{or} \qquad -(3y - 4) = 17$$
$$3y - 4 = 17 \qquad\qquad\qquad -3y + 4 = 17$$
$$3y = 21 \qquad\qquad\qquad -3y = 13$$
$$y = 7 \qquad\qquad\qquad y = -\dfrac{13}{3}$$

8

22. $x = 4$ or -5:

$$4\left|x + \frac{1}{2}\right| = 18$$

$$\left|x + \frac{1}{2}\right| = \frac{18}{4}$$

$$\left|x + \frac{1}{2}\right| = \frac{9}{2}$$

$+\left(x + \frac{1}{2}\right) = \frac{9}{2}$ or $-\left(x + \frac{1}{2}\right) = \frac{9}{2}$

$$x + \frac{1}{2} = \frac{9}{2} \qquad\qquad\qquad -x - \frac{1}{2} = \frac{9}{2}$$

$$x = \frac{8}{2} = 4 \qquad\qquad\qquad -x = \frac{10}{2} = 5$$

$$x = -5$$

23. $x < -3$ or $x > 1$:

$$|x + 1| > 2$$

$+(x + 1) > 2$ $-(x + 1) > 2$

$x + 1 > 2$ $-x - 1 > 2$

$x > 1$ $-x > 3$

                                  $x < -3$

24. $x \leq -12$ or $x \geq 4$:

$$|-x - 4| \geq 8$$

$+(-x - 4) \geq 8$ $-(-x - 4) \geq 8$

$-x - 4 \geq 8$ $x + 4 \geq 8$

$-x \geq 12$ $x \geq 4$

$x \leq -12$

25. $-2 < x < 16$:

$$|x - 7| < 9$$

$+(x - 7) < 9$ $-(x - 7) < 9$

$x - 7 < 9$ $-x + 7 < 9$

$x < 16$ $-x < 2$

                            $x > -2$

8

Chapter Review: Drill Sets

Drill 1

Redraw the given number line on your notepaper, then draw the given equation on the number line.

1. $x > 4$

    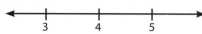

2. $a \geq 3$

3. $y = 2$

    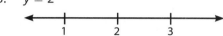

4. $x < 5$

    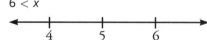

5. $6 < x$

    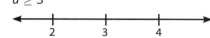

Drill 2

Translate the following into inequality statements.

6. a is less than b.

7. 5 times x is greater than 10.

8. 6 is less than or equal to $4x$.

9. The price of an apple is greater than the price of an orange.

10. The total number of members is at least 19.

Drill 3

Solve the following inequalities.

11. $t - 4 \leq 13$

12. $3b \geq 12$

13. $-5x > 25$

14. $-8 < -4y$

Drill 4

Solve the following inequalities.

15. $2z + 4 \geq -18$

16. $7x + 5 \geq 10x + 14$

17. $d + \dfrac{3}{2} < 8$

18. $\dfrac{2a}{3} > 10 - a$

Drill 5

Solve the following inequalities.

19. $3(x - 7) \geq 9$

20. $\dfrac{x}{3} + 8 < \dfrac{x}{2}$

21. $2x - 1.5 > 7$

22. $\dfrac{6(2x + 8)}{5} \leq 0$

23. $\dfrac{2(3 - x)}{5x} \leq 4$ and $x > 0$

Drill 6

Solve the following inequalities.

24. $4\sqrt{3x - 2} > 20$

25. $\dfrac{2(8 - 3x)}{7} > 4$

26. $0.25x - 3 \leq 1$

27. $2(y + 2)^3 - 5 \geq 49$

28. $\dfrac{4\sqrt[3]{5x - 8}}{3} \geq 4$

Drill 7

Solve the following absolute value equations.

29. $|x| = 5$

30. $|5a| = 15$

31. $|4y + 2| = 18$

32. $|1 - x| = 6$

Drill 8

Solve the following absolute value equations.

33. $3|x - 4| = 18$

34. $2|x + 0.3| = 7$

35. $|6z - 3| = 4z + 11$

36. $\left|\dfrac{x}{4} + 3\right| = 0.5$

Drill 9

Solve each of the following inequalities. Then draw the solution on a number line.

37. $|x + 3| < 1$

38. $5 \geq |2y + 3|$

39. $6 \leq |5b - 9|$

40. $|-12a| < 15$

Drill 10

Solve each of the following inequalities. Then draw the solution on a number line.

41. $|-x| \geq 6$

42. $\dfrac{|x + 4|}{2} > 5$

43. $|z^3| \leq 27$

44. $|0.1x - 3| \geq 1$

45. $\left|\dfrac{3x}{2} + 7\right| \leq 11$

Drill 11

Solve each of the following inequalities. Then draw the solution on a number line.

46. $\left|-\dfrac{x}{5} + \dfrac{2}{3}\right| \leq \dfrac{7}{15}$

47. $|3x - 7| \geq 2x + 12$

48. $|3 + 3x| < -2x$

49. $|-9 - 5x| \leq -4x$

50. $2\left|\dfrac{7y}{4} - 7\right| < \dfrac{3y}{2} + 10$

Drill Sets Solutions

Drill 1

1. $x > 4$

    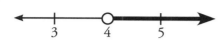

2. $a \geq 3$

3. $y = 2$

4. $x < 5$

    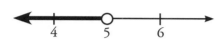

5. $6 < x$

    You can flip inequalities around, moving the left side to the right and vice versa, as long as you flip the sign, too. In this case, $6 < x$ becomes $x > 6$.

    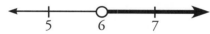

Drill 2

6. **$a < b$:** *a is less than b* translates as $a < b$.

7. **$5x > 10$:** *5 times x is greater than 10* is translated as $5x > 10$.

8. **$6 \leq 4x$:** *6 is less than or equal to 4x* is translated as $6 \leq 4x$.

9. **$a > r$:** Let *a* equal the price of an apple and *r* equal the price of an orange (avoid using the variable *o* for orange, as *o* can be confused with 0 on your scratch paper). *The price of an apple is greater than the price of an orange* is translated as $a > r$.

    Note: In this problem, the variables refer to prices—not the number of apples and oranges.

10. **$m \geq 19$:** Let *m* equal the number of members. *The total number of members is at least 19* is translated as $m \geq 19$.

Drill 3

11. **$t \leq 17$:**

    $$t - 4 \leq 13$$
    $$t \leq 17$$

12. $b \geq 4$:

$$3b \geq 12$$
$$b \geq 4$$

13. $x < -5$:

$$-5x > 25$$
$$x < -5$$

14. $2 > y$:

$$-8 < -4y$$
$$2 > y$$

Drill 4

15. $z \geq -11$:

$$2z + 4 \geq -18$$
$$2z \geq -22$$
$$z \geq -11$$

16. $x \leq -3$: The answer can also be written as $-3 \geq x$. It's more common to write the variable first.

$$7x + 5 \geq 10x + 14$$
$$5 \geq 3x + 14$$
$$-9 \geq 3x$$
$$-3 \geq x$$

17. $d < 6.5$ or $\dfrac{13}{2}$: You can choose to convert to decimals or fractions, as you prefer.

$$d + \frac{3}{2} < 8$$
$$d < 8 - \frac{3}{2}$$
$$d < 8 - 1.5$$
$$d < 6.5$$

18. $a > 6$:

$$\frac{2a}{3} > 10 - a$$
$$2a > 30 - 3a$$
$$5a > 30$$
$$a > 6$$

Drill 5

19. $x \geq 10$:

$$3(x - 7) \geq 9$$
$$x - 7 \geq 3$$
$$x \geq 10$$

20. $x > 48$: The answer can also be written as $48 < x$. It's more common to write the variable first.

$$\frac{x}{3} + 8 < \frac{x}{2}$$
$$6\left(\frac{x}{3} + 8\right) < 6\left(\frac{x}{2}\right) \qquad \text{Multiply by 6 to get rid of fractions.}$$
$$2x + 48 < 3x$$
$$48 < x$$

21. $x > 4.25$:

$$2x - 1.5 > 7$$
$$2x > 8.5$$
$$x > 4.25$$

22. $x \leq -4$: You are allowed to multiply or divide by a constant (a number) even when one side of the equation is 0. That side of the equation just remains 0.

$$\frac{6(2x + 8)}{5} \leq 0$$
$$6(2x + 8) \leq 0$$
$$2x + 8 \leq 0$$
$$2x \leq -8$$
$$x \leq -4$$

23. $x \geq \frac{3}{11}$: The answer can also be written as $\frac{3}{11} \leq x$.

$$\frac{2(3 - x)}{5x} \leq 4 \qquad \begin{array}{l} \text{Since } x > 0 \text{, you can multiply both sides by} \\ 5x \text{ and keep the inequality sign as it is.} \end{array}$$
$$2(3 - x) \leq 20x$$
$$3 - x \leq 10x$$
$$3 \leq 11x$$
$$\frac{3}{11} \leq x$$

Drill 6

24. **$x > 9$:** Both sides of an inequality can be squared as long as both sides are positive. In this case, the square-root side of the equation must be positive because it is greater than the positive number 5, as indicated in the second step of the math.

$$4\sqrt{3x - 2} > 20$$
$$\sqrt{3x - 2} > 5$$
$$3x - 2 > 25$$
$$3x > 27$$
$$x > 9$$

25. **$x < -2$:**

$$\frac{2(8 - 3x)}{7} > 4$$
$$2(8 - 3x) > 28$$
$$8 - 3x > 14$$
$$-3x > 6$$
$$x < -2$$

26. **$x \leq 16$:** If you are more comfortable working with fractions, convert 0.25 to $\frac{1}{4}$ before you solve.

$$0.25x - 3 \leq 1$$
$$0.25x \leq 4$$
$$x \leq 16$$

27. **$y \geq 1$:** When you take a cube root, there is only one solution: The cube has the same sign as the cube root. In this case, the cube root of positive 27 is positive 3.

$$2(y + 2)^3 - 5 \geq 49$$
$$2(y + 2)^3 \geq 54$$
$$(y + 2)^3 \geq 27$$
$$y + 2 \geq 3$$
$$y \geq 1$$

28. **$x \geq 7$:**

$$\frac{4\sqrt[3]{5x - 8}}{3} \geq 4$$
$$4\sqrt[3]{5x - 8} \geq 12$$
$$\sqrt[3]{5x - 8} \geq 3$$
$$\left(\sqrt[3]{5x - 8}\right)^3 \geq (3)^3$$
$$5x - 8 \geq 27$$
$$5x \geq 35$$
$$x \geq 7$$

Drill 7

29. $x = 5$ **or** -5:

$$|x| = 5$$

$$+(x) = 5 \qquad \text{or} \qquad -x = 5$$
$$x = 5 \qquad\qquad\qquad x = -5$$

30. $a = 3$ **or** -3:

$$|5a| = 15$$

$$+(5a) = 15 \qquad \text{or} \qquad -(5a) = 15$$
$$5a = 15 \qquad\qquad\qquad -5a = 15$$
$$a = 3 \qquad\qquad\qquad a = -3$$

31. $y = 4$ **or** -5:

$$|4y + 2| = 18$$

$$+(4y + 2) = 18 \qquad \text{or} \qquad -(4y + 2) = 18$$
$$4y + 2 = 18 \qquad\qquad\qquad -4y - 2 = 18$$
$$4y = 16 \qquad\qquad\qquad -4y = 20$$
$$y = 4 \qquad\qquad\qquad y = -5$$

32. $x = -5$ **or** 7:

$$|1 - x| = 6$$

$$+(1 - x) = 6 \qquad \text{or} \qquad -(1 - x) = 6$$
$$1 - x = 6 \qquad\qquad\qquad -1 + x = 6$$
$$-x = 5 \qquad\qquad\qquad x = 7$$
$$x = -5$$

Drill 8

33. $x = 10$ **or** -2: Isolate the absolute value first. Then split into two equations and solve.

$$3|x - 4| = 18$$
$$|x - 4| = 6$$

$$+(x - 4) = 6 \qquad \text{or} \qquad -(x - 4) = 6$$
$$x - 4 = 6 \qquad\qquad\qquad -x + 4 = 6$$
$$x = 10 \qquad\qquad\qquad -x = 2$$
$$x = -2$$

34. $x = 3.2$ **or** -3.8: Isolate the absolute value first. Then split into two equations and solve.

$$2|x + 0.3| = 7$$
$$|x + 0.3| = 3.5$$

$+(x + 0.3) = 3.5$	or	$-(x + 0.3) = 3.5$
$x + 0.3 = 3.5$		$-x - 0.3 = 3.5$
$x = 3.2$		$-x = 3.8$
		$x = -3.8$

35. $z = 7$ **or** $-\dfrac{4}{5}$:

$$|6z - 3| = 4z + 11$$

$+(6z - 3) = 4z + 11$	or	$-(6z - 3) = 4z + 11$
$6z - 3 = 4z + 11$		$-6z + 3 = 4z + 11$
$2z - 3 = 11$		$3 = 10z + 11$
$2z = 14$		$-8 = 10z$
$z = 7$		$-\dfrac{8}{10} = z$
		$-\dfrac{4}{5} = z$

36. $x = -10$ **or** -14:

$$\left|\frac{x}{4} + 3\right| = 0.5$$

$+\left(\dfrac{x}{4} + 3\right) = 0.5$	or	$-\left(\dfrac{x}{4} + 3\right) = 0.5$
$\dfrac{x}{4} + 3 = 0.5$		$-\dfrac{x}{4} - 3 = 0.5$
$\dfrac{x}{4} = -2.5$		$-\dfrac{x}{4} = 3.5$
$x = -10$		$x = -14$

8

Drill 9

37. $-4 < x < -2$: Because the two solutions overlap, combine them into one inequality: x is greater than -4 *and* less than -2.

$$|x + 3| < 1$$

$+(x + 3) < 1$ $\qquad\qquad$ $-(x + 3) < 1$

$\quad x + 3 < 1$ $\qquad\qquad\qquad$ $-x - 3 < 1$

$\qquad x < -2$ $\qquad\qquad\qquad\qquad$ $-x < 4$

$\qquad\qquad\qquad\qquad\qquad\qquad\qquad$ $x > -4$

$$-4 < x < -2$$

38. $-4 \le y \le 1$: If you prefer to write the variable on the left-hand side, you can flip the inequality, as shown in the last line of the left-hand math, below. If you choose to do this, make sure to flip not just the two sides of math but the inequality sign, as well.

$$5 \ge |2y + 3|$$

$5 \ge +(2y + 3)$ $\qquad\qquad\qquad$ $5 \ge -(2y + 3)$

$5 \ge 2y + 3$ $\qquad\qquad\qquad\qquad$ $5 \ge -2y - 3$

$2 \ge 2y$ $\qquad\qquad\qquad\qquad\qquad$ $8 \ge -2y$

$1 \ge y$ $\qquad\qquad\qquad\qquad\qquad\quad$ $-4 \le y$

$y \le 1$

$$-4 \le y \le 1$$

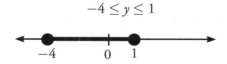

39. $b \ge 3$ or $b \le \frac{3}{5}$:

$$6 \le |5b - 9|$$

$6 \le +(5b - 9)$ $\qquad\qquad\qquad$ $6 \le -(5b - 9)$

$6 \le 5b - 9$ $\qquad\qquad\qquad\qquad$ $6 \le -5b + 9$

$15 \le 5b$ $\qquad\qquad\qquad\qquad\qquad$ $-3 \le -5b$

$3 \le b$ $\qquad\qquad\qquad\qquad\qquad\qquad$ $\frac{3}{5} \ge b$

$$b \ge 3 \ \text{or} \ b \le \frac{3}{5}$$

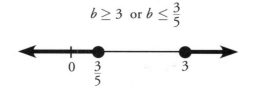

40. $-\frac{5}{4} < a < \frac{5}{4}$:

$$|-12a| < 15$$

$+(-12a) < 15$ $\qquad\qquad\qquad$ $-(-12a) < 15$

$\qquad -12a < 15$ $\qquad\qquad\qquad\qquad$ $12a < 15$

$\qquad\qquad a > -\frac{15}{12}$ $\qquad\qquad\qquad\qquad$ $a < \frac{15}{12}$

$\qquad\qquad a > -\frac{5}{4}$ $\qquad\qquad\qquad\qquad\qquad$ $a < \frac{5}{4}$

$$-\frac{5}{4} < a < \frac{5}{4}$$

Drill 10

41. $x \le -6$ **or** $x \ge 6$:

$$|-x| \ge 6$$

$+(-x) \ge 6$ $\qquad\qquad\qquad\qquad$ $-(-x) \ge 6$

$\qquad -x \ge 6$ $\qquad\qquad\qquad\qquad\qquad$ $x \ge 6$

$\qquad\quad x \le -6$

$$x \le -6 \text{ or } x \ge 6$$

42. $x < -14$ **or** > 6: Isolate the inequality, then split into two equations to solve.

$$\frac{|x + 4|}{2} > 5$$
$$|x + 4| > 10$$

$+(x + 4) > 10$ $\qquad\qquad\qquad\qquad$ $-(x + 4) > 10$

$\qquad x + 4 > 10$ $\qquad\qquad\qquad\qquad$ $-x - 4 > 10$

$\qquad\quad x > 6$ $\qquad\qquad\qquad\qquad\qquad$ $-x > 14$

$\qquad\qquad\qquad\qquad\qquad\qquad\qquad\qquad$ $x < -14$

$$x < -14 \text{ or } x > 6$$

43. $-3 \leq z \leq 3$:

$$|z^3| \leq 27$$

$+\left(z^3\right) \leq 27$ $\qquad\qquad$ $-\left(z^3\right) \leq 27$

$\quad z^3 \leq 27$ $\qquad\qquad\qquad$ $z^3 \geq -27$

$\qquad z \leq 3$ $\qquad\qquad\qquad\quad$ $z \geq -3$

$$-3 \leq z \leq 3$$

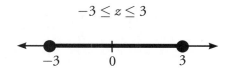

44. $x \leq 20$ **or** $x \geq 40$:

$$|0.1x - 3| \geq 1$$

$+(0.1x - 3) \geq 1$ $\qquad\qquad$ $-(0.1x - 3) \geq 1$

$\quad 0.1x - 3 \geq 1$ $\qquad\qquad\qquad$ $-0.1x + 3 \geq 1$

$\qquad 0.1x \geq 4$ $\qquad\qquad\qquad\quad$ $-0.1x \geq -2$

$\qquad\quad x \geq 40$ $\qquad\qquad\qquad\qquad$ $x \leq 20$

$$x \leq 20 \text{ or } x \geq 40$$

45. $-12 \leq x \leq \dfrac{8}{3}$:

$$\left|\frac{3x}{2} + 7\right| \leq 11$$

$+\left(\dfrac{3x}{2} + 7\right) \leq 11$ $\qquad\qquad$ $-\left(\dfrac{3x}{2} + 7\right) \leq 11$

$\quad \dfrac{3x}{2} + 7 \leq 11$ $\qquad\qquad\qquad$ $-\dfrac{3x}{2} - 7 \leq 11$

$\qquad \dfrac{3x}{2} \leq 4$ $\qquad\qquad\qquad\qquad$ $-\dfrac{3x}{2} \leq 18$

$\qquad 3x \leq 8$ $\qquad\qquad\qquad\qquad$ $-3x \leq 36$

$\qquad\quad x \leq \dfrac{8}{3}$ $\qquad\qquad\qquad\qquad\quad$ $x \geq -12$

$$-12 \leq x \leq \frac{8}{3}$$

Drill 11

46. $1 \leq x \leq \dfrac{17}{3}$: To get rid of all of the fractions at once, multiply an equation or inequality by the common denominator of all of the fractions. The fraction $\dfrac{17}{3}$ can also be written as the mixed fraction $5\dfrac{2}{3}$.

$$\left| -\frac{x}{5} + \frac{2}{3} \right| \leq \frac{7}{15}$$

$$+\left(-\frac{x}{5} + \frac{2}{3} \right) \leq \frac{7}{15} \qquad\qquad\qquad -\left(-\frac{x}{5} + \frac{2}{3} \right) \leq \frac{7}{15}$$

$$-\frac{x}{5} + \frac{2}{3} \leq \frac{7}{15} \qquad\qquad\qquad\qquad \frac{x}{5} - \frac{2}{3} \leq \frac{7}{15}$$

$$15\left(-\frac{x}{5} + \frac{2}{3} \right) \leq \left(\frac{7}{15} \right)15 \qquad\qquad 15\left(\frac{x}{5} - \frac{2}{3} \right) \leq \left(\frac{7}{15} \right)15$$

$$-3x + 10 \leq 7 \qquad\qquad\qquad\qquad 3x - 10 \leq 7$$

$$-3x \leq -3 \qquad\qquad\qquad\qquad\qquad 3x \leq 17$$

$$x \geq 1 \qquad\qquad\qquad\qquad\qquad x \leq \frac{17}{3}$$

$$1 \leq x \leq \frac{17}{3}$$

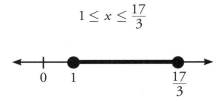

47. $x \leq -1$ **or** $x \geq 19$:

$$|3x - 7| \geq 2x + 12$$

$$+(3x - 7) \geq 2x + 12 \qquad\qquad\qquad -(3x - 7) \geq 2x + 12$$

$$3x - 7 \geq 2x + 12 \qquad\qquad\qquad\qquad -3x + 7 \geq 2x + 12$$

$$x - 7 \geq 12 \qquad\qquad\qquad\qquad\qquad 7 \geq 5x + 12$$

$$x \geq 19 \qquad\qquad\qquad\qquad\qquad\qquad -5 \geq 5x$$

$$\qquad\qquad\qquad\qquad\qquad\qquad\qquad\qquad -1 \geq x$$

$$x \leq -1 \text{ or } x \geq 19$$

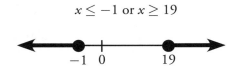

8

48. $-3 < x < -\dfrac{3}{5}$:

$$|3 + 3x| < -2x$$

$+(3 + 3x) < -2x$	$-(3 + 3x) < -2x$
$3 + 3x < -2x$	$-3 - 3x < -2x$
$3 + 5x < 0$	$-3 < x$
$5x < -3$	
$x < -\dfrac{3}{5}$	

$$-3 < x < -\dfrac{3}{5}$$

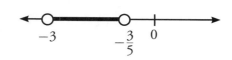

49. $-9 \le x \le -1$:

$$|-9 - 5x| \le -4x$$

$+(-9 - 5x) \le -4x$	$-(-9 - 5x) \le -4x$
$-9 - 5x \le -4x$	$9 + 5x \le -4x$
$-9 \le x$	$9 + 9x \le 0$
	$9x \le -9$
	$x \le -1$

$$-9 \le x \le -1$$

50. $\frac{4}{5} < y < 12$: Isolate the absolute value, and then split into two equations to solve. Multiply the entire inequality by the common denominator of any fractions to get rid of all of the fractions.

$$2\left|\frac{7y}{4} - 7\right| < \frac{3y}{2} + 10$$

$$\left|\frac{7y}{4} - 7\right| < \frac{3y}{4} + 5$$

$$+\left(\frac{7y}{4} - 7\right) < \frac{3y}{4} + 5 \qquad\qquad -\left(\frac{7y}{4} - 7\right) < \frac{3y}{4} + 5$$

$$\frac{7y}{4} - 7 < \frac{3y}{4} + 5 \qquad\qquad -\frac{7y}{4} + 7 < \frac{3y}{4} + 5$$

$$7y - 28 < 3y + 20 \qquad\qquad -7y + 28 < 3y + 20$$

$$4y < 48 \qquad\qquad\qquad 8 < 10y$$

$$y < 12 \qquad\qquad\qquad \frac{8}{10} < y$$

$$\qquad\qquad\qquad\qquad\qquad \frac{4}{5} < y$$

$$\frac{4}{5} < y < 12$$

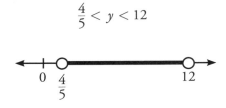

Word Problems

In This Chapter

- Solve Word Problems: Understand, Plan, Solve

- Average = Sum ÷ Number

- Express Revenue as Price × Quantity

- Add Units: Add Apples to Apples

- Multiply and Divide Units: Treat Like Numbers or Variables

- Rate × Time = Distance: Follow the Units

- Rate × Time = Work: Define the Work Unit and Add Rates

In this chapter, you will learn to translate and solve various kinds of math-based stories, or word problems, that may also involve algebra, fractions, and percents. You'll also learn some common formulas that you're expected to know for the exam, such as the average formula and the rate formula.

CHAPTER 9 **Word Problems**

Word problems are everywhere on the GMAT. They come in all shapes and sizes. The stories can be straightforward or they can be confusing and unfamiliar.

Solve Word Problems: Understand, Plan, Solve

Here's an example of a relatively straightforward problem (as straightforward as standardized tests can be, anyway):

> A steel rod 50 meters long is cut into two pieces. If one piece is 14 meters longer than the other, what is the length, in meters, of the shorter piece?
>
> (A) 10
> (B) 18
> (C) 32
> (D) 36
> (E) 40

Use the Understand–Plan–Solve (UPS) process to help you solve the problem. First, Understand what's going on. Second, come up with a Plan. Third, Solve. Here's how the process works and what someone might think (in *italics*) as they use UPS to solve the problem above.

1. **Understand**

    Glance at the entire problem. On story problems, the first thing you'll spot is the Wall of Text (WoT). Slow down a bit and take your time to understand what's going on. On Problem Solving problems, also glance at the answer choices; what form are they in?

    Glance: WoT. Real numbers in the answers and they're pretty spread out. See whether I can eyeball or estimate anything.

    Next, what's the story? Read and jot down info as you go. The problem typically gives you some numbers. It also gives you relationships between things. Since it's a story problem, you'll have to translate that information into some form of math. Jot down this information on your scrap paper, clearly identifying the different parts.

    Read and Jot: There are two pieces of steel, a longer one and a shorter one, and they add up to 50 meters. Jot down: L + S = 50.

    Hmm. The next piece of information is trickier. One piece is 14 meters longer than the other. Logic that out: If you take the shorter piece and add 14, you get the longer piece. Jot down: S + 14 = L.

    Finally, what do they want? Identify what the problem is specifically asking for. Name that thing and write it down. (Don't try to solve yet. That's step 3!)

    They want the shorter piece. Jot down: S = ?

Pro tip: Don't default to x and y for your variables. To minimize careless mistakes, choose variables that tell you what they represent in the problem.

One last thing: If you don't understand what they're telling or asking you, then this is an excellent time to pick your favorite letter and move on to the next problem. Use your business mindset.

2. **Plan**

Reflect for a moment. Do you see any paths to get from what they gave you to what they want you to find? If there's more than one way to get there, which way do you think best matches this problem *and* the way that your brain thinks?

I've got two equations, so I could use algebra to find S. But wait a sec—what's the basic story? There are two parts, a shorter one and a longer one, and they add up to 50. I want the shorter one. Glance at the answers.

The 30-something answers don't make sense! They add up to 50, so something that long would have to be the longer one, not the shorter one.

Take a look at what you've jotted down. Is there some better way to organize the information? This problem is pretty straightforward, but on a different one, maybe you would want to make a table to keep track of multiple variables.

Finally, choose a path. Your plan might involve solving an equation, or doing some estimation, or even trying some numbers to see what works in a problem.

I'll do algebra. I can substitute one equation into the other.

As before, if you don't have a good plan, pick your favorite letter and move on to the next problem.

3. **Solve**

Now that you've got a plan, go for it. Here's the algebraic solution:

$$L + S = 50 \qquad\qquad S + 14 = L$$
$$(S + 14) + S = 50$$
$$2S = 36$$
$$S = 18$$

The shorter one is 18—and this makes sense with your initial logic that the 30-something answers were too big. The correct answer is (B).

One more note on that estimation. The shorter one has to be shorter than the longer one, right? They add up to 50. Hmm. If they were exactly the same length, how long would they be?

They'd each be 25. So the longer one has to be at least a little greater than 25 and the shorter one has to be at least a little less than 25. For a similar future problem, when you know that two items add up to a number and one item is greater than the other, you can use this idea of "half plus a little, half minus a little" to better estimate the possible range of answers for each item.

Here are some common translation terms and expressions that signal addition or subtraction:

Addition

Add, Sum, Total (of parts), More Than:	$+$
The sum of x and y:	$x + y$
The sum of the three funds combined:	$a + b + c$
If Anuj were 50 years older:	$a + 50$
Six pounds heavier than Dave:	$d + 6$
A group of managers and employees:	$m + e$
The cost is marked up by m dollars:	$c + m$

Subtraction

Minus, Difference, Less Than:	$-$
x minus five:	$x - 5$
The positive difference between q and r (if $q > r$):	$q - r$
Four pounds less than expected:	$e - 4$
The profit is the revenue minus the cost:	$P = R - C$

Try another problem:

> There are two trees in the front yard of a school. The trees have a combined height of 60 feet, and the taller tree is three times the height of the shorter tree. What is the height, in feet, of the shorter tree?
>
> (A) 15
> (B) 20
> (C) 45

$$T + S = 60$$
$$3S + S = 60$$

Use the UPS process to solve. First, Understand.

WoT. "Nice" integers in the answers. What do they tell me and what do they want? Jot down:

$$S + T = 60 \qquad 3S = T \qquad S = ?$$

The second piece of information involves multiplication this time, not addition. How can you tell that the equation is $3S = T$ and not $S = 3T$?

Logic it out. You would take the height of the shorter tree and multiply it by 3 in order to get the height of the taller tree, so put the 3 on the same side of the equation as the S.

Make sure to write down a question mark next to what they want. That will help to solidify in your mind that you want the height of the shorter tree—because it's usually the case that the height of the taller tree will be among the answers, as a trap.

Next, Plan.

> *I can solve algebraically or I can try the answers.*

Just like last time, there are at least two ways to solve. Anything else?

9

Yes, actually! Remember ratios? When a story problem tells you a multiplication or division relationship, you can sometimes use ratios to solve. In this case, the taller tree is three times the height of the shorter one; in other words, the ratio of the height of the taller to the height of the shorter is 3 : 1. And the two add up to an actual combined height of 60.

Toss that information into a ratio box and Solve. Put a circle in the box that you want to solve for:

	Taller	Shorter	Total
R	3	1	4
×			
A		◯	60

The unknown multiplier must be 15, so the shorter tree's height is 1 × 15 = 15:

	Taller	Shorter	Total
R	3	1	4
×		× 15	× 15
A		⟨15⟩	60

The correct answer is (A).

Notice that the height of the taller tree, 45, is also among the answers. Don't get so pulled into the story that you forget to use common sense. If the two add up to 60, then the *shorter* tree can't possibly be 45.

Answer (B), 20, is a good trap. They're hoping that you'll see the "three times" figure and think that the answer is $\frac{60}{3} = 20$. Why isn't it?

It's true that 60 is 3 times 20—but 60 represents the combined height, not the height of the taller tree. The 60 is the "whole" in the relationship, while the taller and shorter trees are the two "parts." So set up the ratio of the two parts to find the whole.

Here are some common translation terms and expressions that signal multiplication and division:

Multiplication & Division

Multiply, Product, Times:	×
Quotient, Per, Ratio, Proportion:	÷ or /
m is twice *p*:	$m = 2p$
The number of red marbles times the number of blue marbles:	$r \times b$
One-fifth of *y*:	$\left(\frac{1}{5}\right)y$
n persons have *x* beads each:	total beads $= nx$
A is 4 times the length of *B*:	$A = 4B$
	$A : B = 4 : 1$
The ratio of *x* to *y*:	$\frac{x}{y}$
	$x : y$

Check Your Skills

1. Two conference rooms have a combined capacity of 75 people. Conference Room A can accommodate 15 fewer people than Conference Room B. How many people can Conference Room B accommodate? *45*

2. A local donut shop sold 80 donuts every day for a period of 5 days. In the subsequent 4 days, the shop sold 75 donuts each day. How many total donuts were sold over the entire 9-day period?

Answers can be found on page 349. *700*

Average = Sum ÷ Number

Statistics problems are often in story form on the GMAT. Try this one:

> Over a period of 5 days at a donut shop, the average (arithmetic mean) number of donuts sold per day was 80. In the 4 days after that period, the shop sold a total of 500 donuts. What was the average number of donuts sold per day over the entire 9-day period?

First, Understand. The problem discusses a 9-day period of sales at a donut shop. During the first 5 days, the donut shop averaged 80 donuts sold *per day*. During the next 4 days, the shop sold a *total* of 500 donuts. Jot that information down:

First 5 days: 80/day

Next 4 days: 500 total

Next, the question asks for the average (arithmetic mean) number of donuts sold per day over 9 days. Use the letter *a* to represent this value. (Note: *Arithmetic mean* is a synonym for *average*. The test will often use both terms; all you need to find is the regular average.)

You'll need to memorize the average formula for the GMAT: The average of a group of numbers is the sum of those numbers divided by the number of terms, which is the number of numbers you have:

$$\text{Average} = \frac{\text{Sum}}{\text{\# of terms}}$$

When you see that a problem mentions some concept for which you have memorized a formula, immediately write down the "generic" form of that formula, as shown above.

Include units as well. The average is in donuts *per* day. *Per* means *divided by*, so this unit is donuts/day or $\frac{\text{donuts}}{\text{day}}$. Here is the average formula customized for this problem:

$$\text{Average donuts per day} = \frac{\text{Total donuts}}{\text{Total days}}$$

To find the average, you need both the total number of donuts and the total number of days.

Next, Plan. What do you need to figure out in order to solve for that average?

The problem asked for the average *over the entire 9-day period*, so the total number of days is 9.

However, the problem didn't (directly) give you the total number of donuts. Make up a variable—say, *d* for donuts.

Now rewrite the average formula:

$$a = \frac{d}{9}$$

Now you know that to get a, you need d. How can you get to d?

Take a look at what you jotted down:

> First 5 days: 80/day
>
> Next 4 days: 500 total

The first part and the second part add up to the whole 9 days. This is a part + part = whole relationship all over again. In other words:

$$d = (\text{sum for first 5 days}) + 500$$

As word problems get harder, they're going to make you put more of the pieces together. The problem directly says how many donuts were sold in the last 4 days: 500. However, the number of donuts sold in the first 5 days is still unknown; call that n. The problem gives another average figure here, so plug the information into its own average formula. You don't need to write down the formula on the left, just the one on the right:

$$5\text{-day average} = \frac{\text{Total donuts in first 5 days}}{5} \qquad 80 = \frac{n}{5}$$

Now, find n:

$$80 = \frac{n}{5} \quad \longrightarrow \quad 400 = n$$

The shop sold 400 donuts in the first 5 days. It also sold 500 donuts in the next 4 days, so altogether, it sold $d = 400 + 500 = 900$ donuts over the 9 days.

Plug that total into your first average formula to find a:

$$a = \frac{d}{9} \quad \longrightarrow \quad a = \frac{900}{9} \quad \longrightarrow \quad a = 100$$

The average number of donuts sold per day over the entire 9-day period is 100. That is, on average, 100 donuts were sold each day for 9 days.

As you solve story problems, sometimes you'll notice relationships first. At other times, you'll notice unknowns first. The order is not important. Just keep extracting information from the problem and representing that information on paper.

Turn unknowns into letters and turn relationships into equations. Observe how the equations hook together. In the previous problem, you ultimately wanted the average (a), but you needed the total number of donuts (d) first, and before that you had to find the number of donuts for the first 5 days (n).

9

If you ...	Then you ...	Like this:
Want to solve a story problem: Kelly is three times as old as Bill. In 5 years, Kelly will be twice as old as Bill will be. How old is Bill?	Follow the three steps: 1. Understand (glance, read, jot) 2. Plan (reflect, organize, choose a path) 3. Solve	$B = ?$ $K = 3B$ $K + 5 = 2(B + 5)$ $(3B) + 5 = 2(B + 5)$ $3B + 5 = 2B + 10$ $B = 5$ Bill is 5 years old.

Here are some common translation terms and expressions that signal average or arithmetic mean:

<u>Average & Arithmetic Mean</u>

Average, arithmetic mean: $\qquad$ $\text{Average} = \dfrac{\text{Sum}}{\text{\# of terms}}$

The average of a and b: $\qquad$ $\dfrac{a+b}{2}$

The average salary of the three doctors: $\qquad$ $\dfrac{x+y+z}{3}$

A student's average score on five tests was 87: $\qquad$ $\dfrac{\text{sum}}{5} = 87$ or $\dfrac{a+b+c+d+e}{5} = 87$

Check Your Skills

3. Jan makes a salary of $10,000 per month for 3 months. Then her salary drops to $6,000 per month. After 9 months at $6,000 per month, what will Jan's average (arithmetic mean) monthly pay be for the whole 12-month period?

    Answer can be found on pages 349–350. $\qquad \frac{30}{54} \rightarrow 84 \div 12 = 7,000/mo$

Express Revenue as Price × Quantity

The GMAT expects you to know certain common money relationships. Try this problem:

> At a certain store, 7 shirts cost $63. If each shirt costs the same amount, what is the cost of 3 shirts, in dollars?

First, Understand. The question is asking for the total cost to buy 3 shirts, in dollars (represented by $). This cost for the consumer can also be thought of as revenue for the company selling the shirts.

You are given the cost of 7 shirts ($63); again, this cost to the buyer represents revenue for the seller. You are also told that every shirt has the same price, and you are given another quantity of shirts to care about (3).

9

When every unit of something has the same price, use this equation:

Total revenue = Price × Quantity

This can also be written:

Total cost to consumer = Price × Quantity

Name the variables P for price, Q for quantity, and either C for cost or R for revenue (depending on whether the problem is from the point of view of the customer or the company):

$$C = P \times Q \quad \text{or} \quad R = P \times Q$$

The question asks for the total cost of 3 shirts. In other words, the question asks for the value of C when $Q = 3$.

Next, come up with your Plan. You can write this as an equation; you can also just logic this out on paper:

Equation	Logic
$C = P \times 3 = ?$	7 shirts ⟶ $63
	3 shirts ⟶ $?

To find C, then, you need P, or the price of each shirt. Price is always in dollars *per unit*. It is the cost of one unit of whatever is being sold or bought—in this case, shirts.

The given information (*7 shirts costs $63*) is another instance of Total cost to consumer = Price × Quantity. You have a different quantity (7) and a different total cost ($63), but the problem says that the price per shirt is always the same. So the cost per shirt, P, is the same in both scenarios.

Go ahead and Solve. Set up a second equation or continue to logic this out:

Equation	Logic
$C = P \times 3 = ?$	7 shirts ⟶ $63 ⟶ so 1 shirt $= \dfrac{\$63}{7} = \$9$
$\$63 = P \times 7$	3 shirts ⟶ $?
$\dfrac{63}{7} = P$	
$9 = P$	

Therefore, each shirt costs $9.

Finally, find the total cost of 3 shirts by substituting 9 for P:

Equation	Logic
$C = P \times 3 = ?$	7 shirts ⟶ $63 so 1 shirt $= \dfrac{\$63}{7} = \$9$
$\$63 = P \times 7$	3 shirts ⟶ $3 \times \$9 = \$27$
$\dfrac{63}{7} = P$	
$9 = P$	
$C = P \times 3 = 9 \times 3 = \$27$	

The total cost of 3 shirts is $27.

Here's how the problem would look from the store's point of view:

> A certain store sold 7 shirts for $63. If every shirt has the same price, how much revenue, in dollars, would the store receive from selling 3 shirts?

The answer is the same: $27. Instead of total cost, write total revenue:

> Total revenue = Price × Quantity

Or you can focus on the units. The total on the left is in dollars, while the price is in dollars per shirt:

$$\text{Total money (\$)} = \text{Price}\left(\frac{\$}{\text{shirt}}\right) \times \text{Quantity (shirts)}$$

Try a more complicated problem:

> Five apples and four bananas cost $2.10, while three apples and two bananas cost $1.20. If the cost of each apple is the same, and if the cost of each banana is the same, what is the cost of two apples?

First, Understand. Take this one step at a time.

The cost of five apples and four bananas is $2.10. This $2.10 is a whole made up of two parts: the cost of five apples and the cost of four bananas. Each of *those* costs is a price times a quantity. Set up an equation:

Total cost	=	**Price × Quantity**	+	**Price × Quantity**
		for apples		for bananas

What are you given? You are given the cost of five apples and four bananas ($2.10). You are also given the cost of three apples and two bananas ($1.20).

Furthermore, you are told that the cost (or price) of each apple is the same, and likewise for bananas. These last facts allow you to use the Total cost = Price × Quantity relationship.

The simplest unknowns to name are the price (or cost) of one apple and the price of one banana. Call these A for apple and B for banana. For each total cost you are given, write a separate equation:

Total cost		**Price × Quantity**		**Price × Quantity**		
$2.10	=	$A \times 5$	+	$B \times 4$	=	$5A + 4B$
$1.20	=	$A \times 3$	+	$B \times 2$	=	$3A + 2B$

You are looking for the cost of two apples: $2A = ?$

Next, come up with a Plan. On a math question in school, you'd have to find the values for both A and B, but this question only requires A. What impact does that have on how you Solve?

However you set up the math, you want to make sure that you're solving directly for A. Yes, technically, you can solve for B first and then find A, but why do that extra work?

One option is to combine the equations. Don't start solving quite yet. First, what do you need to do to get the B terms to drop out?

If you multiply the second equation by 2, then both equations will have 4B, so that variable will drop out. Okay, now Solve:

Equation 1:   Equation 2:

$2.10 = 5A + 4B$ $2(1.20) = (3A + 2B)2$

$2.40 = 6A + 4B$

Next, subtract the first equation from the second:

$$2.40 = 6A + 4B$$
$$-(2.10 = 5A + 4B)$$
$$0.30 = A$$

Finally, multiply by 2 to get 2A:

$$0.30 \times 2 = 0.60$$

The cost of two apples is $0.60.

Sometimes the price is known and the quantity is not:

The cost in dollars of x books priced at $12 each is $12x$.

Occasionally, you encounter an up-front **fixed cost**. Cell phone minutes used to be priced this way: You had a fixed cost per call (including the first 2 or 3 minutes), and then you paid for additional minutes at a certain price per minute. The equation looks like this:

Total cost   =   **Fixed cost**   +   **Price** × *Additional* **quantity**

Try this problem:

The charge to reserve a box seat at Colossus Stadium is $1,000, which includes attendance at two games, plus $300 for each additional game attended. If Sani wants to spend no more than $4,000 on a box seat at Colossus Stadium, what is the maximum number of games Sani can attend?

Understand first. To reserve a box seat, Sani first has to pay a flat $1,000 and then an additional $300 each for the third game, the fourth game, the fifth game, and so on. What is the maximum number of games for which Sani can get seats without spending more than $4,000? Call that maximum number of games n.

Sani attends n games total, but two of those games were included in the $1,000 flat fee. So Sani attends $n - 2$ games at the $300 per game ticket price.

Finally, this is an inequality: Sani wants to spend less than or equal to $4,000. Set up an equation:

Fixed cost	+	**Price** × *Additional* **quantity**	≤	**Total cost**
$1,000	+	$300 × $(n - 2)$	≤	$4,000

Next, solve for *n*:

$$1{,}000 + 300(n - 2) \le 4{,}000$$
$$300(n - 2) \le 3{,}000$$
$$n - 2 \le 10$$
$$n \le 12$$

At most, Sani can attend 12 total games.

Last but not least, don't forget this classic relationship:

Profit = Revenues — Costs

This one shows up in business school quite a bit, of course! You also need it on the GMAT, sometimes in combination with the other relationships above.

If you...	Then you...	Like this:
Encounter a money relationship: "The cost of 8 watches is $1,200..."	Write the equation: Total Cost = Price × Quantity, including more items or a fixed cost as necessary	P = Price of one watch $1,200 = P × 8$ $1,200 = 8P$ $150 = P$ One watch costs $150.

Check Your Skills

4. A candy shop sold 50 candy bars in January for $3 each. In February, the shop increased the price of a candy bar by $1, but its revenue from candy bar sales increased by only $10. By how many units did the number of candy bars sold decrease from January to February?

5. At a particular store, seven staplers and five coffee mugs cost $36, while eight staplers and four coffee mugs cost $40. If each stapler costs the same, and if each coffee mug costs the same, how much more does a stapler cost than a coffee mug?

Answers can be found on page 350.

Add Units: Add Apples to Apples

When you solve pure algebra problems, the numbers don't represent anything in particular. In contrast, story problems have a context. Every number has a meaning. That is, every number has a natural **unit** attached.

Up to this point, the units have worked out naturally. For instance, in the steel rod question at the beginning of this chapter, all the lengths were already in meters, so you could ignore the units:

S	+	L	=	50		S	+	14	=	L
meters		meters		meters		meters		meters		meters

What if some of the units were meters and others were feet? In that case, you couldn't add or subtract them. *When you add or subtract quantities, they must have exactly the same units.*

What is $2 plus 45 cents, if $1 = 100 cents? Choose a common unit. It doesn't matter which one you pick, but you must express both quantities in that same unit before you add:

$$\underset{\text{dollars}}{2} \ + \ \underset{\text{dollars}}{0.45} \ = \ \underset{\text{dollars}}{2.45} \qquad\qquad \underset{\text{cents}}{200} \ + \ \underset{\text{cents}}{45} \ = \ \underset{\text{cents}}{245}$$

The result has the same unit as the original quantities. When you add or subtract units, the units do not change. For example:

$$\text{meters} + \text{meters} = \text{meters} \qquad \text{dollars} - \text{dollars} = \text{dollars} \qquad \text{puppies} + \text{puppies} = \text{puppies}$$

Multiply and Divide Units: Treat Like Numbers or Variables

In contrast, when you multiply units, the result has a different unit. For example:

If a room is 6 feet long and 9 feet wide, what is the area of the room?

Area is length times width, so multiply:

Area = 6 feet × 9 feet = 54...

What happens to the feet? Feet times feet equals feet *squared*. Therefore:

Area = 6 feet × 9 feet = 54 feet squared = 54 feet^2 = 54 square feet

If you multiply two quantities that each have units, multiply the units, too. And if you divide, divide the units. For example:

$$3 \text{ books in 1 week} = 3\frac{\text{books}}{\text{week}} \qquad 17 \text{ miles per 1 gallon} = 17\frac{\text{miles}}{\text{gallon}}$$

Not every multiplication in a story problem results in a change in the units. If Alex is twice as old as Brenda, in years, then you can represent that relationship like so:

$$\underset{\text{years}}{A} \quad = \quad 2 \quad \times \quad \underset{\text{years}}{B}$$

The word *twice* has no units (it just means *two times*, or 2 ×).

Some units are naturally ratios of other units. Look for the words *per*, *a*, *for every*, and so on:

$$\text{For every 9 cats, there are 2 dogs} = \frac{C : D}{9 : 2} = \frac{9 \text{ cats}}{2 \text{ dogs}}$$

Prices and averages often have units that are ratios of other units:

$$9 \text{ dollars a shirt} = 9\frac{\text{dollars}}{\text{shirt}} \qquad 100 \text{ donuts per day} = 100\frac{\text{donuts}}{\text{day}}$$

The average formula actually shows the division of units:

$$\text{Average donuts per day} = \frac{\text{Total donuts}}{\text{Total days}}$$

Likewise, the total cost relationship demonstrates that *units cancel in the same way as numbers and variables do*:

$$\text{Total cost} = \text{Price} \times \text{Quantity}$$

$$\text{Dollars} = \frac{\text{dollars}}{\text{shirt}} \times \text{shirts}$$

The units match on both sides of the equation because you can cancel the "shirt" unit just as you would with actual numbers:

$$\text{Dollars} = \frac{\text{dollars}}{\cancel{\text{shirt}}} \times \cancel{\text{shirts}}$$

This cancellation property allows you to convert from a larger to a smaller unit, or vice versa. Try this problem:

How many minutes are in two days?

First, convert 2 days into hours: 1 day = 24 hours. You can probably convert 2 days to 48 hours in your head, but try it on paper with unit cancellation to see how it works:

$$2 \text{ days} \times \frac{24 \text{ hours}}{1 \text{ day}} = 48 \text{ hours} \qquad \cancel{\text{days}} \times \frac{\text{hours}}{\cancel{\text{day}}} = \text{hours}$$

Notice that the "day" units cancel, leaving you with "hours" on top. The fraction you use to multiply is called a *conversion factor*. It's a fancy form of the number 1, because the top (24 hours) equals the bottom (1 day). When you write conversion factors, put the units in place first so that they cancel correctly. Then place the corresponding numbers so that the top equals the bottom.

Keep going to minutes. Set up the conversion factor using 60 minutes = 1 hour:

$$48 \text{ hours} \times \frac{60 \text{ minutes}}{1 \text{ hour}} = 2{,}880 \text{ minutes} \qquad \cancel{\text{hrs}} \times \frac{\text{min}}{\cancel{\text{hr}}} = \text{min}$$

By the way, always write out at least a few letters for every unit. "Hours" can be "hr," and "minutes" can be "min," but never write "h" or "m" alone. You might confuse a single letter for a variable—or you might forget whether *m* stands for meters or minutes.

You can do two or more conversions in one step:

$$2 \text{ days} \times \frac{24 \text{ hours}}{1 \text{ day}} \times \frac{60 \text{ minutes}}{1 \text{ hour}} = 2{,}880 \text{ minutes} \qquad \cancel{\text{days}} \times \frac{\cancel{\text{hrs}}}{\cancel{\text{day}}} \times \frac{\text{min}}{\cancel{\text{hr}}} = \text{min}$$

A common conversion is between miles and kilometers. You don't have to know that 1 mile is approximately 1.6 kilometers; the GMAT will give you this information. However, you will have to be able to use this information to convert between these units. Try this problem:

A distance is 30 miles. What is the approximate distance in kilometers? (1 mile = 1.6 kilometers)

Multiply the given distance by the conversion factor. Set things up to cancel units:

$$30 \text{ miles} \times \frac{1.6 \text{ kilometers}}{1 \text{ mile}} = 48 \text{ kilometers} \qquad \cancel{\text{miles}} \times \frac{\text{km}}{\cancel{\text{miles}}} = \text{km}$$

If you...	Then you...	Like this:
Add or subtract quantities with units	Ensure that the units are the same, converting first if necessary	30 minutes + 2 hours = 30 min + 120 min = 150 min or $= \frac{1}{2}$ hr + 2 hr = $2\frac{1}{2}$ hr
Multiply quantities with units	Multiply the units, canceling as appropriate	$10\frac{bagels}{hr} \times 3$ hr = 30 bagels
Want to convert from one unit to another	Multiply by a conversion factor and cancel	20 min $\times \frac{60 \text{ sec}}{1 \text{ min}} = 1{,}200$ sec

Check Your Skills

6. How many hours are there in two weeks? Do this problem with conversion factors.

7. How long after midnight is 1:04 a.m., in seconds?

Answers can be found on page 351.

Rate × Time = Distance: Follow the Units

A **rate**, or speed, is expressed as a *distance* unit (such as *miles* or *feet*) divided by a *time* unit (such as *hour* or *second*). For example:

$$\text{Rate} = \frac{\text{Distance}}{\text{Time}} \qquad 60 \text{ miles per hour} = \frac{60 \text{ miles}}{1 \text{ hour}}$$

You can rearrange this relationship to isolate distance on one side:

$$\text{Rate} \times \text{Time} = \text{Distance} \qquad 60\ \frac{\text{miles}}{\text{hour}} \times 1 \text{ hour} = 60 \text{ miles}$$

The Rate–Time–Distance formula is commonly called the RTD formula. This version is very similar to the total cost equation:

$$\text{Price} \times \text{Quantity} = \text{Total cost (or revenue)} \qquad 9\ \frac{\text{dollars}}{\text{shirt}} \times 4 \text{ shirts} = 36 \text{ dollars}$$

A price is a kind of rate, too, because it's *per* something. In distance problems, most rates are *per* time, but occasionally, you see a rate *per* something else, such as miles per gallon.

You can also write the formula in the order Distance = Rate × Time and then use an acronym to remember it: DeRT (pronounced "dirt"). The little *e* stands for *equals*.

The exam will sometimes tell you something like "Chipo took 4 hours to travel 60 miles." Avoid expressing rates as time divided by distance, even though that's the order that the sentence seems to give. Instead, *always put time in the denominator*. If you're told that *Chipo took 4 hours to travel 60 miles*, then Chipo's rate was 60 miles ÷ 4 hours, or 15 miles per hour.

9

You can combine the Rate × Time = Distance relationship with other relationships already covered, such as the average formula:

$$\text{Average donuts per day} = \frac{\text{Total donuts}}{\text{Total days}}$$

The same formula can be adapted for rates as well:

$$\text{Average miles per hour} = \frac{\text{Total miles}}{\text{Total hours}}$$

To get the totals on the top and bottom, you often need another relationship you're familiar with:

Part + Part = Whole

Miles for first part of a trip + Miles for second part = Total miles

Hours for first part of a trip + Hours for second part = Total hours

Try this problem:

Gul takes 2 hours to bike 12 kilometers from home to school. If Gul bikes back home by the same route at a rate of 4 kilometers per hour, what is the average rate, in kilometers per hour, for the entire trip?

Be careful. The average rate for a journey is the *total* distance divided by the *total* time. Do not simply take an average of the rates for the two parts of the trip. The answer is not $\frac{6 \text{ km/hr} + 4 \text{ km/hr}}{2} = 5 \text{ km/hr}$.

First, Understand the moving parts of the problem. Gul goes from home to school and back again. The distance between the two points is 12 kilometers. It takes Gul 2 hours to go from H to S, and who knows how long to go back from S to H, but that second part of the trip is done at 4 kilometers per hour.

You are asked for the average rate for the whole trip. Call this *a* and write an equation:

$$\text{Average kilometers per hour} = a \qquad a = \frac{\text{Total kilometers}}{\text{Total hours}}$$

What's the Plan? Find the two missing unknowns: total kilometers and total hours.

Solve. The route from home to school is 12 kilometers, and Gul comes home by the *same* route, so the total kilometers traveled is equal to 12 + 12 = 24:

$$a = \frac{24}{\text{Total hours}}$$

Now, find the total time in hours:

Total hours = Hours spent on the first part of the trip + Hours spent on the second part

Gul spends 2 hours on the first part of the trip. What do you know about the second part of the trip? You know that Gul's rate was 4 kilometers per hour. You also know that this route was the *same* as for the first leg—so the distances are equal. It's easy to miss this information. Whenever the GMAT says "the same," pay attention! It likely represents an equation.

Use Rate × Time = Distance to find the time for the second leg of the journey:

$$\text{Rate} \times \text{Time} = \text{Distance} \qquad 4 \, \frac{\text{kilometers}}{\text{hour}} \times \text{Time (hours)} = 12 \text{ kilometers}$$

Call this time t and solve:

$$4t = 12$$
$$t = 3$$

The total time spent was 2 hours + 3 hours = 5 hours. Plug that figure into the first equation to find the average:

$$a = \frac{24}{\text{Total hours}} = \frac{24}{5} = 4.8$$

Gul's average rate for the whole trip is 4.8 kilometers per hour.

Rate problems can become tricky when you have to use the same relationship repeatedly (Rate × Time = Distance). To keep the various rates, times, and distances straight, you might use subscripts or even whole words:

$$t_1 = 2 \text{ hours} \qquad t_2 = 3 \text{ hours} \qquad \text{Time \#1} = 2 \text{ hrs} \qquad \text{Time \#2} = 3 \text{ hrs}$$

If you have more than one time represented in the problem, then using t everywhere for every time is likely to lead to mistakes. Tables or grids can also help keep quantities straight.

If you...	Then you...	Like this:
See a rate problem	Use Rate × Time = Distance, putting in units to keep the math correct	$7\,\dfrac{\text{miles}}{\text{hr}} \times 3 \text{ hrs} = 21 \text{ miles}$

Check Your Skills

8. Alaitz ran 24 miles at a rate of 3 miles per hour, then took 4 hours to run an additional 6 miles. What was Alaitz's average speed, in miles per hour, for the entire run?

Answer can be found on page 351.

Rate × Time = Work: Define the Work Unit and Add Rates

Work problems are very similar to Rate–Time–Distance problems. The main difference is that work takes the place of distance:

$$\text{Rate} \times \text{Time} = \text{Distance} \qquad\qquad \text{Rate} \times \text{Time} = \text{Work}$$

$$20\,\frac{\text{miles}}{\text{hour}} \times 3 \text{ hours} = 60 \text{ miles} \qquad 20\,\frac{\text{chairs}}{\text{hour}} \times 3 \text{ hours} = 60 \text{ chairs}$$

Define work by the task done. It could be building chairs, painting houses, manufacturing soda cans, etc. One unit of output (1 chair built, 1 house painted, 1 can produced) is one unit of work.

If the "job" is to paint a house or fill a warehouse, then doing the full job once is one unit of work. Occasionally, it can be helpful to invent small units of work ("widgets") so that you avoid dealing with fractions. If the problem references "half a warehouse per day," you could say that a warehouse contains 10 boxes. The rate then becomes 5 boxes per day.

As before, always put time in the denominator. Translate *it takes Scout 3 minutes to build a chair* as 1 chair per 3 minutes. As a rate, this translates to $\frac{1}{3}$ of a chair per minute:

$$\text{Rate} = 1 \text{ chair per 3 minutes} \qquad \text{Rate} = \frac{1 \text{ chair}}{3 \text{ minutes}} = \frac{1}{3} \text{ chair per minute}$$

If two people or machines work at the same time side by side, you can *add* their rates. Try this problem:

> Jay can build a chair in 3 hours. Kay can build a chair in 5 hours. How long will it take both of them, working together, to build 8 chairs?

First, focus on Jay. What is Jay's rate of work? Put time in the denominator:

$$\text{Jay's rate} = \frac{1 \text{ chair}}{3 \text{ hours}} = \frac{1}{3} \text{ chair per hour} \qquad\qquad \text{T } \frac{1}{3} = A \quad B = \frac{1}{5} \text{ T}$$

In 1 hour, Jay can build $\frac{1}{3}$ of a chair.

Now, figure out Kay's rate of work:

$$\text{Kay's rate} = \frac{1 \text{ chair}}{5 \text{ hours}} = \frac{1}{5} \text{ chair per hour}$$

In 1 hour, Kay can build $\frac{1}{5}$ of a chair.

Together, then, Jay and Kay can build $\frac{1}{3} + \frac{1}{5}$ of a chair in an hour. This is the "adding rates" principle in action.

Find the sum of the fractions:

$$\overset{5}{\underset{3}{\frac{1}{}}} + \overset{}{\underset{5}{\frac{1}{}}}^{3} = \frac{8}{15} \text{ chair per hour}$$

Together, Jay and Kay build $\frac{8}{15}$ of a chair in 1 hour. That's an annoying fraction; what easier fraction is it close to? It's about one-half, so they can build a little bit more than half of a chair in one hour.

How long will it take them to build 8 chairs? Since they can build a bit more than half of a chair in an hour, then in 2 hours they can build 1 whole chair and even get started on the next one. To build 8 chairs, then, will take somewhat less than 16 hours. Depending on your answers, that estimate might be enough already.

Use the full Rate × Time = Work equation to find the exact value (if necessary). The work is 8 chairs. The time is unknown. Therefore:

$$R \times T = W$$
$$\frac{8}{15} \times T = 8$$
$$T = 8\left(\frac{15}{8}\right)$$
$$T = 15$$

It takes Jay and Kay, working together, 15 hours to build 8 chairs.

As with Rate–Time–Distance problems, keep the various quantities separate. If you need to use the same equation more than once, distinguish the different cases clearly. For instance, there could be three cases in this problem: Jay working alone, Kay working alone, or the two of them working together.

If you...	Then you...	Like this:
See a work problem	Use Rate × Time = Work, choosing work units and adding rates when appropriate	$7\,\dfrac{goblets}{hr} \times 3\,hrs = 21\,goblets$

Check Your Skills

9. It takes Alpha 6 hours to build a shelf. Beta can do the same work twice as fast. How many shelves can Alpha and Beta, working together, build in a 24-hour period?

$\frac{1}{6}T \qquad \frac{1}{3}r$

Answer can be found on page 351.

$8 + 4 = 12$

Here are some common translation terms and expressions that signal rates or work:

Rates & Work

Five dollars every two weeks: $\dfrac{5\ dollars}{2\ weeks} \longrightarrow 2.5$ dollars a week

The technician takes 3 hours to complete the job: $Rate = \dfrac{1\ job}{3\ hours} = \dfrac{1}{3}$ job per hour

Machine A and Machine B work simultaneously: Let A = rate of Machine A and B = rate of Machine B: Combined rate $= A + B$

Check Your Skills Answer Key

1. **45:** Translate the text into math.

$$\text{Eq. 1: } a + b = 75$$
$$\text{Eq. 2: } a = b - 15$$

The second translation is trickier. Which room can hold more people, A or B? Room B is bigger, so subtract 15 from room B to get the capacity in room A. (Alternatively, you could add 15 people to room A to get room B: $a + 15 = b$.)

The second equation already isolates a. Substitute it into the first equation, and solve for b.

$$(b - 15) + b = 75$$
$$2b - 15 = 75$$
$$2b = 90$$
$$b = 45$$

2. **700:** Calculate how many donuts were sold over the first 5 days, then add that to the number sold in the last 4 days.

$$\text{First 5 days: } 80 \times 5 = 400 \text{ donuts}$$
$$\text{Next 4 days: } 75 \times 4 = 300 \text{ donuts}$$
$$\text{Total donuts: } 400 + 300 = 700 \text{ donuts}$$

3. **\$7,000:** First, Understand. When the problem mentions something for which you know a standard formula (in this case, average), write down that standard formula. Also, when the problem has multiple time periods, organize the information carefully.

First 3 months:

$$\text{Average monthly pay} = \frac{\text{Total pay for 3 months}}{\text{Months}}$$

$$\$10,000 = \frac{\text{Total pay for 3 months}}{3}$$

The total pay for the first 3 months is $\$10,000 \times 3 = \$30,000$.

Last 9 months:

$$\text{Average monthly pay} = \frac{\text{Total pay for 9 months}}{\text{Months}}$$

$$\$6,000 = \frac{\text{Total pay for 9 months}}{9 \text{ months}}$$

The total pay for the last 9 months is $\$6,000 \times 9 = \$54,000$.

Next, find the average for the overall 12-month period.

$$\text{Average monthly pay} = \frac{\text{Total pay for all months}}{\text{\# of months}}$$

$$\text{Average monthly pay} = \frac{30{,}000 + 54{,}000}{3 + 9}$$

$$\text{Average} = \frac{\overset{7}{\cancel{8}4{,}000}}{\cancel{12}} = 7{,}000$$

4. **10:** The candy shop sold 50 candy bars in January for \$3 each.

$$\text{January revenue} = \text{Price} \times \text{Quantity} = \$3 \text{ per bar} \times 50 \text{ bars} = \$150$$

In February, the shop made \$10 more, or \$160. The candy bars cost \$1 more, or \$4 each. Solve for the number of candy bars sold, b.

$$\$160 = \$40 \times b$$
$$40 = b$$

The shop sold 50 bars in January but only 40 bars in February, so the shop sold $50 - 40 = 10$ fewer bars in February.

5. **\$4:** The first sentence provides two different equations. Translate both.

$$7S + 5C = \$36 \qquad\qquad 8S + 4C = \$40$$

Your first instinct might be to simplify the second equation by dividing everything by 4. It's great to notice something like that, but don't jump to solving yet—you're still in the Understand phase. (As a general rule on this test, don't do work until you have to.)

The question stem indicates that a stapler costs more than a coffee mug. Let S equal the price of one stapler, and let C equal the price of one coffee mug. The difference in price is $S - C$, so the question is $S - C = ?$

When you're asked for a combo, or combination of variables, examine the math to see whether you can solve directly for the entire combo (in this case, $S - C$). If you can, that will be faster than finding S and C individually.

How can you combine the two equations to leave you with one positive S and one negative C? Subtract the first equation from the second equation.

$$
\begin{array}{r}
8S + 4C = 40 \\
-(7S + 5C = 36) \\
\hline
S - C = 4
\end{array}
$$

A stapler costs \$4 more than a coffee mug. You might be thinking that it would never occur to you to do this yourself. Maybe that was true—before! Now that you know it's possible, though, you can train yourself to think about it. In the future, take a moment during your Plan phase to see whether this type of shortcut exists; it often does. If not, you can solve for the variables individually, with substitution or elimination.

6. **336:** 1 week = 7 days and 1 day = 24 hours.

$$2 \text{ weeks} \times \frac{7 \text{ days}}{1 \text{ week}} \times \frac{24 \text{ hours}}{1 \text{ day}} = 336 \text{ hours}$$

To multiply 14 and 24 without a calculator, break the calculation into two parts:

$$14 \times 24 = \begin{matrix} 10 & \times & 24 & = & 240 \\ 4 & \times & 24 & = & 96 \end{matrix} = 240 + 96 = 336$$

7. **3,840:** Convert 1 hour to seconds and 4 minutes to seconds.

$$1 \text{ hours} \times \frac{60 \text{ minutes}}{1 \text{ hour}} \times \frac{60 \text{ seconds}}{1 \text{ minute}} = 3,600 \text{ seconds}$$

$$4 \text{ minutes} \times \frac{60 \text{ seconds}}{1 \text{ minute}} = 240 \text{ seconds}$$

Therefore, 3,600 seconds + 240 seconds = 3,840 seconds.

8. **2.5:** To find Alaitz's average speed, find the total distance and the total time.

Total distance = 24 miles + 6 miles = 30 miles

Total time = time for first part of trip + 4 hours

Now, find the time Alaitz took for the first part of the trip.

Rate × Time = Distance

3 miles per hour × t = 24 miles

t = 8 hours

The total time is 8 + 4 = 12 hours. Finally, calculate Alaitz's average speed.

$$\text{Average} = \frac{\text{Total distance}}{\text{Total time}}$$

$$a = \frac{30 \text{ miles}}{12 \text{ hours}} = 2.5 \text{ miles per hour}$$

9. **12:** First, find Alpha's rate, using the formula Rate × Time = Work.

$R \times 6$ hours = 1 shelf

$$R = \frac{1 \text{ shelf}}{6 \text{ hours}}$$

Alpha's rate is $\frac{1}{6}$ shelf per hour. Beta can work twice as fast, so Beta's rate is $2 \times \frac{1}{6} = \frac{1}{3}$ shelf per hour.

Add the individual rates to find their combined rate.

$$\frac{1}{6} + \frac{1}{3} = \frac{1}{6} + \frac{2}{6} = \frac{3}{6} = \frac{1}{2}$$

Together, they can build $\frac{1}{2}$ of a shelf per hour. Finally, calculate how much work is completed in 24 hours at their combined rate:

$$\frac{1}{2} \text{ shelf per hour} \times 24 \text{ hours} = 12 \text{ shelves}$$

Chapter Review: Drill Sets

Drill 1

Translate and solve the following problems.

1. If −5 is 7 more than *z*, what is $\frac{z}{4}$?

2. The total weight of two jugs of milk is 2.2 kilograms. The lighter jug weighs 0.4 kilogram less than the heavier jug. What is the weight of the lighter jug?

3. Norman is 12 years older than Mica. In 6 years, he will be twice as old as Mica. How old is Norman now?

4. Three lawyers each earn an average (arithmetic mean) of $300 per hour. How much money have they earned in total after each has worked 4 hours?

5. A clothing store bought a container of 100 shirts for $20. If the store sold all of the shirts at $0.50 per shirt, what is the store's gross profit on the entire container?

6. The average of 2, 13, and *x* is 10. What is *x* ?

    (A)   7
    (B)   11
    (C)   15

7. Four children collect 33 candies that fall out of a piñata. If 3 of the children pick up the same number of candies and the fourth child picks up three fewer candies than each of the other children, how many candies does the fourth child collect?

    (A)   6
    (B)   9
    (C)   12

Drill 2

Translate and solve the following problems.

8. To put on a concert, a band pays $10,000 to rent a venue and another $15,000 for security; the band has no other costs. Attendees at the concert pay an average price of $40 for a ticket to the concert. If everyone who attends the concert must purchase a ticket, how many tickets must the band sell to make a gross profit of $7,000 on the concert?

9. Toshi is 7 years older than Kado, who is twice as old as Juni. If Juni is 8 years old, how old is Toshi?

10. A plane left Chicago in the morning and made 3 flights before returning to Chicago. The plane traveled twice as far on the first flight as on the second flight, and the plane traveled three times as far on the second flight as on the third flight. If the third flight was 45 miles, how many miles was the first flight?

11. It costs a certain bicycle factory $10,000 to operate for one month, plus $300 for each bicycle produced during the month. Each of the bicycles sells for a retail price of $700. The gross profit of the factory is measured by total income from sales minus the production costs of the bicycles and the factory operation cost. If 50 bicycles are produced and sold during the month, what is the factory's gross profit?

12. Arnaldo earns $11 for each ticket that he sells, plus a bonus of $2 per ticket for each ticket he sells over the first 100 tickets. If Arnaldo was paid $2,400, how many tickets did he sell?

13. If the average (arithmetic mean) of the five numbers *x* − 3, *x*, *x* + 3, *x* + 4, and *x* + 11 is 45, what is the value of *x* ?

14. Ten years ago, Sana was half as old as Byron. If Byron is now 15 years older than Sana, how old will Sana be in 7 years?

    (A) 25

    (B) 32

    (C) 40

15. Jordan buys 5 books with an average (arithmetic mean) price of $9. If Jordan then buys another book with a price of $15, what is the average price of the 6 books?

    (A) $10

    (B) $14

    (C) $18

16. Alicia is producing a magazine issue that costs $3 per magazine to print. In addition, she has to pay $10,500 to her staff to design the issue. If Alicia sells each magazine for $10, how many magazines must she sell to break even?

Drill 3

Translate and solve the following unit conversion problems.

17. What is the temperature in Fahrenheit when it is 30 degrees Celsius? $\left[C = \frac{5}{9}(F - 32) \right]$

18. How many minutes are there in 10 days?

19. On her bicycle, Miriam travels 50 yards in 10 seconds. How many feet does she travel in 2 minutes? (1 yard = 3 feet)

    (A) 500

    (B) 900

    (C) 1,800

20. A recipe calls for 1.6 cups of sugar and 2 quarts of flour. How many gallons does the resulting mixture of sugar and flour measure? (1 gallon = 4 quarts; 1 quart = 4 cups) Leave your answer in decimal form. (For you chefs: Ignore the difference between dry measures and liquid measures.)

21. How many 1-inch-square tiles would it take to cover the floor of a closet that has dimensions 5 feet by 2 feet? (1 foot = 12 inches)

Drill 4

Translate and solve the following rate problems.

22. A pool has sprung a leak and is losing water at a rate of 5 milliliters per second. How many liters of water is this pool losing per hour? (1 liter = 1,000 milliliters)

23. Jiang drove away from Marksville at a constant speed of 64 miles per hour. How many miles was Jiang from Marksville after 2 hours and 15 minutes of driving?

    (A) 128

    (B) 144

    (C) 176

24. Tyrone began the drive from Billington to Camville at 7:30. He drove at a constant speed of 40 miles per hour for the first hour and 50 miles per hour after that. If the distance from Billington to Camville is 160 miles, at what time did Tyrone arrive in Camville?

    (A) 10:30

    (B) 10:54

    (C) 11:00

25. If Roger took 2 hours to walk to a store that is 3 miles away, and then ran home along the same path in 1 hour, what was Roger's average rate, in miles per hour, for the round trip?

26. Sigi and Rob began running a 10-mile path around a lake at the same time. Sigi ran at a constant rate of 8 miles per hour. Rob ran at a constant rate of $6\frac{2}{3}$ miles per hour. Sigi finished running the 10-mile path how many minutes sooner than Rob did?

9

27. Svetlana ran the first 5 kilometers of a 10-kilometer race at a speed of 12 kilometers per hour. At what speed, in kilometers per hour, will she have to run the last 5 kilometers of the race if she wants to complete the 10-kilometer race in 55 minutes?

Drill 5

Translate and solve the following work problems.

28. A standard machine can fill 15 gallons of paint per hour. A deluxe machine fills gallons of paint at twice the rate of a standard machine. How many hours will it take a deluxe machine to fill 150 gallons of paint?

29. Machine A produces 15 widgets per minute. Machine B produces 18 widgets per minute. How many widgets will the machines produce together in 20 minutes?

30. At 2:00 p.m., a hose was placed into an empty pool and turned on. The pool, which holds 680 gallons of water, reached its capacity at 5:24 p.m. How many gallons of water per hour did the hose add to the pool?

31. Machine X, working alone at a constant rate, can produce a certain number of chocolates in 5 hours. Machine Y, working alone at a constant rate, can produce the same number of chocolates in 2 hours. If Machine X produces 180 chocolates per hour, how many chocolates does Machine Y produce per hour?

    (A)   72
    (B)   450
    (C)   900

Drill 6

Translate and solve the following word problems.

32. If Kayin's salary were 20% higher, it would be 20% less than Lorena's. If Lorena's salary is $60,000, what is Kayin's salary?

33. A $10 shirt is marked up by 30%, then by an additional 50%. What is the new price of the shirt?

34. A share of Stock Q increased in value by 20%, then decreased in value by 10%. The new value of a share of Stock Q is what percent of its initial value?

    (A)   108%
    (B)   110%
    (C)   118%

35. Akira currently weighs 160 pounds. If he must lose 8 pounds in order to qualify for a certain sporting event, what percent of his body weight must he lose in order to qualify?

    (A)   0.05%
    (B)   0.5%
    (C)   5%

Drill 7

Translate and solve the following word problems.

36. At a birthday party, children can choose 1 of the following 3 flavors of ice cream: chocolate, vanilla, or strawberry. If $\frac{1}{2}$ of the children choose chocolate, 20% of the children choose vanilla, and the remaining 15 children choose strawberry, how many children are at the party?

37. Lily stayed up last night to watch a meteor shower from her roof. Of the meteors visible from her roof, 10% were exceptionally bright, and of these, 80% inspired Lily to write a haiku. If Lily was inspired to write a haiku by 20 exceptionally bright meteors, how many meteors were visible from her roof last night?

38. An African elephant can lift 6% of its body weight using its trunk alone. If an African elephant weighs 1,000 times as much as a white-handed gibbon, how many gibbons can an African elephant lift at once with its trunk?

39. Last year, Country X received $\frac{7}{4}$ as much precipitation as Country Y, which received $\frac{2}{3}$ as much precipitation as Country Z. If Country X received 280 centimeters of precipitation, how much precipitation, in centimeters, did Country Z receive?

    (A) 160
    (B) 240
    (C) 420

40. In Farrah's workday playlist, $\frac{1}{3}$ of the songs are jazz, $\frac{1}{4}$ are R & B, $\frac{1}{6}$ are rock, $\frac{1}{12}$ are country, and the remainder are world music. What fraction of the songs in Farrah's playlist are world music?

41. At a music convention, $\frac{2}{5}$ of the attendees play no musical instrument, $\frac{1}{4}$ play exactly one musical instrument, $\frac{3}{10}$ play exactly two musical instruments, and the remaining 8 attendees play three or more musical instruments. How many people are attending the convention?

42. Every junior at Central High School studies exactly one language: 75 percent of the juniors study Gaelic, one-sixth of the juniors study Spanish, and the other 7 juniors study Tagalog. How many juniors are there at Central High School?

Drill 8

Translate and solve the following word problems.

43. Yemi wants to buy a blue umbrella for no more than $25. At a certain store, $\frac{3}{5}$ of the umbrellas cost more than $25, and of the remaining umbrellas, $\frac{7}{8}$ are not blue. If the store has 400 umbrellas, how many of the umbrellas meet Yemi's requirements?

    (A) 20
    (B) 140
    (C) 210

44. To make one serving of her signature punch, Mariko mixes $\frac{1}{2}$ cup of grape juice, $\frac{3}{4}$ cup of passion fruit juice, and $\frac{1}{8}$ cup of sparkling water. If Mariko makes 22 cups of punch, how many servings of punch will there be?

    (A) 16
    (B) 20
    (C) 30

45. Of the movies in Santosh's collection, $\frac{1}{3}$ are animated features, $\frac{1}{4}$ are live-action features, and the remainder are documentaries. If $\frac{2}{5}$ of the documentaries are depressing, what fraction of the films in Santosh's collection are depressing documentaries?

46. Of all the homes on Park Avenue, $\frac{1}{3}$ are termite-ridden, and $\frac{3}{5}$ of these are collapsing. What fraction of the homes are termite-ridden but NOT collapsing?

47. A bag contains only red and green marbles. Of these, $\frac{3}{4}$ of the marbles in the bag are green. Of the green marbles, $\frac{1}{3}$ are cracked. If there are 6,000 red marbles in the bag, how many cracked green marbles are there?

9

Drill Sets Solutions

Drill 1

1. **−3:** Check the relationship logically. The question says −5 is more than z, so add 7 to z when translating.

$$-5 = z + 7$$
$$-12 = z$$
$$\frac{z}{4} = \frac{-12}{4} = -3$$

2. **0.9 kg:** Let H = the weight of the heavier jug and L = the weight of the lighter jug. First, translate the two equations given in the problem and note down the question.

$$H + L = 2.2 \qquad L = H - 0.4 \qquad L = ?$$

Then, decide how to solve. Since the question asks for L, one option is to isolate H and substitute into the other equation.

$$H + L = 2.2 \leftarrow L + 0.4 = H$$
$$(L + 0.4) + L = 2.2$$
$$2L + 0.4 = 2.2$$
$$2L = 1.8$$
$$L = 0.9$$

3. **18:** Make a table to keep track of the two people and the two points in time. Put a big circle around the element you need to solve for.

	Now	+6 yrs
Norman	N	$N + 6$
Mica	M	$M + 6$

Add translations from the text to your table. For the +6 years column, make sure to use $N + 6$ and $M + 6$, not just N and M.

	Now	+6 yrs
Norman	$N = M + 12$	$N + 6 = 2(M + 6)$
Mica	M	$M + 6$

Solve for N, Norman's age now. First, pull the two equations from the table; in one of them, isolate M (the variable you *don't* want).

$$(N + 6) = 2(M + 6)$$
$$N - 12 = M$$

Then substitute and solve for N.

$$N + 6 = 2\big((N - 12) + 6\big)$$
$$N + 6 = 2(N - 6)$$
$$N + 6 = 2N - 12$$
$$18 = N$$

One more thing: Want to see a neat shortcut? This shortcut works whenever the problem tells you these two facts: first, that the two people are a certain number of years apart (e.g., A is 5 years older than B), and second, that one person is twice as old as the other at a certain point in time (e.g., 4 years ago, A was twice as old as B).

In this problem, Norman is 12 years older than Mica. In 6 years, he will be twice as old as Mica. At that point in time, the younger person's age is equal to the difference in their ages and the older person is exactly twice that. In this case, in 6 years, Mica will be 12 (the difference in their ages) and Norman will be 24 (twice Mica's age). Today, six years earlier, Norman is only $24 - 6 = 18$.

4. **$3,600:** The lawyers each earn the same amount, so calculate what one lawyer earns in 4 hours, then multiply by 3.

In 4 hours, one lawyer earns: $\quad 4 \times \$300 = \$1,200$

The three lawyers together earn: $\quad \$1,200 \times 3 = \$3,600$

5. **$30:** Pretend this is your store and logic it out. How would you figure out how much gross profit you made today?

Profit equals revenue minus cost. You sold 100 shirts for $0.50 each, so your revenue was $(100)(0.5) = \$50$. Your direct cost was $20. Therefore, your profit was $\$50 - \$20 = \$30$.

6. **(C) 15:** The three values are 2, 13, and x. If you just had 13 and 2, the average would be halfway between, or 7.5. The actual average is 10, so the value of x must be greater than 7.5 in order to make the average higher. In fact, it must be greater than the actual average, 10, in order to pull the average up to 10. Eliminate answer (A).

To find the exact number, use the average formula.

$$\text{Average} = \frac{\text{Sum}}{\text{\# of terms}}$$
$$10 = \frac{2 + 13 + x}{3}$$
$$30 = 2 + 13 + x$$
$$30 = 15 + x$$
$$15 = x$$

By the way, you can take the logic-it-out approach even further, if you're comfortable thinking about averages. The 2 value is 8 "under" the average of 10. The 13 value is 3 "over" the average of 10. Averages have an interesting pattern: The "over–under" for any set of data points in an average has to be equal—the sum amount "under" the average has to balance with the sum amount "over" the average. Since the 2 value is 8 under, while the 13 value is only 3 over, the third value, x, has to be 5 over the average to balance out. In other words, the value of x must be $10 + 5 = 15$. This pattern always holds for averages.

7. **(A) 6:** Let *c* stand for each of the three children who pick up the same number of candies. Let *f* stand for the fourth child. Therefore:

$$\text{Eq. 1: } c + c + c + f = 33 \quad \text{or} \quad 3c + f = 33$$

$$\text{Eq. 2: } f = c - 3$$

If you were to substitute the second equation into the first, you would be solving for *c*, not *f*. Instead, rearrange the second equation to isolate *c*, then solve directly for *f*. The second equation becomes $f + 3 = c$.

$$3c + f = 33$$
$$3(f + 3) + f = 33$$
$$3f + 9 + f = 33$$
$$4f = 24$$
$$f = 6$$

Drill 2

8. **800:** Let *n* = the number of tickets sold. Recall that Profit = Revenue − Cost. The problem indicates that the band wants a profit of $7,000. What are the revenues and the costs?

Revenue: $40n$

Costs: $10{,}000 + 15{,}000 = 25{,}000$

Profit $= 7{,}000$

Profit $=$ Revenue $-$ Cost

$7{,}000 = 40n - 25{,}000$

$32{,}000 = 40n$

$\dfrac{32{,}000}{40} = n$ Divide top and bottom of fraction by 10.

$\dfrac{3{,}200}{4} = n$ Divide top and bottom of fraction by 4.

$800 = n$

9. **23:** Let $T =$ Toshi's age, $K =$ Kado's age, and $J =$ Juni's age.

$$T = K + 7$$
$$K = 2J$$
$$J = 8$$

Before solving, check that the logical relationships make sense. Toshi is older than Kado, and Kado is older than Juni. The equations agree with that logic, so move forward with the computations. Substitute and solve.

$$K = 2 \times J = 2 \times (8) = 16$$
$$T = K + 7 = (16) + 7 = 23$$

10. **270:** Let $F =$ the distance of the first flight, $S =$ the distance of the second flight, and $T =$ the distance of the third flight. What is F?

$$F = 2S$$
$$S = 3T$$
$$T = 45$$

Substitute and solve.

$$S = 3 \times (45) = 135$$
$$F = 2 \times (135) = 270$$

11. **$10,000:** Recall that Profit = Revenue − Cost. Set up equations to find the revenues and costs.

$$\text{Revenue} = 50 \times 700 = 35{,}000$$
$$\text{Cost} = 10{,}000 + (50 \times 300) = 10{,}000 + 15{,}000 = 25{,}000$$
$$\text{Profit} = \text{Revenue} - \text{Cost} = 35{,}000 - 25{,}000 = 10{,}000$$

12. **200:** Let x equal the total number of tickets sold. Therefore, $(x - 100)$ is equal to the number of tickets Arnoldo sold beyond the first 100.

$$11x + 2(x - 100) = 2{,}400$$
$$11x + 2x - 200 = 2{,}400$$
$$13x = 2{,}600$$
$$x = 200$$

13. **42:** You can use the multiply-by-5 shortcut to find 45×5, but don't do so immediately. In general, don't do math until you have to; you may be able to simplify before you have to multiply. Use the average formula to set up the math.

$$\frac{(x - 3) + (x) + (x + 3) + (x + 4) + (x + 11)}{5} = 45$$
$$\frac{5x + 15}{5} = 45$$
$$x + 3 = 45$$
$$x = 42$$

14. **(B) 32:** Let $S =$ Sana's age and $B =$ Byron's age. What is $S + 7$? Set up a table.

	−10 yrs	Now	+7 yrs
Sana	$S - 10$	S	$S + 7$
Byron	$B - 10$	B	$B + 7$

$$S - 10 = \frac{B - 10}{2}$$
$$B = S + 15$$

Substitute the second equation into the first, and solve for S.

$$S - 10 = \frac{(S + 15) - 10}{2}$$
$$2S - 20 = S + 5$$
$$S = 25$$

You're not quite done! The question asks for $S + 7$: $(25) + 7 = 32$.

Have you already done problem 3? This one has the same shortcut, but in disguise: Ten years ago, S was half as old as B, so B was twice as old as S. The two are 15 years apart. So 10 years ago, S was 15 (the difference in their ages) and B was 30. Today, then, S is 25 years old, and in 7 years, S will be 32 years old.

15. **(A) $10:** The average for 5 books is $9, so purchasing one more book for $15, should increase the average. However, the new average can't be greater than the cost of that one additional book, $15, so answer (C) can't be correct. In fact, since 5 books average $9 and only one averages $15, the new average should be closer to the original $9 than to $15. Answer (B) is too big, as well. The answer must be (A).

To find the exact value for the new average, find the sum Jordan spent on all 6 books. First, find the cost of the original 5 books.

$$\frac{\text{Sum}}{\text{\# of terms}} = \text{Average}$$
$$\text{Sum} = \text{Average} \times \text{\# of terms}$$
$$= 9 \times 5$$
$$= 45$$

Next, find the sum of the cost for all 6 books: $45 + $15 = $60. Then, find the new average.

$$\text{New average} = \frac{\$60}{6} = \$10$$

16. **1,500:** Let m = the number of magazines sold.

$$\text{Total cost} = 3m + 10,500$$
$$\text{Total revenue} = 10m$$

Breaking even occurs when total revenue equals total cost. In other words, profit is $0. Therefore, set cost equal to revenue and solve for m.

$$3m + 10,500 = 10m$$
$$10,500 = 7m$$
$$1,500 = m$$

Drill 3

17. **86 degrees:** Use the conversion formula, replacing the variable C with the temperature in Celsius.

$$C = \frac{5}{9}(F - 32)$$

$$30 = \frac{5}{9}(F - 32)$$

$$\overset{6}{\cancel{30}} \times \frac{9}{\cancel{5}} = F - 32$$

$$54 = F - 32$$

$$86 = F$$

18. **14,400:** One day has 24 hours. Each of those hours has 60 minutes. Set up conversion ratios to solve.

$$10 \ \cancel{\text{days}} \times \frac{24 \ \cancel{\text{hours}}}{1 \ \cancel{\text{day}}} \times \frac{60 \ \text{minutes}}{1 \ \cancel{\text{hour}}} = 14,400 \ \text{minutes}$$

To make the math a little easier, first multiply 24 by 6, then add two zeros. One more note: $24 \times 6 = 12 \times 2 \times 6 = 12 \times 12$. You can always rearrange multiplication in this manner to make it easier. And $12 \times 12 = 144$ is one of your perfect squares to memorize.

19. **(C) 1,800:** You have answer choices, so logic it out before diving into the computations. If Miriam can travel 50 yards in 10 seconds, she goes 150 feet in 10 seconds. Therefore, she goes 1,500 feet in 100 seconds. But 100 seconds isn't 2 minutes yet, so she has farther to go. Glance at the answers; only choice (C) is possible.

Here are the exact calculations.

$$\frac{50 \ \cancel{\text{yards}}}{10 \ \cancel{\text{seconds}}} \times \frac{3 \ \text{feet}}{1 \ \cancel{\text{yard}}} \times \frac{60 \ \cancel{\text{seconds}}}{1 \ \text{minute}} = 50 \times 3 \times 6 = 900 \frac{\text{feet}}{\text{min}}$$

Don't stop yet! Answer (B) is a trap. The question asks how far Miriam can travel in 2 minutes, not just 1 minute. If she can travel 900 feet in 1 minute, then she can travel $900 \times 2 = 1,800$ feet in 2 minutes.

20. **0.6:** Convert both 1.6 cups and 2 quarts into gallons using conversion ratios:

$$1.6 \ \cancel{\text{cups}} \times \frac{1 \ \cancel{\text{quart}}}{4 \ \cancel{\text{cups}}} \times \frac{1 \ \text{gallon}}{4 \ \cancel{\text{quarts}}} = \frac{1.6}{16} \ \text{gallons} = 0.1 \ \text{gallons}$$

$$2 \ \cancel{\text{quarts}} \times \frac{1 \ \text{gallon}}{4 \ \cancel{\text{quarts}}} = \frac{2}{4} \ \text{gallons} = 0.5 \ \text{gallons}$$

$$0.1 \ \text{gallons} + 0.5 \ \text{gallons} = 0.6 \ \text{gallons}$$

21. **1,440:** There is a hidden trap in this question. The dimensions of this room are in square feet, not feet (because 5 feet × 2 feet = 10 square feet). To avoid this trap, convert the dimensions to inches first, then multiply:

$$5 \text{ feet} \times \frac{12 \text{ inches}}{1 \text{ foot}} = 60 \text{ inches}$$

$$2 \text{ feet} \times \frac{12 \text{ inches}}{1 \text{ foot}} = 24 \text{ inches}$$

The dimensions of the closet in inches are 60 inches by 24 inches, or 60 × 24 = 1,440 square inches. Each tile is 1 square inch, so it will take 1,440 tiles to cover the floor. Note: To break 60 × 24 into a more manageable calculation, try (60 × 20) + (60 × 4) = 1,200 + 240 = 1,440.

Drill 4

22. **18:** There is no mandatory order for processing the conversions. Start with 5 milliliters per second, and make the appropriate conversions.

$$\frac{5 \text{ milliliters}}{\text{second}} \times \frac{60 \text{ seconds}}{1 \text{ minute}} \times \frac{60 \text{ minutes}}{1 \text{ hour}} = \frac{5 \times 60 \times 60 \text{ milliliters}}{\text{hour}}$$

Don't multiply yet. Continue with the conversion to see whether you can simplify before you multiply.

$$\frac{5 \times 60 \times 60 \text{ milliliters}}{\text{hour}} \times \frac{1 \text{ liter}}{1,000 \text{ milliliters}}$$

$$= \frac{\overset{1}{5} \times \overset{3}{6} \times 6 \text{ liters}}{\underset{1}{\cancel{\underset{\cancel{2}}{10}}} \text{ hours}}$$

$$= \frac{18 \text{ liters}}{\text{hour}}$$

23. **(B) 144:** The answers are decently far apart, so begin by estimating. At a rate of 64 miles per hour, Jiang would go 64 miles in 1 hour, 64 × 2 = 128 miles in 2 hours, and 64 × 3 = 192 in 3 hours. The correct answer is not (A), 128, because Jiang actually drove for 2 hours and 15 minutes, not 2 hours.

Answer choice (B) is closer to 2 hours than 3, and answer (C) is closer to 3 hours. Since you're looking for an answer that is just 15 minutes past 2 hours, the answer must be (B).

If you had to solve exactly, you could use the $D = RT$ formula. But you could also continue to logic it out. 15 minutes is one-quarter of a full hour (60 minutes). At a rate of 64 miles per hour, then, Jiang can go $\frac{64}{4} = 16$ miles. In 2 hours and 15 minutes, Jiang travels 128 + 16 = 144 miles.

24. **(B) 10:54:** Logic this out. Tyrone had to cover a total of 160 miles. For the first hour only, Tyrone drove 40 miles per hour (mph), so he drove a total of 40 miles in that first hour. During the second hour, he drove 50 mph and covered another 50 miles, for a total of 90 miles so far. During the third hour, he added another 50 miles, for a total of 140 miles. What time is it now? He started at 7:30 and it's 3 hours later, so it's 10:30 and he hasn't covered the necessary 160 miles. Eliminate answer (A).

    Glance at the answers. The next "nice" number is 11:00, or another half an hour. At 50 mph, Tyrone goes 25 miles in half an hour, so at 11:00 he will have gone a total of $140 + 25 = 165$ miles. That's too far! The answer can't be (C) either; it must be (B).

25. **2:** Find the average rate by dividing the total distance traveled by the total time spent traveling. In this case, Roger traveled 3 miles to the store and 3 miles back, covering a total of 6 miles in 3 hours.

$$R = \frac{(3+3)\text{ miles}}{(2+1)\text{ hours}} = \frac{6\text{ miles}}{3\text{ hours}} = 2\frac{\text{miles}}{\text{hour}}$$

26. **15:** Use the $D = RT$ equation to calculate how long it took each individual to run the path.

    Sigi ran the 10-mile path at a rate of 8 miles per hour. Set up an equation and solve for Sigi's time.

D (mi)	$=$	R (mi/hr)	$\times$	T (hr)
10	$=$	8	$\times$	t

$$10 = 8t$$

$$\frac{5}{4} = t$$

Therefore, Sigi completed the path in 1 hour and 15 minutes.

Rob ran the 10-mile path at a rate of $6\frac{2}{3}$ mi/hr $= \frac{20}{3}$ mi/hr. Set up an equation and solve for Rob's time.

D (mi)	$=$	R (mi/hr)	$\times$	T (hr)
10	$=$	$\frac{20}{3}$	$\times$	t

$$10 = \frac{20}{3}t$$

$$\frac{3}{\underset{2}{\cancel{20}}} \times \cancel{10} = t$$

$$\frac{3}{2} = t$$

Therefore, Rob ran the path in 1 hour and 30 minutes.

Finally, subtract Sigi's time from Rob's time to calculate how much sooner Sigi finished.

$$\text{1 hour 30 min} - \text{1 hour 15 min} = \text{15 min}$$

27. **10:** In order to calculate Svetlana's speed during the second half of the race, first calculate how long it took her to run the first half of the race. The first half of the race is 5 kilometers in length, and Svetlana ran at a speed of 12 kilometers per hour. Set up an equation and solve for her time.

D (km)	=	R (km/hr)	×	T (hr)
5	=	12	×	t

$$5 = 12t$$

$$t = \frac{5}{12} \text{ hr}$$

To calculate the time Svetlana has available to run the second half of the race, subtract her time for the first half of the race from her goal time for the entire race. To do this calculation, first convert her goal time from minutes to hours.

$$55 \text{ min } \times \frac{1 \text{ hr}}{60 \text{ min}} = \frac{55}{60} \text{ hr} = \frac{11}{12} \text{ hr}$$

Then subtract Svetlana's time for the first half of the race from this value.

$$\frac{11}{12} \text{ hr} - \frac{5}{12} \text{ hr} = \frac{6}{12} \text{ hr} = \frac{1}{2} \text{ hr} = 0.5 \text{ hr}$$

Svetlana must complete the second 5 kilometers in 0.5 hours. Logic it out: If Svetlana covers 5 kilometers in half an hour, then she would cover 10 kilometers in one hour; that is, her rate is 10 km/h.

Drill 5

28. **5:** Logic this out. The standard machine fills 15 gallons of paint in 1 hour, so to fill 150 gallons would take 10 hours. The deluxe machine works twice as fast, so it will take half as long, or 5 hours.

Alternatively, calculate the rate of the deluxe machine by multiplying the rate of the standard machine by 2. If the standard machine can fill 15 gallons per hour, then the deluxe machine can fill 30 gallons per hour.

Use the $W = RT$ formula to solve for the amount of time the machine takes to fill 150 gallons of paint.

W (gal)	=	R (gal/hr)	×	T (hr)
150	=	30	×	t

$$150 = 30t$$

$$t = \frac{15\cancel{0}}{3\cancel{0}} = \frac{15}{3} = 5$$

The "textbook math" approach takes a lot longer. If you don't feel comfortable using logic, go back and try some easier problems the logical way until you start feeling comfortable thinking your way through the math. This will help you not only on the test but also in school and even at work.

29. **660:** Since the two machines are working together, add their rates. Together, they produce $15 + 18 = 33$ widgets per minute. In 10 minutes, they produce 330 widgets. In 20 minutes, they produce twice that, or 660 widgets.

30. **200:** First, find the time it took to fill the pool.

$$5{:}24 \text{ p.m.} - 2{:}00 \text{ p.m.} = 3 \text{ hours } 24 \text{ minutes}$$

Next, convert the minutes portion of this time to hours.

$$24 \text{ min} \times \left(\frac{1 \text{ hr}}{60 \text{ min}}\right) = \frac{24}{60} \text{ hr} = \frac{2}{5} \text{ hr} = 0.4 \text{ hr}$$

It took 3.4 hours to fill the pool, and the capacity of the pool is 680 gallons. Use the $W = RT$ equation to solve for the rate.

W (gal)	$=$	R (gal/hr)	$\times$	T (hr)
680	$=$	r	$\times$	3.4

$$680 = 3.4r$$

$$r = \frac{680}{3.4} = \frac{6800}{34} = \frac{\overset{2}{\cancel{68}}\,00}{\cancel{34}} = 200 \frac{\text{gallons}}{\text{hour}}$$

31. **(B) 450:** Do a logic check first. Machine X is slower, taking 5 hours to do the same job as Machine Y can do in 2 hours. Therefore, Machine Y can produce more than 180 chocolates in an hour. Eliminate answer choice (A). Further, Machine Y takes less than half the time, so it is more than twice as fast. The answer should be more than $2 \times 180 = 360$.

To compute exactly, use the $W = RT$ formula to determine how many chocolates Machine X produces in 5 hours.

W (choc)	$=$	R (choc/hr)	$\times$	T (hr)
W	$=$	180	$\times$	5

$$W = 180 \times 5$$
$$W = 900$$

First, answer (C) is a trap. Machine Y is capable of producing the same number, 900 chocolates, but in 2 hours, not in 1 hour. Since you've already eliminated answer (A), the answer must be (B). It's quick to check in this case: If Machine Y produces 900 chocolates in 2 hours, then it must produce half that, or 450 chocolates, in 1 hour.

9

Drill 6

32. **$40,000:** *20% higher* is equal to 120%. Similarly, *20% less than* is equal to 80%.

One approach is to logic this out. L = 60,000. Start with this part of the story: *It would be 20% less than* 60,000. Since you need to multiply, change 80% to a fraction.

$$\frac{4}{5}(60,000) = \frac{4}{\cancel{5}}\left(\overset{12,000}{\cancel{60,000}}\right) = 48,000$$

Next, address this part of the story: *If Kayin's salary were 20% higher, it would be* 48,000. That is, if Kayin's actual salary were multiplied by 120%, it would equal 48,000.

$$\frac{6}{5}K = 48,000$$

$$K = 48,000\left(\frac{5}{6}\right)$$

$$K = \underset{8,000}{\cancel{48,000}}\left(\frac{5}{\cancel{6}}\right) = 40,000$$

Therefore, Kayin's actual salary is $40,000.

Alternatively, translate everything into one equation and solve algebraically.

$$\frac{6}{5}K = \frac{4}{5}(60,000)$$

$$K = \left(\frac{4}{5}\right)\left(\frac{5}{6}\right)(60,000)$$

$$K = \left(\frac{4}{\cancel{5}}\right)\left(\frac{\cancel{5}}{\cancel{6}}\right)\left(\overset{10,000}{\cancel{60,000}}\right)$$

$$K = 40,000$$

33. **$19.50:** When you are asked to multiply percents, it is usually best to use a fractional representation because you can then simplify before you multiply.

$$\$10 \times \frac{13\cancel{0}}{10\cancel{0}} \times \frac{15\cancel{0}}{10\cancel{0}}$$

$$= \$\cancel{10} \times \frac{13}{\cancel{10}} \times \frac{15}{10}$$

$$= \$13 \times \frac{3}{2}$$

$$= \$19.50$$

Or do the calculations one step at a time. First, since $3 is 30% of $10, the shirt is marked up to $10 + $3 = $13. Next, the shirt is marked up another 50%. That's the equivalent of taking half of the price and adding it to the original price: $13 + $6.50 = $19.50.

34. **(A) 108%:** Imagine that the initial price was \$100. When the stock increased by 20%, the resulting value was \$100 + \$20 = \$120. If the stock then went down 10%, the final value was \$120 − \$12 = \$108. The final percentage was $\frac{108}{100} = 108\%$. The advantage of choosing 100 as your starting number is not having to bother to do that last calculation; the dollar amount equals the new percentage.

Alternatively, set up an equation and solve.

$\frac{120}{100}$	×	$\frac{90}{100}$	×	Q	=	$\frac{x}{100}$	×	Q
120 percent	of	90 percent	of	Q	is	what percent	of	Q ?

$$\frac{120}{100} \times \frac{90}{100} \times Q = \frac{x}{100} \times Q \qquad \text{Divide both sides by Q and simplify.}$$

$$\frac{6}{5} \times \frac{9}{10} = \frac{x}{100}$$

$$100 \times \frac{6}{5} \times \frac{9}{10} = x$$

$$\overset{2}{\cancel{100}} \times \frac{6}{\cancel{5}} \times \frac{9}{\cancel{10}} = x$$

$$108 = x$$

35. **(C) 5%:** Use the percent change formula or calculate directly.
Direct calculation:

$160 = 100\%$	This is the starting fact.
$16 = 10\%$	Move the decimal one to the left to get 10.
$8 = 5\%$	Divide by 2 to get 5%.

$$\text{Percent change: } \% \text{ change} = \frac{\text{Change}}{\text{Original}}$$

$$\% \text{ change} = \frac{\overset{1}{\cancel{8}}}{\underset{2}{\cancel{160}}} = \frac{1}{20}$$

This value is on the to-memorize list from the Fractions chapter: $\frac{1}{20} = 5\%$. If you forget the conversion, you can also find the percentage by making the fraction *of 100* or *per cent*: $\frac{1}{20} = \frac{5}{100} = 5\%$.

Drill 7

36. **50:** Logic this out. If 50% chose chocolate and 20% chose vanilla, then 30% chose strawberry. Benchmark in reverse to find 100%.

$$30\% = 15$$
$$10\% = 5$$
$$100\% = 50$$

Alternatively, create variables to represent the number of kids who chose each flavor and relate them to the total number of children.

$$\text{Total kids} = t$$
$$\text{Kids who like chocolate} = c = 0.5t$$
$$\text{Kids who like vanilla} = v = 0.2t$$
$$\text{Kids who like strawberry} = s = 15$$

$$c + v + s = t$$
$$0.5t + 0.2t + 15 = t$$
$$0.7t + 15 = t$$
$$15 = 0.3t$$

Change that decimal to a fraction to make the math easier.

$$15 = \frac{3}{10}t$$
$$\left(\frac{10}{3}\right)\overset{5}{15} = t$$
$$50 = t$$

37. **250:** Let m equal the number of meteors visible from Lily's roof. The 20 meteors that inspired Lily represent 10% of 80% of the visible meteors.

$$20 = \frac{10}{100}\left(\frac{80}{100}\right)m$$
$$20 = \frac{1}{10}\left(\frac{4}{5}\right)m$$
$$\left(\frac{50}{4}\right)\left(\overset{5}{20}\right) = m$$
$$250 = m$$

38. **60:** The elephant weighs the same as 1,000 gibbons. Therefore, the elephant can lift 6% of 1,000 gibbons with its trunk.

$$\frac{6}{100} \times 1,000 \text{ gibbons} = 6 \times 10 \text{ gibbons} = 60 \text{ gibbons}$$

39. **(B) 240:** Let x equal the precipitation in Country X, y equal the precipitation in Country Y, and z equal the precipitation in Country Z. Translate the equations.

$$x = \frac{7}{4}y$$
$$y = \frac{2}{3}z$$

Since the question tells you the value of x and you are solving for the value of z, substitute the second equation into the first to simplify to a single equation with x and z.

$$x = \frac{7}{4}\left(\frac{2}{3}z\right)$$
$$x = \frac{7}{6}z$$

Now plug in for x and solve for z.

$$280 = \frac{7}{6}z$$
$$\left(\frac{6}{7}\right)\left(\overset{40}{\cancel{280}}\right) = z$$
$$(6)(40) = z$$
$$240 = z$$

40. $\frac{1}{6}$: In order to determine the fraction of the songs that are world music, first figure out what fraction of the songs are *not* world music. Add the fractions for the other types of music.

$$\text{Not world} = \frac{1}{3} + \frac{1}{4} + \frac{1}{6} + \frac{1}{12}$$
$$= \frac{4}{12} + \frac{3}{12} + \frac{2}{12} + \frac{1}{12}$$
$$= \frac{10}{12}$$
$$= \frac{5}{6}$$

Of the songs in Farrah's playlist, $\frac{5}{6}$ are *not* world music, so the remaining $\frac{1}{6}$ of the songs are world music.

41. **160:** Let $p =$ the total number of people attending the convention. In order to find the total number, determine what fraction of the total the 8 people (who play three or more instruments) represent, and then use that to find the total.

You could add up the three fractions and subtract from 1—but wait! Adding fractions is annoying because you have to find a common denominator. Take a look at those fractions again: They are all on the to-memorize conversion list. Convert them to percentages.

$$\frac{2}{5} = 40\%$$

$$\frac{1}{4} = 25\%$$

$$\frac{3}{10} = 30\%$$

The sum is $40\% + 25\% + 30\% = 95\%$. The 8 people who play three or more instruments must represent the remaining 5%. Use this fact to find 100%, or the total number of attendees.

$5\% = 8$	This is the starting fact.
$10\% = 16$	Double the number to get 10%.
$100\% = 160$	Add a 0 to get 100%.

42. **84:** Define variables and translate the information. Let $G =$ the number of juniors studying Gaelic, $S =$ the number of juniors studying Spanish, $T =$ the number of juniors studying Tagalog, and $J =$ the total number of juniors.

$$J = G + S + T = ?$$

One-sixth doesn't convert nicely into a percentage (it's not an integer), so instead translate 75% into a fraction. If $\frac{3}{4}$ of the juniors study G and $\frac{1}{6}$ study S, what fraction represents the 7 who study T?

$$\frac{3}{4} + \frac{1}{6} = \frac{9}{12} + \frac{2}{12} = \frac{11}{12}$$

Since $\frac{11}{12}$ study G or S, the remaining 7 studying T must represent $\frac{1}{12}$ of the juniors. Use that to determine the total number of juniors.

$$\frac{1}{12} = 7$$
$$\frac{12}{12} = 7 \times 12 = 84$$

Drill 8

43. **(A) 20:** First, determine how many umbrellas fit Yemi's price range. There are 400 umbrellas total, but $\frac{3}{5}$ cost more than the $25 that Yemi is willing to pay.

$$400\left(1 - \frac{3}{5}\right) = \text{umbrellas under \$25}$$

$$\overset{80}{\cancel{400}}\left(\frac{2}{\cancel{5}}\right) = \text{umbrellas under \$25}$$

$$160 = \text{umbrellas under \$25}$$

There are 160 *remaining* umbrellas, so answer (C) is out. According to the problem, $\frac{7}{8}$ of these are not blue, so only $\frac{1}{8}$ are under $25 and blue. Glance at the answers; only answer (A) makes sense.

$$\overset{2}{\cancel{16}}0\left(\frac{1}{\cancel{8}}\right) = \text{blue umbrellas under \$25}$$

$$20 = \text{blue umbrellas under \$25}$$

44. **(A) 16:** First, figure out how many cups of liquid are in one serving of Mariko's punch.

$$\frac{1}{2} + \frac{3}{4} + \frac{1}{8} = \frac{4}{8} + \frac{6}{8} + \frac{1}{8} = \frac{11}{8}$$

Each serving of punch consists of $\frac{11}{8}$ cups of liquid. (This is a weird number. Don't try to turn it into a mixed fraction yet. Glance at the answers: Something must cancel out later to make a "nicer" number, so just keep going.)

Next, $\frac{11}{8}$ times the number of servings equals 22 cups. Let *s* equal the number of servings and solve:

$$\frac{11}{8}s = 22$$

$$s = 22\left(\frac{8}{11}\right)$$

$$s = 2(8) = 16$$

45. $\frac{1}{6}$: Begin by finding the fraction of the movies that are documentaries. Let *M* equal the total number of movies and *d* equal the number of documentaries. You can simplify the math by multiplying both sides by the common denominator, 12.

$$M = \frac{1}{3}M + \frac{1}{4}M + d$$

$$(12)(M) = \left(\frac{1}{3}M + \frac{1}{4}M + d\right)(12)$$

$$12M = 4M + 3M + 12d$$

$$5M = 12d$$

$$\frac{5}{12}M = d$$

Of the movies, $\frac{5}{12}$ are documentaries. Of these, $\frac{2}{5}$ are depressing. Therefore:

$$\frac{\cancel{5}}{12}M \times \frac{2}{\cancel{5}} = \frac{2}{12}M = \frac{1}{6}M$$

Of the movies in Santosh's collection, $\frac{1}{6}$ are depressing documentaries.

Alternatively, because the problem discusses only fractions (never any real numbers), you can work through the problem using a real number. Choose a number that is a multiple of all of the denominators of the fractions in the problem. In this case, choose $3 \times 4 \times 5 = 60$.

Assuming that there are 60 total movies, then $\frac{1}{3}$, or 20 movies, are animated, and $\frac{1}{4}$, or 15 movies, are live-action. The remainder, or $60 - 20 - 15 = 25$, are documentaries.

Of the documentaries, $\frac{2}{5}$, or $25\left(\frac{2}{5}\right) = 10$, are depressing. These 10 depressing documentaries represent $\frac{10}{60}$, or $\frac{1}{6}$, of all of the movies.

46. $\frac{2}{15}$: Let h = the total number of homes. You know that $\frac{1}{3}$ of the homes are termite-ridden. If $\frac{3}{5}$ of the termite-ridden homes are collapsing, then $1 - \frac{3}{5} = \frac{2}{5}$ of the termite-ridden homes are NOT collapsing. Therefore:

$$\left(h \times \frac{1}{3}\right) \times \frac{2}{5} = \frac{2}{15}h$$

Alternatively, because the problem only mentions fractions (no real numbers), you can choose your own real number to solve. Choose a number that is a multiple of all of the denominators in the problem. In this case, choose $3 \times 5 = 15$.

Assuming that there are 15 homes total, then $\frac{1}{3}$, or 5 homes, are termite-ridden. Of these 5 homes, $\frac{3}{5}$, or 3 homes, are collapsing. That leaves 2 homes that are both termite-ridden and NOT collapsing; this represents $\frac{2}{15}$ of the total homes.

47. **6,000:** If $\frac{3}{4}$ of the marbles are green, then $\frac{1}{4}$ are red. The ratio of green to red marbles is 3 to 1, so if there are 6,000 red marbles, then there are 3 times as many green marbles. That is, there are $6,000 \times 3 = 18,000$ green marbles. Determine how many of these are cracked.

$$\overset{6,000}{\cancel{18,000}} \times \frac{1}{\cancel{3}} = 6,000$$

It may seem odd that you're back at 6,000. It is occasionally the case that the answer is the same as another number in the problem!

CHAPTER 10

Geometry

In This Chapter

- Circles
    - Know One Thing about a Circle: Know Everything Else
    - Sector: Slice of Pizza
- Triangles
    - Sum of Any Two Sides > Third Side
    - Sum of the Three Angles = 180°
    - Same Sides = Same Angles, and Vice Versa
    - Perimeter: Sum of Sides
    - Apply Area Formula: Any Side Can Be the Base
    - Know Two Sides of a Right Triangle: Find the Third Side
- Quadrilaterals
    - Parallelogram: Cut into Triangles OR Drop Height
    - Rectangles = Parallelogram + Four Right Angles
    - Squares = Rectangle + Four Equal Sides
- Geometry: Word Problems with Pictures
- Coordinate Plane—Position Is a Pair of Numbers
    - Know Just One Coordinate = Find a Line
    - Know a Range = Shade a Region
    - Read a Graph = Drop a Line to the Axes
    - Plot a Relationship: Given an x, Find a y
    - Lines in the Plane: Use Slope and y-intercept to Plot

In this chapter, you will learn about all of the key facts, rules, and formulas for the major shapes tested by the exam, including triangles, squares, rectangles, and circles. You'll also learn the necessary rules for the coordinate plane.

CHAPTER 10 Geometry

For many students, geometry brings to mind complicated shapes and the need to memorize lots of formulas. It's true that you will need to memorize some rules and formulas, but you can generally get away with limiting your studies to triangles, squares, rectangles, circles, and the coordinate plane.

On occasion, you may see a more complicated shape, such as a cylinder or a rhombus, but these shapes don't often appear on the GMAT. If you like, you can choose not to study the "weird" shapes and just make a guess if one happens to pop up on your test.

Circles

A circle is a set of points that are all the same distance from a central point. By definition, every circle has a center, usually labeled *O*, which is not itself a point on the circle. The **radius** of a circle is the distance between the center of the circle and a point on the circle. *Any* line segment connecting the center and *any* point on the circle is a radius (usually labeled *r*). All radii in the same circle have the same length. For example:

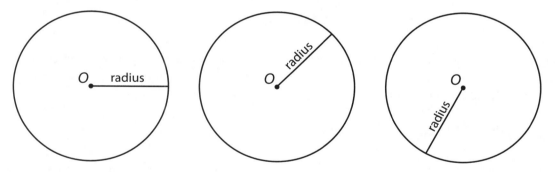

Know One Thing about a Circle: Know Everything Else

Now imagine a circle of radius 7. What else can you figure out about that circle?

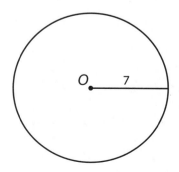

The next easiest thing to figure out is the **diameter** (usually labeled *d*), which passes through the center of a circle and connects two opposite points on the circle:

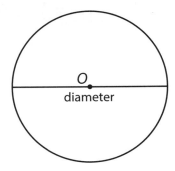

A diameter is two radii laid end to end, so it is always exactly twice the length of the radius. This relationship can be expressed as $d = 2r$. A circle with radius 7 has a diameter of 14.

The **circumference** (usually referred to as *C*) is a measure of the distance around a circle. The circumference is essentially the perimeter of a circle:

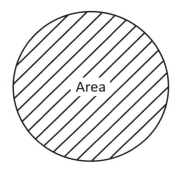

For any circle, the circumference and the diameter have a consistent relationship. If you divide the circumference by the diameter, you always get the same number: 3.14 … This number has decimals that continue forever, and it is indicated by the Greek letter π (pi). To recap:

$$\frac{\text{Circumference}}{\text{Diameter}} = \pi \quad \text{or} \quad \pi d = C$$

In a circle with a diameter of 14, the circumference is $\pi(14)$, or 14π. Most of the time, you will not approximate this as 43.96 (which is 14×3.14). Instead, keep it as 14π.

You can relate the circumference directly to the radius, since the diameter is twice the radius. This relationship is commonly expressed as $C = 2\pi r$. Be comfortable using either equation.

Finally, the **area** (usually labeled *A*) is the space inside the circle:

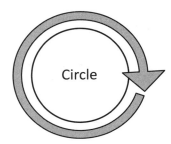

The area of a circle and its radius always have the same relationship. If you know the radius of the circle, then you can find the area using the formula $A = \pi r^2$. For a circle of radius 7, the area is $\pi(7)^2$, or 49π. *Once you know the radius, you can find the diameter, the circumference, and the area:*

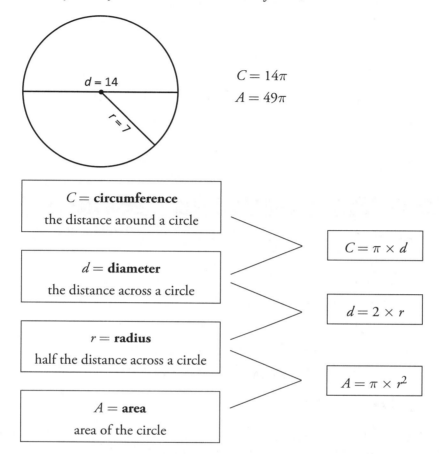

These relationships are true of any circle. What's more, *if you know any one of these values, you can determine the rest,* since they're all connected.

Say that the area of a circle is 36π. How do you find the other measures? Start with the formula for the area, which involves the radius:

$$36\pi = \pi r^2$$

Solve for the radius by isolating r:

$36\pi = \pi r^2$	Divide by π.
$36 = r^2$	Take the square root of both sides.
$6 = r$	

Now that you know the radius, multiply it by 2 to get the diameter, which is 12. Finally, to find the circumference, multiply the diameter by π. The circumference is 12π. You can fill in the measurements on your circle:

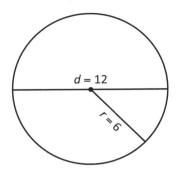

If you...	Then you...	Like this:
Know one thing about a circle	Can find out everything else about the circle by using the standard formulas	If $r = 4$, then $d = 8$, $C = 8\pi$, and $A = 16\pi$

Check Your Skills

1. The radius of a circle is 9. What is the area?

2. The circumference of a circle is 17π. What is the diameter?

3. The area of a circle is 25π. What is the circumference?

Answers can be found on page 420.

Sector: Slice of Pizza

Imagine that you have a circle with an area of 36π. Now cut it in half and make it a semicircle. Any fractional portion of a circle is known as a **sector**. Think of a sector as a slice of pizza:

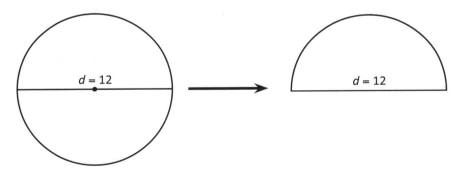

What effect does cutting the circle in half have on the basic elements of the circle? The diameter stays the same, as does the radius. But the area and the circumference are cut in half. The area of the semicircle is 18π, and the curved part of the semicircle's perimeter is 6π. When you deal with sectors, you call the remaining portion of the circumference the **arc length**. For this sector, the arc length is 6π.

10

If, instead of cutting the circle in half, you cut it into quarters, each piece of the circle would have $\frac{1}{4}$ the area of the entire circle and $\frac{1}{4}$ the circumference:

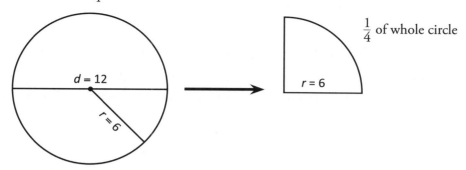

$\frac{1}{4}$ of whole circle

Now, on the GMAT, you're unlikely to be told that you have one-quarter of a circle. Rather, you would be told something about the **central angle**, which is the degree measure between two given radii. Take a look at the quarter circle. Normally, there are 360° in a full circle. What is the degree measure of the angle between the two radii? The same thing that happens to area and circumference happens to the central angle. It is now $\frac{1}{4}$ of 360°, which is 90° (you illustrate a 90° angle with a box at the angle). For example:

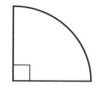

$$\frac{1}{4} = \frac{90°}{360°}$$

Let's see how you can use the central angle to determine sector area and arc length. Imagine that the original circle still has area 36π, but now the sector, or slice of pie, has a central angle of 60°:

What fractional amount of the circle remains if the central angle is 60°? The whole is 360°, and the part is 60°, so $\frac{60}{360}$ is the relevant fraction, which reduces to $\frac{1}{6}$. In other words, a sector with a central angle of 60° is $\frac{1}{6}$ of the entire circle. Here are the formulas for sector area and arc length:

Sector area = Fraction × Entire area = $\frac{1}{6} \times (36\pi) = 6\pi$

Arc length = Fraction × Entire circumference = $\frac{1}{6} \times (12\pi) = 2\pi$

$$\frac{1}{6} = \frac{60°}{360°} = \frac{\text{Sector area}}{\text{Circle area}} = \frac{\text{Arc length}}{\text{Circumference}}$$

10

In the last example, the central angle indicated the fractional amount that the sector represented. But any of the three properties of a sector, namely central angle, arc length, and area, could be used to convey that same information, since all three measures are related.

Try this problem:

> A sector has a radius of 9 and an area of 27π. What is the central angle of the sector?

You still need to determine the fraction of the circle that the sector represents. This time, however, use the radius to figure out the area of the whole circle. From that, you can figure out the fractional amount the sector represents:

$$\text{Area of the whole circle} = \pi r^2 = \pi(9)^2 = 81\pi$$

$$\frac{\text{Sector area}}{\text{Circle area}} = \frac{27\pi}{81\pi} = \frac{1}{3}$$

The sector is $\frac{1}{3}$ of the entire circle. The full circle has an angle of 360°, so multiply that by $\frac{1}{3}$ to find the angle of the sector:

$$\frac{1}{3} \times 360 = 120$$

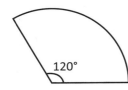

$$\frac{1}{3} = \frac{120°}{360°} = \frac{27\pi \; (\text{Sector area})}{81\pi \; (\text{Circle area})}$$

Every question about sectors will provide you with enough information to calculate one of the following fractions, which represent the sector as a fraction of the circle:

$$\frac{\text{Central angle}}{360} \qquad \frac{\text{Sector area}}{\text{Circle area}} \qquad \frac{\text{Arc length}}{\text{Circumference}}$$

All of these fractions have the same value for the same sector of a circle (in other words, each fraction is equal). Once you know this value, you can find any measure of the sector of the original circle.

If you...	Then you...	Like this:
Encounter a sector	Figure out the fraction of the circle that the sector represents	If central angle = 45° and radius = 5, then fraction = $\frac{45}{360} = \frac{1}{8}$, and area = $\frac{1}{8}\pi r^2 = \frac{1}{8}\pi(5)^2 = \frac{25}{8}\pi$

Check Your Skills

4. A sector has a central angle of 270° and a radius of 2. What is the area of the sector?

5. A sector has an arc length of 4π and a radius of 3. What is the central angle of the sector? $6\pi = C$

6. A sector has an area of 40π and a radius of 10. What is the arc length of the sector? $\frac{4\pi}{6\pi} = \frac{240}{360}$

Answers can be found on page 420.

Triangles

Triangles are relatively common in GMAT Geometry problems. You'll often find them hiding in problems that seem to be about rectangles or other shapes.

Sum of Any Two Sides $>$ Third Side

The sum of any two side lengths of a triangle is always greater than the third side length.

Grab your pen and draw a straight line. Now use that straight line as one side of a triangle, and draw two more lines to create the full triangle. The other two lines added together have to be longer than the straight line that you started from, since the straight line is the shortest possible distance between the two endpoints.

A related idea is that any side is greater than the *difference* of the other two side lengths. Otherwise, you can't even connect the dots and draw a complete triangle. The pictures below illustrate these points:

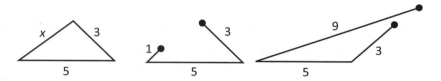

In the first triangle, what is the greatest number x could be? What's the smallest? Could x be 9? 1?

x must be less than $3 + 5 = 8$

x must be greater than $5 - 3 = 2$

$2 < x < 8$

If you...	Then you...	Like this:
Want to know how long the third side of a triangle could be	Find the sum and the difference of the other two sides; the length of the third side must be less than the sum, but more than the difference	First side $= 6$ Second side $= 4$ Third side must be less than $6 + 4 = 10$ and greater than $6 - 4 = 2$

Check Your Skills

7. Two sides of a triangle have lengths 5 and 19. Can the third side have a length of 13 ?

8. Two sides of a triangle have lengths 8 and 17. What is the range of possible values of the length of the third side?

Answers can be found on page 421.

Sum of the Three Angles = 180°

The internal angles of a triangle must sum to 180°. As a result, if you know two angles in the triangle, you can find the third angle. Take a look at this triangle:

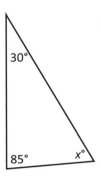

The three internal angles must sum to 180°, so $30 + 85 + x = 180$, so $x = 65$. The third angle is 65°.

The GMAT can also test you in more complicated ways. Consider this triangle:

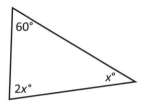

You only know one of the angles, and the other two are both given in terms of x. Again, the three angles must sum to 180°. Solve for x:

$$60 + x + 2x = 180$$

$$3x = 120$$

$$x = 40$$

The angle labeled x has a measure of 40°, and the angle labeled $2x$ has a measure of 80°.

By the way, a straight line also has a measure of 180°:

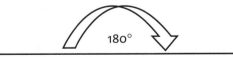

The GMAT does not always draw triangles to scale. On Problem Solving questions (one of the two types of GMAT Quant questions), a figure will be drawn to scale unless there is a note saying that it is not drawn to scale. On Data Sufficiency questions (the other type of GMAT Quant question), the figures will not say whether they are drawn to scale, so assume they are not.

10

If you...	Then you...	Like this:
Know two angles of a triangle or can represent all three in terms of a single variable	Can find all angles using the "sum to 180" principle	First angle = 3x Second angle = 4x Third angle = 40° 3x + 4x + 40 = 180 7x = 140 or x = 20

Check Your Skills

Find the missing angle(s).

9.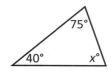

10.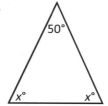

11.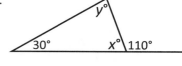

Answers can be found on pages 421–422.

Same Sides = Same Angles, and Vice Versa

Internal angles of a triangle are important on the GMAT for another reason: *Sides correspond to their opposite angles.* That is, the longest side is opposite the greatest angle, and the shortest side is opposite the smallest angle. Think about an alligator opening its mouth. As the angle between its upper and lower jaws increases, the distance between its top and bottom teeth increases. For example:

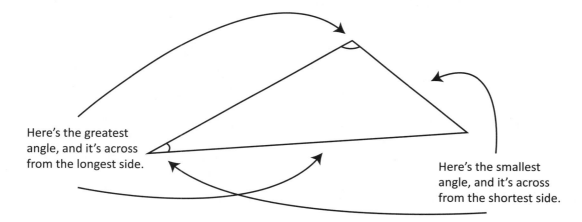

Here's the greatest angle, and it's across from the longest side.

Here's the smallest angle, and it's across from the shortest side.

This relationship works both ways. If you know the sides of the triangle, you can make inferences about the angles. If you know the angles, you can make inferences about the sides. (Note: In the figure below, a two-letter designation, such as *AC*, refers to the line that lies between the points *A* and *C*. The three-letter designation, ∠*ABC,* refers to the angle traced by starting from the letter *A*, going through *B*, and ending on *C*.) For example:

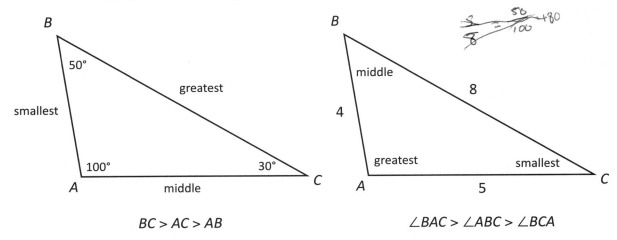

$$BC > AC > AB$$ $$\angle BAC > \angle ABC > \angle BCA$$

Lots of triangles have two or even three equal sides. These triangles also have two or three equal angles, respectively. You can classify triangles by the number of equal sides or angles that they have:

- A triangle that has two equal angles and two equal sides is an **isosceles** triangle.
- A triangle that has three equal angles (all 60°) and three equal sides is an **equilateral** triangle.

The relationship between equal angles and equal sides works in both directions. Take a look at these isosceles triangles, and think about what additional information you can infer from them:

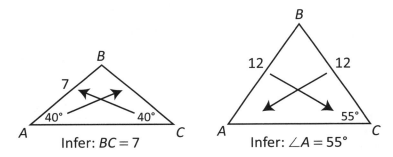

The GMAT loves isosceles triangles. Examine this challenging example:

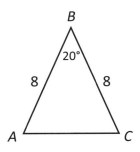

Take a look at the triangle to see what other information you can fill in. Specifically, do you know the degree measure of either angle *BAC* or angle *BCA*?

Because side *AB* is the same length as side *BC*, angle *BAC* must have the same degree measure as angle *BCA*. Label each of those angles $x°$:

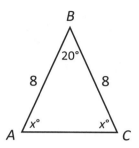

The three internal angles sum to 180. So $20 + x + x = 180$, and $x = 80$. Therefore, the two angles, *BAC* and *BCA*, each equal 80°. You can't find the side length *AC* without more advanced math that the GMAT doesn't test.

If you...	Then you...	Like this:
See two equal sides in a triangle	Set the angles opposite each side equal	Two sides both equal 8, so the angles opposite those sides are equal
See two equal angles in a triangle	Set the sides opposite each angle equal	Two angles equal 30°, so the sides opposite those angles are equal

Check Your Skills

Find the value of *x*.

12.

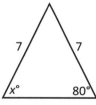

13.

14.

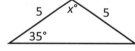

Answers can be found on page 422.

Perimeter: Sum of Sides

The **perimeter** of a triangle is the sum of the lengths of all three sides. Look at the triangle below:

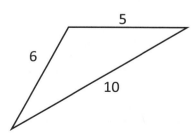

In this triangle, the perimeter is $5 + 6 + 10 = 21$. This is a relatively simple property of a triangle, so often it will be used in combination with another property. Try this next problem. What is the perimeter of triangle *PQR* ?

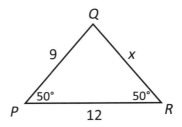

To solve for the perimeter, you need to determine the value of *x*. Because angles *QPR* and *QRP* are both 50°, their opposite sides (*QR* and *PQ,* respectively) have equal lengths. Therefore, side *QR* also has a length of 9. The perimeter of triangle *PQR* is therefore $9 + 9 + 12 = 30$.

Check Your Skills

Find the perimeter of each triangle.

15.
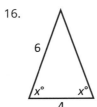

Note: Figure not drawn to scale.

16.
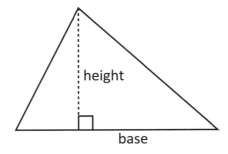

Answers can be found on page 423.

Apply Area Formula: Any Side Can Be the Base

The area of a triangle equals $\frac{1}{2} \times$ (base) $\times$ (height). In area formulas for any shape, be clear about the relationship between the base and the height. *The base and the height must be perpendicular to each other.*

In a triangle, one side of the triangle is the base. Any of the sides can be the base, but it is most common to make the bottom side the base.

The height is formed by dropping a line from the opposite point of the triangle straight down toward the base so that it forms a 90° angle with the base. The small square located where the height and base meet is used to denote a right angle (see the figure below). You can also say that the height is **perpendicular** to the base, or vice versa. For example:

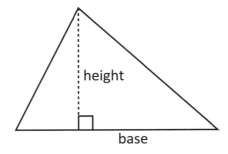

The GMAT often asks you about familiar shapes while presenting them in unfamiliar orientations. In particular, the triangle may be oriented in a way that makes it difficult to call the bottom side the base. The three triangles below are all the same triangle, but the same side is not always used as the base. The figures also show the corresponding height:

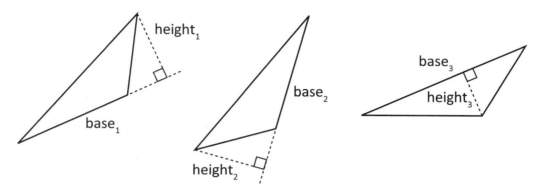

As in the first two examples, the height can be *outside* the triangle! You can extend the base farther in order to create the 90° angle. As long as there is a right angle between the base and the height, any orientation is permitted.

If you...	Then you...	Like this:
Need the area of a triangle	Apply the area formula, using *any* convenient side as the base	$A = \frac{1}{2} (\text{base})(\text{height})$ Use any side as the base; draw the right-angle height from the corner opposite the base

Check Your Skills

Find the area of each triangle.

17.

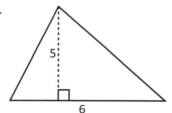

18.

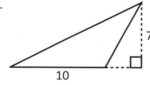

Answers can be found on page 423.

Know Two Sides of a Right Triangle: Find the Third Side

Right triangles are very common on the GMAT. A right triangle is any triangle in which one of the angles is a right angle (90°). You might be asked this on the test:

What is the perimeter of triangle *ABC* ?

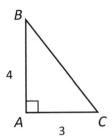

With only two sides of the triangle, how do you get the perimeter? Because this is a right triangle, you can use the Pythagorean theorem, which applies only to right triangles. According to the theorem, the lengths of the three sides of a right triangle are related by the equation $a^2 + b^2 = c^2$, where a and b are the lengths of the sides that form the right angle, also known as **legs**, and c is the length of the side opposite the right angle, also known as the **hypotenuse**.

In the given triangle, sides *AB* and *AC* are a and b (it doesn't matter which is which), and side *BC* is c. Therefore:

$$a^2 + b^2 = c^2$$
$$(3)^2 + (4)^2 = (BC)^2$$
$$9 + 16 = (BC)^2$$
$$25 = (BC)^2$$
$$5 = BC$$

The triangle looks like this:

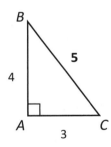

The perimeter is equal to $3 + 4 + 5 = 12$.

Often, you can take a shortcut around using the Pythagorean theorem. The GMAT favors a subset of right triangles, called **Pythagorean triples**. The triangle above is an example. The side lengths are 3, 4, and 5—all integers—and this triple is called a 3–4–5 triangle.

10

While there are quite a few of these triples, only a couple are useful to commit to memory for the GMAT; if you memorize them, you can save yourself from having to calculate them on the test. For each triple, the first two numbers are the lengths of the sides that form the right angle, and the third (and greatest) number is the length of the hypotenuse. For example:

3–4–5, or its "double" 6–8–10

5–12–13

Note that you can double, triple, or otherwise apply a common multiplier to these lengths; 3–4–5 is really a ratio of 3 : 4 : 5.

Try this problem:

What is the area of triangle *DEF*?

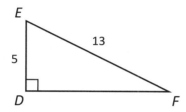

What do you need in order to find the area of triangle *DEF*? The area formula is $\frac{1}{2} \times$ (base) $\times$ (height).

This is a right triangle, so sides *DE* and *DF* are already perpendicular to each other. Treat one of them as the base and the other as the height.

How do you find the length of side *DF*? First, you can always use the Pythagorean theorem to find the length of the third side of a right triangle if you know the lengths of the other two sides. In this case, the formula would look like this: $(DE)^2 + (DF)^2 = (EF)^2$.

But don't follow through on that calculation! If you have memorized the Pythagorean triples, then when you see a right triangle in which one of the legs has a length of 5 and the hypotenuse has a length of 13, you will know that the length of the other leg must be 12:

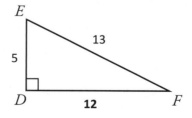

Now you have what you need to find the area of triangle *DEF*:

Area $= \frac{1}{2} \times (12) \times (5) = (6)(5) = 30$

10

If you...	Then you...	Like this:
Know two sides of a right triangle	Can find the third side, either by recognizing a triple or by using the full Pythagorean theorem	A leg of a right triangle has length 18, while the hypotenuse has length 30. How long is the third side? $18^2 + x^2 = 30^2$ and solve for x, or recognize that this is a multiple of the $3 : 4 : 5$ triple (each side is multiplied by 6). $x = 24$

Check Your Skills

Solve the following problems.

19.

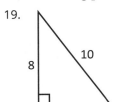

What is the length of the third side of the triangle in the figure above?

20.

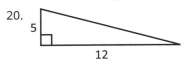

What is the length of the third side of the triangle in the figure above?

21.

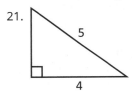

What is the area of the triangle in the figure above?

Answers can be found on pages 423–424.

Quadrilaterals

A quadrilateral is any figure with *four* sides. Quadrilaterals can always be cut up into two triangles by slicing across the middle to connect opposite corners. Therefore, what you know about triangles could apply in a problem involving quadrilaterals. In many cases, you won't want to cut up the quadrilateral that way, but it's good to know you could.

Parallelogram: Cut into Triangles OR Drop Height

The GMAT frequently deals with **parallelograms**. A parallelogram is any four-sided figure in which the opposite sides are parallel and equal. Opposite angles are also equal, and adjacent angles (angles that are next to each other without another angle in between) add up to 180°. For example:

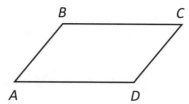

In the parallelogram above, sides *AB* and *CD* are parallel and have equal lengths. Sides *AD* and *BC* are parallel and have equal lengths. Angles *ADC* and *ABC* are equal. Angles *BAD* and *BCD* are equal. This is shown below:

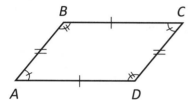 Use hash marks to indicate equal lengths or equal angles.

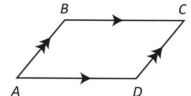 Use arrows to indicate parallel lines.

In any parallelogram, the diagonal divides the parallelogram into two equal triangles:

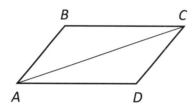 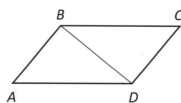 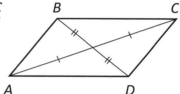

Triangle *ABC* = Triangle *ACD*   Triangle *ABD* = Triangle *BCD*   The diagonals also cut each other in half (bisect each other).

For any parallelogram, the perimeter is the sum of the lengths of all the sides and the area is equal to (base) × (height). With parallelograms, as with triangles, remember that the base and the height *must* be perpendicular to one another:

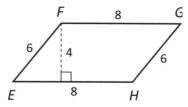

In the parallelogram above, what is the perimeter and what is the area? The perimeter is the sum of the sides: $6 + 8 + 6 + 8 = 28$.

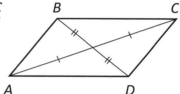

Alternatively, you can use one of the properties of parallelograms to calculate the perimeter in a different way. Parallelograms always have two sets of equal sides. In this parallelogram, two of the sides have a length of 6 and two of the sides have a length of 8. Therefore, the perimeter equals $(2)(6) + (2)(8)$. You can factor out a 2 and say that the perimeter equals $2(6 + 8) = 28$.

To calculate the area, you need a base and a height. It might be tempting to say that the area is $6 \times 8 = 48$. But the two sides of this parallelogram are not perpendicular to each other. The dotted line drawn into the figure, however, is perpendicular to side *EH*. You need to "drop a height," or draw a perpendicular line, to the base. The area of parallelogram *EFGH* is $8 \times 4 = 32$.

If you...	Then you...	Like this:
Want the perimeter or area of a parallelogram	Find all sides (for the perimeter) or drop a height (for the area)	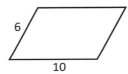 Perimeter $= 2(6 + 8) = 28$ Area $= 8 \times 4 = 32$

Check Your Skills

22. What is the perimeter of the parallelogram?

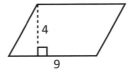

23. What is the area of the parallelogram?

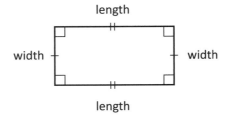

Answers can be found on page 424.

Rectangles = Parallelogram + FOUR Right Angles

Rectangles are a specific type of parallelogram. Rectangles have all the properties of parallelograms, plus one more: *All four internal angles of a rectangle are right angles.* With rectangles, you refer to one pair of sides as the length and one pair of sides as the width. It doesn't matter which is which, though traditionally the shorter side is the width and the longer side is the length. For example:

length

width   width

length

The formula for the perimeter of a rectangle is the same as for the perimeter of a parallelogram. You can sum the lengths of the four sides, or you can double the sum of the length and the width: $2(l + w)$.

The formula for the area of a rectangle is also the same as for the area of a parallelogram. But for any rectangle, the length and width are by definition perpendicular to each other, so you don't need to find a separate height. For this reason, the area of a rectangle is commonly expressed as (length) × (width), or $A = lw$.

For the following rectangle, find the perimeter and the area:

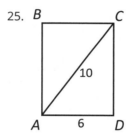

Start with the perimeter. The perimeter is $2(5 + 7) = 24$. Alternatively, $5 + 5 + 7 + 7 = 24$.

Now find the area. The formula for area is (length) × (width), or $(7)(5) = 35$.

Finally, the diagonal of a rectangle cuts the rectangle into two equal *right* triangles, with all the properties you expect of right triangles.

Check Your Skills

Find the area and perimeter of each rectangle.

24.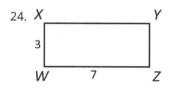

25.

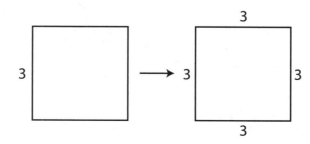

Answers can be found on pages 424–425.

Squares = Rectangle + Four Equal Sides

The most special type of rectangle is a square. *A square is a rectangle in which all four sides are equal.* Knowing only one side of a square is enough to determine the perimeter and area of a square.

For instance, if the length of the side of a square is 3, then all four sides have a length of 3:

The perimeter of the square is $3 + 3 + 3 + 3$, which equals 12. Alternatively, once you know the length of one side of a square, multiply that length by 4 to find the perimeter: $4s = (4)(3) = 12$.

To find the area, use the same formula as for a rectangle: Area = length × width.

But because the shape is a square, the length and the width are equal. Therefore, the area of a square is Area = $(\text{side})^2$, which is the side length squared.

In this case, the area is equal to s^2, or $(3)^2 = 9$.

Squares are like circles: If you know one measure, you can find everything. This is because they are both "regular" figures. All circles look like each other, and all squares look like each other. For circles, the most fundamental measure is the radius, and then you can calculate everything else. For squares, the most fundamental measure is the side length.

Geometry: Word Problems with Pictures

Now that you know various properties of shapes, such as perimeter and area, how do you use these properties to answer GMAT Geometry questions, especially ones with more than one figure? Try this problem:

> Rectangles *EFGH* and *ABCD*, shown below, have equal areas. The length of side *AB* is 5. What is the length of diagonal *AC* ?

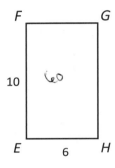
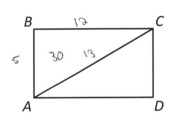

First, *draw your own copies of the shapes, and fill in everything you know.* For this problem, redraw both rectangles. Label side *AB* with a length of 5. Also, make note of what you're looking for—in this case, you want the length of diagonal *AC*. Label that diagonal with a question mark:

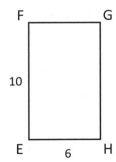
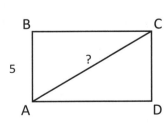

Now turn to the question. Many geometry questions are similar to the story problems discussed in Chapter 9. Both types of problems provide you with information that may be disguised. The information is related through common mathematical relationships, which also may be disguised or implied. In geometry, the information can be presented in words or visually.

So has the question above provided you with any information that can be expressed mathematically? Can you create equations?

Yes. The two rectangles have equal areas: Area$_{ABCD}$ = Area$_{EFGH}$. You can do even better than that.

The formula for area of a rectangle is Area = (length) × (width). The equation can be rewritten as (length$_{ABCD}$) × (width$_{ABCD}$) = (length$_{EFGH}$) × (width$_{EFGH}$).

The length and width of rectangle *EFGH* are 6 and 10, and the length of *AB* is 5. Plug the values into the equation and solve:

$$(5) \times (\text{width}_{ABCD}) = (6) \times (10)$$
$$(5) \times (\text{width}_{ABCD}) = 60$$
$$\text{width}_{ABCD} = 12$$

Any time you learn a new piece of information, add that information to your picture:

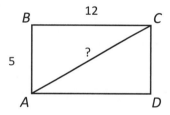

To recap, you've redrawn the shapes and filled in all the given information (such as side lengths, angles, etc.). You've made note of what the question wants. Just as you start any story problem by identifying unknowns, creating variables, and writing down givens, the first steps for geometry problems are to draw or redraw figures, add all of the given information, and confirm what you're being asked to find.

Next, you made use of additional information provided. The problem stated that the two rectangles had equal areas. You created an equation to express this relationship, and then you plugged in the values you knew to solve for the width of rectangle *ABCD*. This process is identical to the process used to solve word problems—you identify relationships and create equations. After that, you solve the equations for the missing value (in this case, the width of *ABCD*).

In some ways, all you have done so far is set up the problem. In fact, aside from noting that you need to find the length of diagonal *AC*, nothing you have done so far seems to have directly helped you actually solve for that value. So far, you have found that the width of rectangle *ABCD* is 12.

So why bother solving for the width of rectangle *ABCD* when you're not even sure why you'd need it? You are *likely* to need that missing value. On the vast majority of GMAT problems, two general principles hold:

1. Intermediate steps are required to solve for the value you want.

2. The GMAT almost never provides extraneous information.

10

As a result, something that you *can* solve for is likely to be a stepping stone on the way to the answer.

This doesn't mean that you should calculate everything you possibly could—you could find the diagonal of rectangle *EFGH*, for instance—or calculate quantities at random. Rather, as you practice these problems, you'll gain a sense of the kinds of stepping stones that the GMAT prefers.

Now that you know the width of *ABCD*, what can you figure out that you couldn't before? Take another look at the value you're looking for: the length of *AC*.

You've already identified a relationship mentioned in the question—that both rectangles have equal areas. But for many geometry problems, there are additional relationships that aren't as obvious.

The key to this problem is to recognize that *AC* is not only the diagonal of rectangle *ABCD*, but also the hypotenuse of a right triangle, because all four interior angles of a rectangle are right angles:

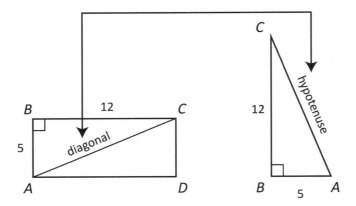

Now that you know *AC* is the hypotenuse of a right triangle, you can use two sides to find the third. One way to get the number is through the Pythagorean theorem.

This is a Pythagorean triple, though! The hypotenuse is *AC* = 13.

If you do want to use the Pythagorean theorem, sides *BC* and *AB* are the legs of the triangle, and *AC* is the hypotenuse. Therefore:

$$(BC)^2 + (AB)^2 = (AC)^2$$
$$(12)^2 + (5)^2 = (AC)^2$$
$$144 + 25 = (AC)^2$$
$$169 = (AC)^2$$
$$13 = AC$$

Let's recap what happened in the last portion of this question. You needed an insight that wasn't obvious: that the diagonal of rectangle *ABCD* is also the hypotenuse of right triangle *ABC*. Once you had that insight, you could apply "right triangle" thinking to get that unknown side. The last part of this problem required you to make inferences from the figures.

Sometimes you need to make a jump from one shape to another through a common element. For instance, you needed to see *AC* as both a diagonal of a rectangle and as a hypotenuse of a right triangle. Here, *AC* was common to both a rectangle and a right triangle, playing a different role in each.

These inferences can also make you think about what information you need to find another value.

10

Putting this all together, we recommend a 4-step process that you can apply to any geometry problem:

Step 1: *Draw or redraw figures, fill in all given information, and identify the target.* Fill in all known angles and lengths, and make note of any equal sides or angles.

Step 2: *Identify relationships and create equations.* Start with relationships that are explicitly stated somewhere.

Step 3: *Solve the equations for the missing value.* If you can solve for a value, you will almost always need that value to answer the question.

Step 4: *Make inferences from the figures.* You often need to use relationships that are not explicitly stated.

Try this problem:

> Rectangle *PQRS* is inscribed in circle *O* pictured below. The center of circle *O* is also the center of rectangle *PQRS*. If the circumference of circle *O* is 5π, what is the area of rectangle *PQRS*?

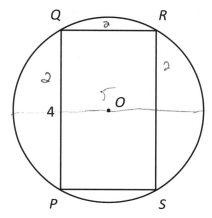

Inscribed means that the inside shape is just large enough to exactly touch the edges of the outside shape. The four corners of the rectangle just touch the circle.

First, redraw the figure on your paper and fill in all of the given information. The question didn't explicitly give you the value of any side lengths or angles, but it did say that *PQRS* is a rectangle. That means all four internal angles are right angles. This is how the GMAT tests what you know about the key properties of different shapes. Also, identify what you're looking for: the rectangle's area. Here's what your picture should look like:

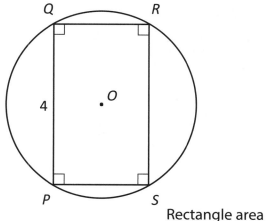

Rectangle area = ?

Now identify relationships and create equations. The question stated that the circumference of circle O is 5π. Find the radius of the circle:

$$C = 2\pi r$$
$$5\pi = 2\pi r$$
$$5 = 2r$$
$$2.5 = r$$

If the radius is 2.5, then the diameter of circle O is 5.

Why do you find the radius and diameter? *This is how you will make a connection between the circle and the rectangle.* Now is the time to make inferences from the figures.

Ultimately, this question is asking for the area of rectangle $PQRS$. What information do you need to find that value? You have the length of PQ. If you can find the length of either QR or PS, you can find the area of the rectangle.

What is the connection between the rectangle and the radius or diameter? Put in a diameter:

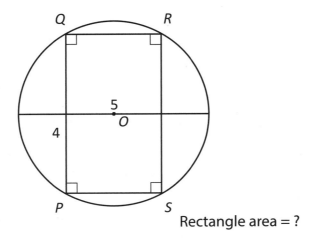

Rectangle area = ?

That didn't help much. What if you drew the diameter so that it passed through the center but touched the circle at points P and R? The center of the circle is also the center of the rectangle, so the diameter of the circle is also the diagonal of the rectangle:

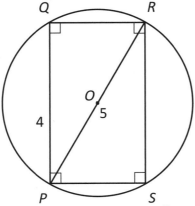

Rectangle area = ?

Connection made! *PR* is the "bridge" between the two figures. (You could also draw the diagonal *QS*.)

Where do you go from here? You still need the length of either *QR* or *PS* (which are the same, because this is a rectangle). Can you get either one of those values? Yes. *PQR* is a right triangle. Maybe it's not oriented the way you are used to, but all the elements are there: It's a triangle and one of its internal angles is a right angle. You also know the lengths of two of the sides: *PQ* and *PR*.

Triangle *PQR* is a 3–4–5 Pythagorean triple, so the length of *QR* is 3:

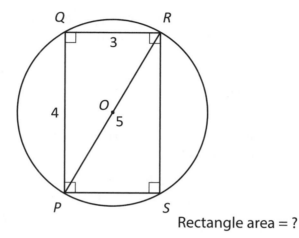

Rectangle area = ?

At last, you have what you need to find the area of rectangle *PQRS*. Since the area is equal to (length) × (width), the area is (4) × (3) = 12.

The key insight in this problem was to realize that you could draw a diameter that would also act as the diagonal of the rectangle, linking the two figures together. You also had to recognize that *PQR* was a right triangle, even though it may have been hard to see. These kinds of insights will be crucial to success on the GMAT.

If you...	Then you...	Like this:
Face a geometry problem with more than one figure	Follow the basic 4-step process to solve, finding intermediate unknowns and looking for bridges or links between the shapes	1. Redraw, fill in, label target 2. Spot relationships and write equations 3. Solve for what you can 4. Make inferences

Check Your Skills

26.

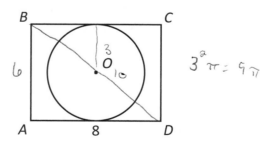

In rectangle *ABCD*, the distance from *B* to *D* is 10. What is the area of the circle that is tangent to sides *AD* and *BC* of the rectangle?

The answer can be found on page 425.

Coordinate Plane—Position Is a Pair of Numbers

A **coordinate plane** is a more advanced version of a number line. Here are some examples:

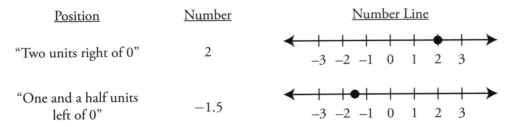

Position	Number	Number Line
"Two units right of 0"	2	
"One and a half units left of 0"	−1.5	

A **number line** is a ruler or measuring stick that goes as far as you want in both directions. With the number line, you can say where something is with a single number. In other words, you can link a position with a number.

You use either positive or negative numbers to indicate the position of a **point** either right or left of 0. When you are dealing with the number line, a point and a number mean the same thing:

If you're shown where the point is on the number line, you can tell the number:

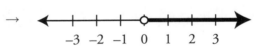

→ *The point is at −2.*

If you're told the number, you can show where the point is on the number line:

The point is at 0. →

This works even if you have only partial information about the point. If you are told something about where the point is, you can tell *something* about the number, and vice versa.

For instance, if the number is positive, then the point lies somewhere to the right of 0 on the number line. Even though you don't know the exact location of the point, you do know a range of potential values:

The number is positive.
In other words, the number is greater than (>) 0.

→

This also works in reverse. If you are given a range of potential positions on a number line, you can tell what that range is for the number:

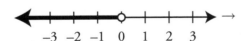

→ The number is less than (<) 0.

10

How does this get more complicated? What if you want to be able to locate a point that's not on a straight line, but on a page?

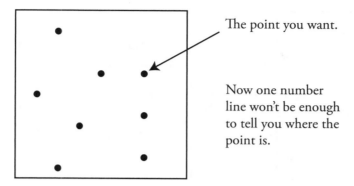

The point you want.

Now one number line won't be enough to tell you where the point is.

Begin by inserting the number line into the picture. This will help you determine how far to the right or left of 0 the point is:

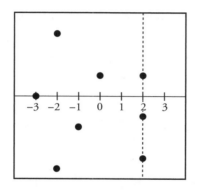

The point is 2 units to the right of 0.

But all three points that touch the dotted line are two units to the right of 0. You don't have enough information to determine the unique location of the point.

To locate the point, you also need to know how far up or down the dotted line the point is. For that, you'll need another number line. This number line, however, is going to be vertical. Using this vertical number line, you can measure how far above or below 0 a point is:

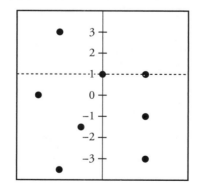

The point is 1 unit above 0.

Notice that this number line by itself also does not provide enough information to determine the unique location of the point.

If you combine the information from the two number lines, you can determine both how far left or right *and* how far up or down the point is:

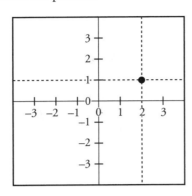

The point is 2 units to the right of 0.

and

The point is 1 unit above 0.

Now you have a unique description of the point's position. Only one point on the page is *both* 2 units to the right of 0 *and* 1 unit above 0. On a page, you need two numbers to indicate position.

As with the number line, information can travel in either direction. If you're told the two numbers that indicate a point's location, you can place that point on the page:

The point is 3 units to the left of 0.

and

The point is 2 units below 0.

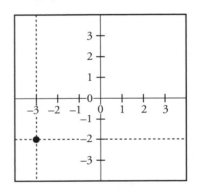

If, on the other hand, you see a point on the page, you can identify its location and extract the two numbers:

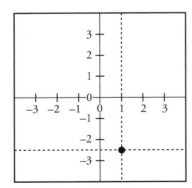

The point is 1 unit to the right of 0.

and

The point is 2.5 units below 0.

10

Now that you have two pieces of information for each point, you need to keep straight which number is which. In other words, you need to know which number gives the left–right position and which number gives the up–down position:

The **x-coordinate** is the left–right number.

> Numbers to the right of 0 are positive.
> Numbers to the left of 0 are negative.

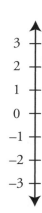

$$-3 \quad -2 \quad -1 \quad 0 \quad 1 \quad 2 \quad 3$$

This number line is the x-axis.

The **y-coordinate** is the up–down number.

> Numbers above 0 are positive.
> Numbers below 0 are negative.

This number line is the y-axis.

The point where the x-axis and the y-axis cross is called the **origin**. This is always 0 on both axes.

Now when describing the location of a point, you can use the technical terms: x-coordinate and y-coordinate:

The x-coordinate of the point is 1 and the y-coordinate of the point is 0.

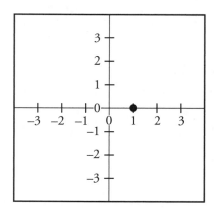

In short, you can say that, for this point, $x = 1$ and $y = 0$. In fact, you can go even farther. You can say that the point is at $(1, 0)$. This shorthand always has the same basic layout. The first number in the parentheses is the x-coordinate and the second number is the y-coordinate: (x, y). One easy way to remember this is that x comes before y in the alphabet. The origin has coordinates $(0, 0)$. For example:

The point is at $(-3, -1)$.

OR

The point has an x-coordinate of -3 and a y-coordinate of -1.

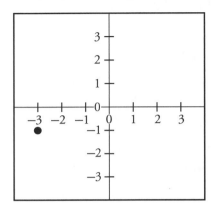

Now you have a fully functioning **coordinate plane**: an x-axis and a y-axis drawn on a page. The coordinate plane allows you to determine the unique position of any point on a **plane** (essentially, a big, flat sheet of paper).

And in case you were ever curious about what **one-dimensional** and **two-dimensional** mean, now you know. A line is one-dimensional, because you only need one number to identify a point's location. A plane is two-dimensional, because you need two numbers to identify a point's location.

If you...	Then you...	Like this:
Want to plot a point on the coordinate plane	Use the x-coordinate for right–left of $(0, 0)$, and use the y-coordinate for up–down from $(0, 0)$	$(3, 2)$ is 3 units right of and 2 units up from the origin

Check Your Skills

27. Draw a coordinate plane and plot the given points.

   1. $(3, 1)$
   2. $(-2, 3.5)$
   3. $(0, -4.5)$
   4. $(1, 0)$

10

28.

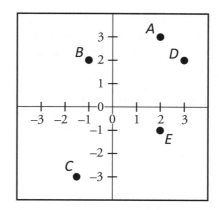

Match each coordinate with the appropriate point on the coordinate plane.

1. $(2, -1)$
2. $(-1.5, -3)$
3. $(-1, 2)$
4. $(3, 2)$

Answers can be found on page 426.

Know Just One Coordinate = Find a Line

You need to know both the *x*-coordinate and the *y*-coordinate to plot a point exactly on the coordinate plane. If you only know one coordinate, you can't tell precisely where the point is, but you can narrow down the possibilities.

Let's say that all you know is that the point is 4 units to the right of 0:

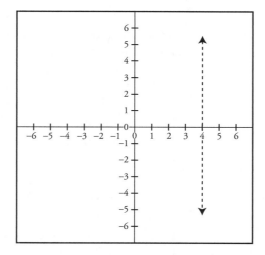

The *x*-coordinate still indicates left–right position. If you fix the left–right position but not the up–down position, then the point can only move up and down—forming a vertical line.

10

In this case, any point along the vertical dotted line is 4 units to the right of 0. In other words, every point on the dotted line has an *x*-coordinate of 4. You could shorten that and say $x = 4$. You don't know anything about the *y*-coordinate, which could be any number. All the points along the dotted line have different *y*-coordinates but the same *x*-coordinate, which equals 4.

So if you know that $x = 4$, then the point can be anywhere along a vertical line that crosses the *x*-axis at $(4, 0)$. For example:

If you know that $x = -3$…

Then you know:

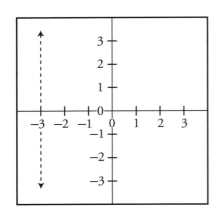

Every point on the dotted line has an *x*-coordinate of –3.

Points on the dotted line include $(-3, 1)$, $(-3, -7)$, $(-3, 100)$, and so on. In general, if you know the *x*-coordinate of a point and not the *y*-coordinate, then all you can say about the point is that it lies on a vertical line.

Now imagine that all you know is the *y*-coordinate of a number. Say you know that $y = -2$. How could you represent this on the coordinate plane? In other words, what are all the points for which $y = -2$?

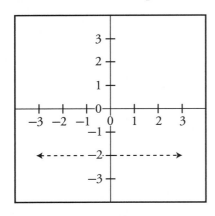

Every point 2 units below 0 fits this condition. These points form a horizontal line. You don't know anything about the *x*-coordinate, which could be any number. All the points along the horizontal dotted line have different *x*-coordinates but the same *y*-coordinate, which equals –2. For instance, $(-3, -2)$, $(-2, -2)$, and $(50, -2)$ are all on the line.

For example:

If you know that $y = 1$...

Then you know:

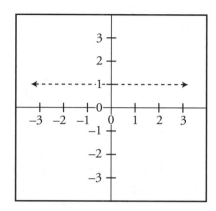

Every point on the dotted line has a y-coordinate of 1.

If you know the y-coordinate but not the x-coordinate, then you know the point lies somewhere on a horizontal line.

If you...	Then you...	Like this:
Know just one coordinate	Have either a horizontal or a vertical line	If $x = 3$, then a vertical line runs through the number 3 on the x-axis. The y-coordinate could be anything.

Check Your Skills

Draw a coordinate plane and plot the following lines.

29. $x = 6$
30. $y = -3$
31. $x = 0$

Answers can be found on page 427.

Know a Range = Shade a Region

What do you do if all you know is that $x > 0$? To answer that, return to the number line for a moment. As you saw earlier, if $x > 0$, then the target is anywhere to the right of 0:

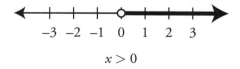

$$x > 0$$

Now look at the coordinate plane. All you know is that *x* is greater than 0. And you don't know *anything* about *y*, which could be any number.

How do you show all the possible points? You can shade in part of the coordinate plane: the part to the right of 0.

If you know that $x > 0$…

Then you know:

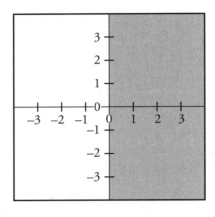

Every point in the shaded region has an *x*-coordinate greater than 0.

What if you know that $y < 0$? Then you can shade in the bottom half of the coordinate plane—where the *y*-coordinate is less than 0. The *x*-coordinate can be anything.

If you know that $y < 0$…

Then you know:

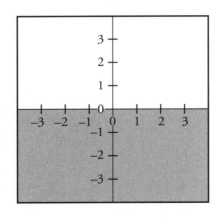

Every point in the shaded region has a *y*-coordinate less than 0.

Finally, if you know information about both *x* and *y*, then you can narrow down the shaded region.

If you know that $x > 0$ and $y < 0$...

Then you know:

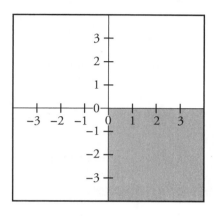

The only place where x is greater than 0 and y is less than 0 is the bottom right quarter of the plane. So you know that the point lies somewhere in the bottom right quarter of the coordinate plane.

The four quarters of the coordinate plane are called **quadrants**. Each quadrant corresponds to a different combination of signs of x and y. The quadrants are numbered I, II, III, or IV, as shown below, starting with the top right quadrant and moving counterclockwise:

In quadrant:	The x-coordinate is:	The y-coordinate is:
I	positive	positive
II	negative	positive
III	negative	negative
IV	positive	negative

If you...	Then you...	Like this:
Only know ranges for one or both coordinates	Can plot a shaded region corresponding to the proper range	If $x < 0$, then shade all of the points to the left of the y-axis

10

Check Your Skills

32. In which quadrant do the following points lie?

    1. $(1, -2)$
    2. $(-4.6, 7)$
    3. $(-1, -2.5)$
    4. $(3, 3)$

33. Which quadrant or quadrants are indicated by the following?

    1. $x < 0, y > 0$
    2. $x < 0, y < 0$
    3. $y > 0$
    4. $x < 0$

Answers can be found on page 428.

Read a Graph = Drop a Line to the Axes

If you see a point on a coordinate plane, how do you determine its coordinates? To find an *x*-coordinate, drop an imaginary line down to the *x*-axis (or up to it) and find the corresponding number:

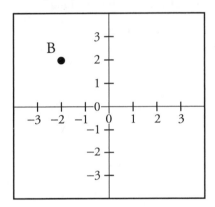 →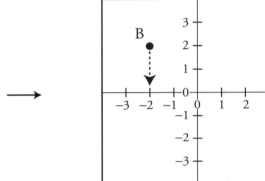

The line hits the *x*-axis at -2, so the *x*-coordinate of the point is -2. Now to find the *y*-coordinate, employ a similar technique. This time, draw a horizontal line instead of a vertical line:

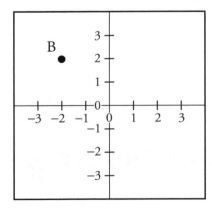 →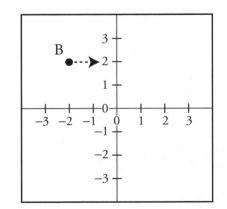

The line touches the *y*-axis at 2, which means the *y*-coordinate of the point is 2. Thus, the coordinates of point *B* are $(-2, 2)$.

Now suppose that you know that the point is on a slanted line in the plane. Try this problem:

On the line shown, what is the *y*-coordinate of the point that has an *x*-coordinate of -4?

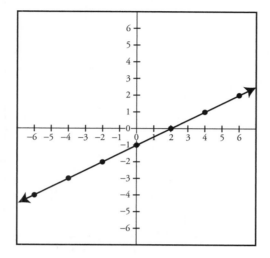

Go from the axis that you know (here, the *x*-axis) to the line that contains the point, and then to the *y*-axis (the axis you don't know):

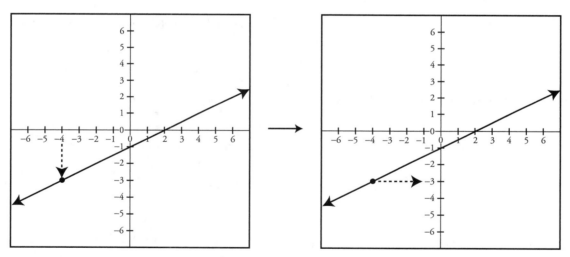

So the point on the line that has an *x*-coordinate of -4 has a *y*-coordinate of -3.

Check Your Skills

34. On the following graph, what is the *y*-coordinate of the point on the line that has an *x*-coordinate of −3 ?

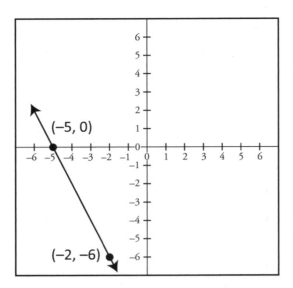

Answer can be found on page 428.

Plot a Relationship: Given an *x*, Find a *y*

The most frequent use of the coordinate plane is to display a relationship between *x* and *y*. Often, this relationship is expressed this way: If you tell me *x*, I can tell you *y*.

As an equation, this sort of relationship looks like this:

$$y = \text{some expression involving } x$$

Another way of saying this is that you have *y* "in terms of" *x*.

Examples:

$$y = 2x + 1$$
$$y = x^2 - 3x + 2$$
$$y = \frac{x}{x + 2}$$

If you plug a number in for *x* in any of these equations, you can calculate a value for *y*.

Take $y = 2x + 1$. You can generate a set of *y*'s by plugging in various values of *x*. Start by making a table:

x	y = 2x + 1
−1	$y = 2(-1) + 1 = -1$
0	$y = 2(0) + 1 = 1$
1	$y = 2(1) + 1 = 3$
2	$y = 2(2) + 1 = 5$

Now that you have some values, see what you can do with them. You can say that when *x* equals 0, *y* equals 1. These two values form a pair. You express this connection by plotting the point (0, 1) on the coordinate plane. Similarly, you can plot all the other points that represent an *x*–*y* pair from the table:

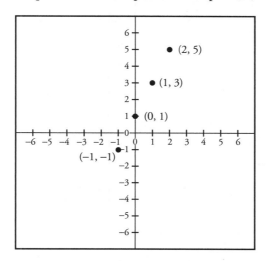

You might notice that these points seem to lie on a straight line. You're right—they do. In fact, any point that you can generate using the relationship $y = 2x + 1$ also lies on the line.

This line is the graphical representation of $y = 2x + 1$:

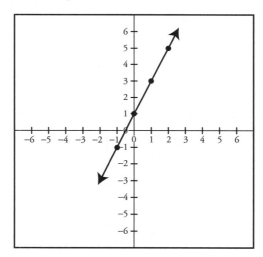

So now you can talk about equations in visual terms. In fact, that's what lines and curves on the coordinate plane are—they represent all the *x–y* pairs that make an equation true. Take a look at the following example:

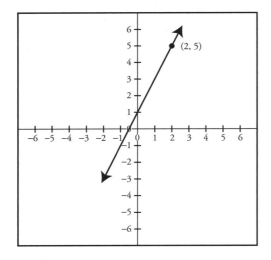

$$y = 2x + 1$$

$$5 = 2(2) + 1$$

The point (2, 5) lies on the line
$y = 2x + 1$.

If you plug in 2 for *x* in
$y = 2x + 1$, you get 5 for *y*.

You can even speak more generally, using variables:

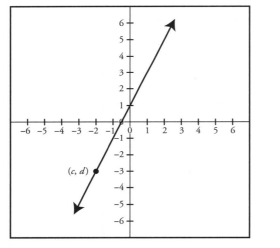

$$y = 2x + 1$$

$$d = 2(c) + 1$$

The point (*c*, *d*) lies on the line
$y = 2x + 1$.

If you plug in *c* for *x* in
$y = 2x + 1$, you get *d* for *y*.

If you...	Then you...	Like this:
Want to plot a relationship in the *x–y* plane	Input values of *x* into the relationship, get values of *y* back, then plot the (*x*, *y*) pairs	Plot $y = 4 - x$ If $x = 0$, then $y = 4$, etc. Then plot (0, 4), etc.

10

Check Your Skills

35. True or False? The point (9, 21) is on the line $y = 2x + 1$.

36. True or False? The point (4, 14) is on the curve $y = x^2 - 2$.

Answers can be found on pages 428–429.

Lines in the Plane: Use Slope and *y*-Intercept to Plot

The relationship $y = 2x + 1$ formed a line in the coordinate plane. You can generalize this relationship. Any relationship of the following form represents a line:

$$y = mx + b \qquad \text{m and b represent numbers (positive or negative).}$$

For instance, in the equation $y = 2x + 1$, $m = 2$ and $b = 1$.

A **linear equation** is one that forms a straight line when it is plotted on a coordinate plane. A linear equation does not contain any exponents and, if you put x and y on the same side of the equation, the two variables are not multiplied together. For example:

Lines		Not Lines
$y = 3x - 2$	$m = 3, b = -2$	$y = x^2$
$y = -x + 4$	$m = -1, b = 4$	$y = \dfrac{1}{x}$
$y = \dfrac{1}{2}x$	$m = \dfrac{1}{2}, b = 0$	$y = x(x + 3)$
These are called linear equations.		These equations are not linear.

The variables m and b have special meanings when you are dealing with linear equations. The variable m represents the **slope** of the line. The slope tells you how steep the line is and whether the line is rising or falling:

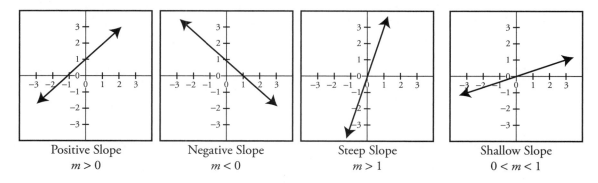

Positive Slope	Negative Slope	Steep Slope	Shallow Slope
$m > 0$	$m < 0$	$m > 1$	$0 < m < 1$

Imagine someone walking along the line from left to right. When the person is walking uphill, the slope is positive. If the person is walking downhill, the slope is negative. If the slope is very gradual, it is called a shallow slope. If it would be very hard to walk uphill, the slope is considered steep. (And if you'd be in danger of tumbling down while walking downhill, that's also very steep!)

The variable *b* represents the ***y*-intercept**. The *y*-intercept indicates where the line crosses the *y*-axis. Any line or curve always crosses the *y*-axis when $x = 0$. To find the *y*-intercept, plug in 0 for *x* in the equation and find *y*. For example:

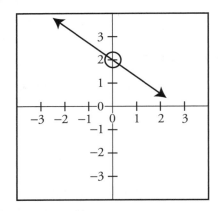 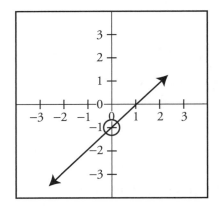

In the first image above, the *y*-intercept is 2, because the line crosses the *y*-axis at $y = 2$. In the second image, the *y*-intercept is -1.

As you saw earlier, you can graph an equation by solving the equation for various values of *x* to obtain a value of *y*, and then plot and connect those points. However, you can use *m* and *b* in a linear equation to plot a line more quickly than by plotting several points on the line. For example:

Plot the line $y = \frac{1}{2}x - 2$:

Begin with the *y*-intercept: $b = -2$, so the line crosses the *y*-axis at $y = -2$. Plot that point:

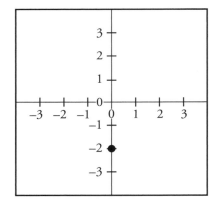

Now you're going to use the slope in order to finish drawing the line. Every slope, whether an integer or a fraction, can be thought of as a fraction. In this equation, $m = \frac{1}{2}$:

$$\frac{1}{2} \rightarrow \frac{\text{Numerator}}{\text{Denominator}} \rightarrow \frac{\text{Rise}}{\text{Run}} \rightarrow \frac{\text{Change in } y}{\text{Change in } x}$$

10

The numerator of the fraction indicates how many units you want to move in the y direction—in other words, how far up or down you want to move. The denominator indicates how many units you want to move in the x direction—in other words, how far left or right you want to move. For this particular equation, the slope is $\frac{1}{2}$, which means you want to move up 1 unit and right 2 units from the y-intercept:

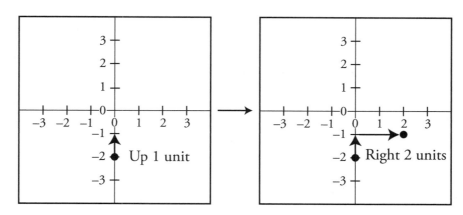

If you start from the point $(0, -2)$ and go 1 unit up and 2 units to the right, you'll end up at the point $(2, -1)$. The point $(2, -1)$ is also a point on the line and a solution to the equation $y = \frac{1}{2}x - 2$. In fact, you can plug in the x value and solve for y to check that you did this correctly:

$$y = \frac{1}{2}x - 2$$
$$y = \frac{1}{2}(2) - 2$$
$$y = -1$$

If you go up another 1 unit and right another 2 units, you will end up with another point that appears on the line. Although you could keep doing this indefinitely, in reality you need only two points to draw a line. Connect the dots and you're done! Here's what the equation looks like graphically:

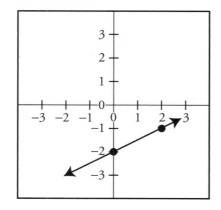

$$y = \frac{1}{2}x - 2$$

Try this problem:

Graph the equation $y = -\frac{3}{2}x + 4$.

10

First, plot the *y*-intercept. In this equation, $b = 4$, so the line crosses the *y*-axis at the point $(0, 4)$:

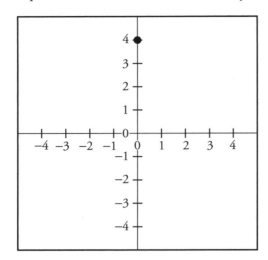

Now use the slope to find a second point. This time, the slope is $-\frac{3}{2}$, which is a negative slope. Associate the negative sign with the numerator: -3. To find the next point, go *down* 3 units and right 2 units:

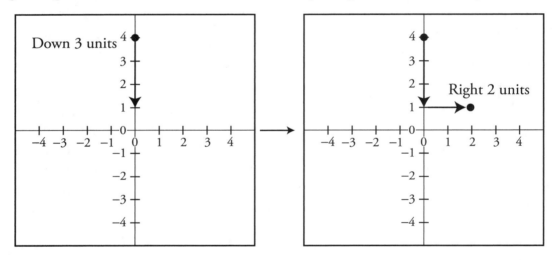

Therefore, $(2, 1)$ is another point on the line. Now that you have two points, you can draw the line:

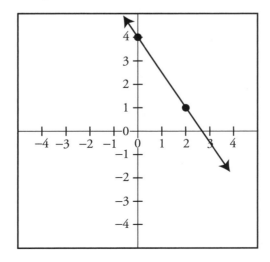

$$y = -\frac{3}{2}x + 4$$

If you...	Then you...	Like this:
Want to plot a linear equation in the *x–y* plane	Put the equation in the form of $y = mx + b$, then use m and b to draw the line	Plot $y = 4 - x$ Rearrange the equation: $y = -1x + 4$ Slope is -1, and y-intercept is $(0, 4)$

Check Your Skills

For questions 37 and 38, what are the slope and *y*-intercept of the given lines?

37. $y = 3x + 4$

38. $2y = 5x - 12$

39. Draw a coordinate plane and graph the line $y = 2x - 4$. Identify the slope and the *y*-intercept.

Answers can be found on page 429.

Check Your Skills Answer Key

1. **81π:** The formula for area is $A = \pi r^2$. The radius is 9, so the area is equal to $\pi(9)^2 = 81\pi$.

2. **17:** The circumference of a circle can be written $C = 2\pi r$ or $C = \pi d$. The question asks for the diameter, so use the latter formula.

$$17\pi = \pi d$$
$$17 = d$$

3. **10π:** Both the area and circumference of a circle are defined in terms of the radius. Use the given area of 25π to find the radius.

$$A = \pi r^2$$
$$25\pi = \pi r^2$$
$$25 = r^2$$
$$5 = r$$

Now use the radius of 5 to find the circumference.

$$C = 2\pi r$$
$$C = 2\pi(5)$$
$$C = 10\pi$$

4. **3π:** If the central angle of the sector is $270°$, then it is $\frac{3}{4}$ of the full circle, because $\frac{270°}{360°} = \frac{3}{4}$.

Since the radius is 2, the area of the full circle is $\pi(2)^2 = 4\pi$. The area of the sector is $\frac{3}{4} \times 4\pi = 3\pi$.

5. **$240°$:** Because you are given the arc length, find the circumference of the full circle and use that to determine what fraction the sector represents.

$$C = 2\pi r = 2\pi(3) = 6\pi$$

$$\text{Fraction of total} = \frac{4\pi}{6\pi} = \frac{2}{3}$$

The central angle is $\frac{2}{3} \times 360° = 240°$.

6. **8π:** Begin by finding the area of the whole circle; use that information to determine what fraction the sector represents.

$$\text{Area} = \pi(10)^2 = 100\pi$$

$$\text{Fraction of sector} = \frac{40\pi}{100\pi} = \frac{4}{10} = \frac{2}{5}$$

The circumference of the whole circle is $2\pi r = 20\pi$. Use this, coupled with the fraction of the sector, to determine the arc length.

$$\text{Arc length} = \frac{2}{5} \times 20\pi = 8\pi$$

7. **No:** The two known sides of the triangle are 5 and 19. The third side of the triangle must be greater than $19 - 5$ but less than $19 + 5$. In other words, the third side must be between 14 and 24. The number 13 is less than 14, so 13 cannot be the length of the third side.

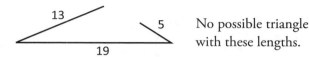

No possible triangle with these lengths.

8. **$9 <$ third side < 25:** If the two known sides of the triangle are 8 and 17, then find the range for the third side by taking the sum and the difference of the known sides.

$$\text{Sum} = 8 + 17 = 25$$
$$\text{Difference} = 17 - 8 = 9$$

The third side must be greater than 9 but less than 25.

9. **$65°$:** The internal angles of a triangle sum to $180°$.

$$40 + 75 + x = 180$$
$$115 + x = 180$$
$$x = 65$$

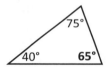

10. **$65°$:** The three angles of the triangle sum to $180°$, so $50 + x + x = 180$.

$$50 + x + x = 180$$
$$2x = 130$$
$$x = 65$$

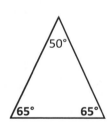

11. $x = 70, y = 80°$: In order to determine the missing angles of the triangle, you'll need to do a little work with the picture. Start with x. Straight lines have a degree measure of $180°$, so $110 + x = 180$, and $x = 70$.

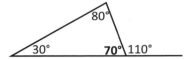

Now find y.

$$30 + 70 + y = 180$$
$$y = 80$$

12. **80°**: In this triangle, two sides have the same length (so this triangle is isosceles). Therefore, the two angles opposite the two equal sides are also equal: x must be $80°$.

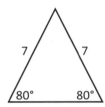

13. **4**: In this triangle, two angles are equal (so this triangle is isosceles). The two sides opposite the equal angles must also be equal, so x must equal 4.

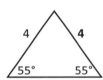

14. **110°**: Two sides have the same length, so the angles opposite the equal sides must also be equal. (This is an isosceles triangle.)

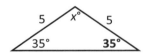

Now solve for x.

$$35 + 35 + x = 180$$
$$x = 110$$

15. **25:**

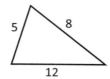

The perimeter of a triangle equals the sum of the three sides, or $5 + 8 + 12 = 25$.

16. **16:** The two angles labeled x are equal, so the sides opposite the equal angles must also be equal.

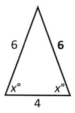

The perimeter is $6 + 6 + 4 = 16$.

17. **15:** In the triangle shown, the base is 6 and the height is 5. Therefore, you can find the area.

$$A = \frac{1}{2}bh$$
$$A = \frac{1}{2}(6)(5)$$
$$A = 15$$

18. **35:** In this triangle, the base is 10 and the height is 7. Remember that the height must be perpendicular to the base, but the height doesn't need to lie within the triangle. Now find the area.

$$A = \frac{1}{2}bh$$
$$A = \frac{1}{2}(10)(7)$$
$$A = 35$$

When you see a problem like this one on the GMAT, you will probably not be given the dotted lines; you'll likely have to realize yourself that the height can be drawn outside of the triangle.

19. **6:** This is a right triangle, so you can use the Pythagorean theorem to solve for the length of the third side. First, though, check to see whether this is one of the Pythagorean triples. It is a multiple of the 3–4–5 triangle; the sides here are 6–8–10. The third side must equal 6.

If you don't recognize the triple, you need to use the Pythagorean theorem to solve.

$$a^2 + 8^2 = 10^2$$
$$a^2 + 64 = 100$$
$$a^2 = 36$$
$$a = 6$$

20. **13:** This triangle is one of the Pythagorean triples, a 5–12–13 triangle. The hypotenuse equals 13.

    Alternatively, use the Pythagorean theorem to solve.

$$5^2 + 12^2 = c^2$$
$$25 + 144 = c^2$$
$$169 = c^2$$
$$13 = c$$

21. **6:** This is a 3–4–5 triple, so the length of the third side is 3.

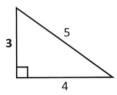

    The base of the triangle is 4 and the height is 3. Plug this into the area formula.

$$A = \frac{1}{2}bh$$
$$A = \frac{1}{2}(4)(3)$$
$$A = 6$$

22. **32:** In parallelograms, opposite sides have equal lengths, so two of the sides of the parallelogram have a length of 6 and two sides have a length of 10.

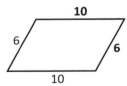

    The perimeter is $2(6 + 10) = 32$.

23. **36:** The area of a parallelogram is $b \times h$. In this parallelogram, the base is 9 and the height is 4, so the area is $(9) \times (4) = 36$.

24. **Area = 21, perimeter = 20:** In rectangles, opposite sides have equal lengths, so the rectangle looks like this:

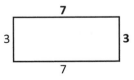

    The perimeter is $2(l + w) = 2(3 + 7) = 20$. The area is $l \times w = (7)(3) = 21$.

25. **Area = 48, perimeter = 28:** The diagonal of the rectangle creates a right triangle, so you can use the Pythagorean theorem to find the length of side *CD*. Alternatively, triangle *ACD* is a 6–8–10 triangle, so the length of side *CD* is 8.

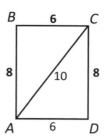

The perimeter of the rectangle is $2(l + w) = 2(8 + 6) = 28$. The area is $l \times w = (8)(6) = 48$.

26. **9π:** Redraw the figure *without* the circle so that you can focus on the rectangle. Add in the diagonal *BD*, as well as its length.

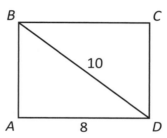

Look at right triangle *ABD*, where *BD* functions not only as the diagonal of rectangle *ABCD* but also as the hypotenuse of the right triangle. Find the third side of triangle *ABD*, either using the Pythagorean theorem or by recognizing a Pythagorean triple (6–8–10).

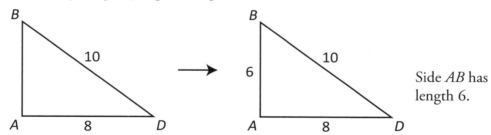

Side *AB* has length 6.

Now redraw the figure *with* the circle but without the diagonal, since you've gotten what you need from that: the other side of the rectangle.

10

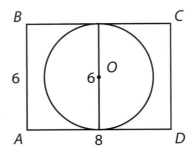

Since the circle touches both AD and BC, its diameter must be 6.

Finally, find the radius and compute the area:

$$d = 6 = 2r$$
$$3 = r$$

$$\text{Area} = \pi r^2$$
$$\text{Area} = \pi(3)^2$$
$$\text{Area} = 9\pi$$

27.

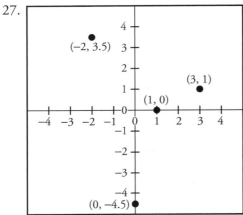

28. 1. $(2, -1)$: **E**

    2. $(-1.5, -3)$: **C**

    3. $(-1, 2)$: **B**

    4. $(3, 2)$: **D**

10

29.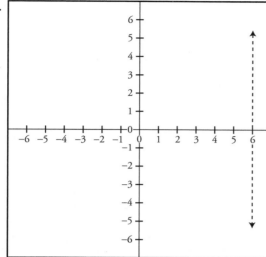

$x = 6$

30.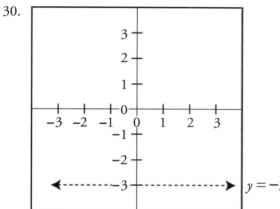

$y = -3$

31.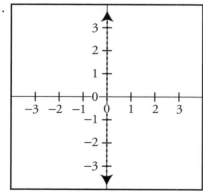

$x = 0$ is the y-axis.

32. 1. $(1, -2)$ is in **Quadrant IV**.

    2. $(-4.6, 7)$ is in **Quadrant II**.

    3. $(-1, -2.5)$ is in **Quadrant III**.

    4. $(3, 3)$ is in **Quadrant I**.

$x < 0$ $y > 0$ II	$x > 0$ $y > 0$ I
III $x < 0$ $y < 0$	IV $x > 0$ $y < 0$

33. 1. $x < 0, y > 0$ indicates **Quadrant II**.

    2. $x < 0, y < 0$ indicates **Quadrant III**.

    3. $y > 0$ indicates **Quadrants I and II**.

    4. $x < 0$ indicates **Quadrants II and III**.

34. **−4:** The point on the line with $x = -3$ has a y-coordinate of -4.

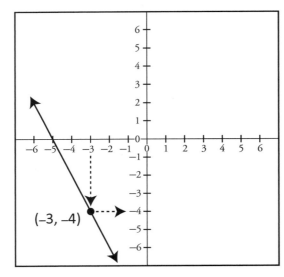

(−3, −4)

35. **False:** Test the point $(9, 21)$ in the equation $y = 2x + 1$. Plug in 9 for x to see whether you get 21 for y.

$$y = 2(9) + 1 = 19$$

Because 19 does not equal 21, the point $(9, 21)$ does not lie on the line.

10

36. **True:** Test the point $(4, 14)$ in the equation $y = x^2 - 2$. Plug in 4 for x to see whether you get 14 for y.

$$y = (4)^2 - 2 = 14$$

Because 14 is the desired number, the point $(4, 14)$ does lie on the curve defined by the equation $y = x^2 - 2$.

37. **Slope = 3, *y*-intercept = 4:** The equation $y = 3x + 4$ is already in $y = mx + b$ form, so you can directly find the slope and *y*-intercept. The slope, *m*, is 3, and the *y*-intercept, *b*, is 4.

38. **Slope = 2.5 or $\frac{5}{2}$, *y*-intercept = −6:** To find the slope and *y*-intercept of a line, put the equation in $y = mx + b$ form. Divide the original equation by 2 to make that happen. Therefore, $2y = 5x - 12$ becomes $y = 2.5x - 6$. The slope, *m*, is 2.5 (or $\frac{5}{2}$) and the *y*-intercept, *b*, is −6.

39.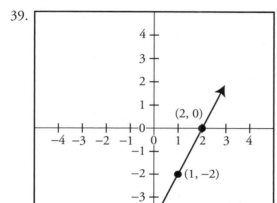

    $y = 2x - 4$

    slope = 2

    y-intercept = −4

    Think of the slope, 2, as a fraction: $\frac{2}{1}$. Go up 2 and to the right 1.

Chapter Review: Drill Sets

Drill 1

1. The radius of a circle is 4. What is its area?

2. The diameter of a circle is 7. What is its circumference?

3. The radius of a circle is 3. What is its circumference?

4. The area of a circle is 25π. What is its diameter?

5. The circumference of a circle is 18π. What is its area?

Drill 2

6. The area of a circle is 100π. What is its circumference?

7. The diameter of a circle is 16. Calculate its radius, circumference, and area.

8. Circle M has a circumference of 6π and circle N has an area of 8π. Which circle has the greater area?

9. Circle X has a diameter of 10 and circle Y has a circumference of 12π. Which circle has the greater area?

10. Circle A has an area of 64π and circle B has an area of 16π. The radius of circle A is how many times the radius of circle B ?

Drill 3

11. A circle has a radius of 8 and a sector with a central angle of 90°. What is the area of the sector?

12. A sector has a central angle of 45°. If the radius of the circle is 12, what is the arc length of the sector?

13. A sector has an arc length of 7π and the diameter of the circle is 14. What is the central angle of the sector?

14. A sector of a circle has a central angle of 270°. If the circle has a radius of 4, what is the area of the sector?

15. A circle has a radius of 12. If a sector of the circle has area of 24π, what is the central angle of the sector?

Drill 4

16. The area of a sector is $\frac{1}{10}$ of the area of the full circle. What is the central angle of the sector?

17. A circle has a radius of 18. What is the perimeter of a sector of the circle with a central angle of 60° ?

18. A circle has a radius of 8 and a sector with an area of 8π. What is the arc length of the sector?

19. A sector of a circle has an arc length of $\frac{\pi}{2}$ and a central angle of 45°. What is the radius of the circle?

20. Circle A has a radius of 4 and a sector with a central angle of 90°. Circle B has a radius of 6 and a sector with a central angle of 45°. Which sector has the greater area?

Drill 5

Note: Figures are not necessarily drawn to scale.

21. A triangle has two sides with lengths of 5 and 11, respectively. What is the range of values for the length of the third side?

22. The length of the hypotenuse of a right triangle is 10, and the length of one of the legs is 6. What is the length of the other leg?

23.

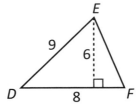

What is the area of triangle *DEF* ?

24.

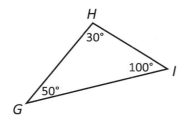

Which side of triangle *GHI* has the longest length? (Note: Figure not drawn to scale.)

25.

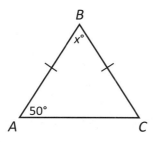

What is the value of *x* ?

Drill 6

Note: Figures are not necessarily drawn to scale.

26. Two sides of a triangle have lengths 4 and 8. Which of the following are possible side lengths of the third side? (More than one answer may apply.)

(A) 2

(B) 4

(C) 6

(D) 8

27.

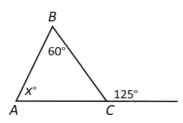

What is the value of *x* ?

28. Isosceles triangle *ABC* has two sides with lengths 3 and 9. What is the length of the third side?

29.

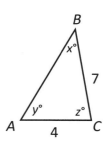

Which of the following could be the length of side *AB*, if *x < y < z* ?

(A) 6

(B) 10

(C) 14

30.

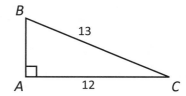

What is the area of right triangle *ABC* ?

Drill 7

Note: Figures are not necessarily drawn to scale.

31.

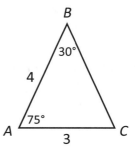

What is the perimeter of triangle *ABC* ?

32.

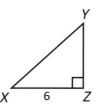

The area of right triangle *XYZ* is 12. What is the length of its hypotenuse?

10

33.

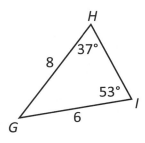

What is the length of side *HI* ?

34.

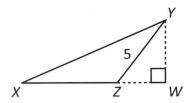

Which triangle has the greater perimeter?

35.

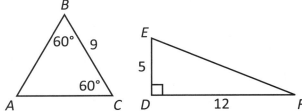

ZW has a length of 3 and *XZ* has a length of 6. What is the area of triangle *XYZ* ?

Drill 8

Note: Figures are not necessarily drawn to scale.

36.

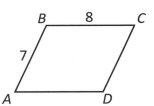

What is the perimeter of parallelogram *ABCD* ?

37.

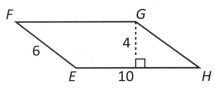

What is the area of parallelogram *EFGH* ?

38.

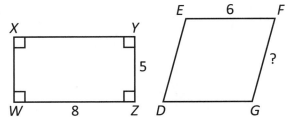

The rectangle and the parallelogram pictured have the same perimeters. What is the length of side *FG* ?

39.

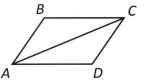

In parallelogram *ABCD*, triangle *ABC* has an area of 12. What is the area of triangle *ACD* ?

40.

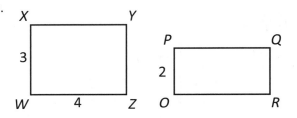

Rectangle *WXYZ* and rectangle *OPQR* have equal areas. What is the length of side *PQ* ?

Drill 9

Note: Figures are not necessarily drawn to scale.

41.

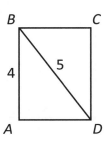

What is the area of rectangle *ABCD* ?

42.

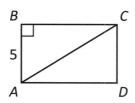

In rectangle *ABCD*, the area of triangle *ABC* is 30. What is the length of diagonal *AC* ?

43.

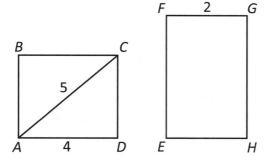

Rectangle *EFGH* has an area three times that of rectangle *ABCD*. What is the length of side *GH* ?

44. A rectangle has an area of 22 and a perimeter of 26. What are the length and width of the rectangle?

45.

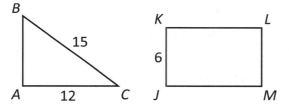

Right triangle *ABC* and rectangle *JKLM* have equal areas. What is the perimeter of rectangle *JKLM* ?

Drill 10

Note: Figures are not necessarily drawn to scale.

46. What is the perimeter of a square with an area of 25 ?

47. A rectangle and a square have the same area. The square has a perimeter of 32, and the rectangle has a width of 4. What is the length of the rectangle?

48.

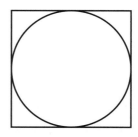

A circle is inscribed in a square, as shown in the figure above. If the area of the circle is 16π, what is the perimeter of the square?

49.

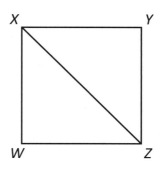

Square *WXYZ* has an area of 36. What is the length of diagonal *XZ* ?

50.

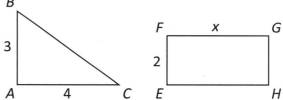

Right triangle *ABC* and rectangle *EFGH* have the same perimeter. What is the value of *x* ?

Drill 11

51. Draw a coordinate plane and plot the following points.

1. (2, 3)
2. (−2, −1)
3. (−5, −6)
4. (4, −2.5)

10

52.

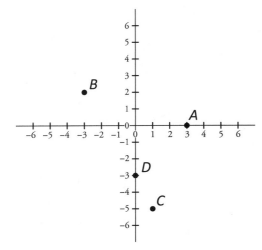

What are the *x*- and *y*-coordinates of the following points?

53.

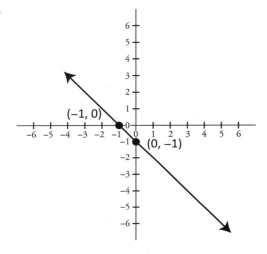

What is the *y*-coordinate of the point on the line that has an *x*-coordinate of 3 ?

54.

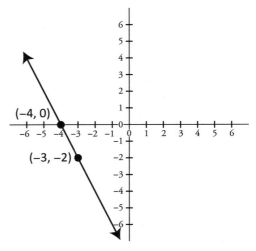

What is the *x*-coordinate of the point on the line that has a *y*-coordinate of −4?

55. Does the point (−3, 5) lie on the line
$y = -2x - 1$?

Drill 12

56. For the line $y = 3x - 4$, what is the *y*-coordinate when $x = 2$?

57. What is the *y*-intercept of the line
$y = -2x - 7$?

58. Graph the line $y = \frac{1}{2}x + 3$.

59. Graph the line $\frac{1}{2}y = -\frac{1}{2}x + 1$.

Drill Sets Solutions

Drill 1

1. **16π:**

$$A = \pi r^2$$
$$A = \pi(4)^2$$
$$A = 16\pi$$

2. **7π:**

$$C = \pi d$$
$$C = \pi(7) = 7\pi$$

3. **6π:**

$$C = 2\pi r$$
$$C = 2\pi(3)$$
$$C = 6\pi$$

4. **10:**

$$A = \pi r^2$$
$$25\pi = \pi r^2$$
$$25 = r^2$$
$$5 = r$$
$$d = 2r = 2(5) = 10$$

5. **81π:** The connection between circumference and area is radius. Use the circumference to solve for the radius.

$$C = 2\pi r$$
$$18\pi = 2\pi r$$
$$18 = 2r$$
$$9 = r$$

Use the radius to find the area.

$$A = \pi r^2$$
$$A = \pi(9)^2$$
$$A = 81\pi$$

Drill 2

6. **20π:**

$$A = \pi r^2$$
$$100\pi = \pi r^2$$
$$100 = r^2$$
$$10 = r$$

Use the radius to find the circumference.

$$C = 2\pi r$$
$$C = 2\pi(10)$$
$$C = 20\pi$$

7. **$r = 8$, $C = 16\pi$, and $A = 64\pi$:**

$$d = 2r$$
$$16 = 2r$$
$$8 = r$$

$$C = 2\pi r$$
$$C = 2\pi(8)$$
$$C = 16\pi$$

$$A = \pi r^2$$
$$A = \pi(8)^2$$
$$A = 64\pi$$

8. **Circle M:** One option is to find the radius of both circles, as the circle with the greater radius also has the greater area. Another option is to find the radius of circle M and then calculate its area. The second option is shown.

$$C_M = 2\pi r_M$$
$$6\pi = 2\pi r_M$$
$$3 = r_M$$
$$A_M = \pi r_M^2$$
$$A_M = \pi(3)^2$$
$$A_M = 9\pi$$

Because 9π is greater than 8π, circle M has the greater area.

9. **Circle Y:** You could find the area of each circle and compare. Note, though, that it is sufficient to find the radii of both circles. The one with the greater radius also has the greater area. Don't do more work than you're asked to do! You only need to tell which circle has the greater area; you don't need to be able to say what that area is.

$$d_X = 10 \qquad\qquad C_Y = 2\pi r_Y$$
$$r_X = 5 \qquad\qquad 12\pi = 2\pi r_Y$$
$$6 = r_Y$$

Since circle Y has the greater radius, it has the greater area.

10. **Two times:** The radius of circle A is two times the radius of circle B.

Careful: Don't use the areas to determine the ratio of the two circles' radii. Find the radii.

$$A_A = \pi r_A{}^2 \qquad\qquad A_B = \pi r_B{}^2$$
$$64\pi = \pi r_A{}^2 \qquad\qquad 16\pi = \pi r_B{}^2$$
$$64 = r_A{}^2 \qquad\qquad 16 = r_B{}^2$$
$$8 = r_A \qquad\qquad 4 = r_B{}^2$$

Thus, 8 is 2 times 4.

Drill 3

11. **16π:** If the sector has a central angle of 90°, then the sector is $\frac{90}{360} = \frac{1}{4}$ of the circle. Next, find the area of the entire circle.

$$A = \pi r^2$$
$$A = \pi (8)^2$$
$$A = 64\pi$$

The area of the sector is $\frac{1}{4}$ of the total area: $64\pi \times \frac{1}{4} = 16\pi$.

12. **3π:** A central angle of 45° corresponds to $\frac{45}{360} = \frac{1}{8}$ of the circle. Next, find the circumference of the entire circle.

$$C = 2\pi r$$
$$C = 2\pi (12)$$
$$C = 24\pi$$

The arc length of the sector is $\frac{1}{8}$ of the total circumference: $24\pi \times \frac{1}{8} = 3\pi$.

13. **180°:** To find the central angle of the sector, determine what fraction of the full circle the sector represents. Use the arc length to find the circumference of the circle, and use those two numbers to determine the fraction.

$$C = \pi d$$
$$C = \pi(14) = 14\pi$$
$$\frac{\text{Arc length}}{\text{Circumference}} = \frac{7\pi}{14\pi} = \frac{1}{2}$$

The arc length represents $\frac{1}{2}$ of the circle. The central angle, therefore, also represents $\frac{1}{2}$ of the total 360°: $360° \times \frac{1}{2} = 180°$.

14. **12π:** The sector represents $\frac{270°}{360°} = \frac{3}{4}$ of the circle. Next, find the area of the whole circle.

$$A = \pi r^2$$
$$A = \pi(4)^2$$
$$A = 16\pi$$

The area of the sector is $\frac{3}{4}$ of the total area: $16\pi \times \frac{3}{4} = 12\pi$.

15. **60°:** First, find the area of the whole circle.

$$A = \pi r^2$$
$$A = \pi(12)^2$$
$$A = 144\pi$$

The sector represents $\frac{24\pi}{144\pi} = \frac{1}{6}$ of the circle. The central angle is also $\frac{1}{6}$ of the circle: $360° \times \frac{1}{6} = 60°$.

Drill 4

16. **36°:** If the area of the sector is $\frac{1}{10}$ of the area of the full circle, then the central angle is also $\frac{1}{10}$ of the degree measure of the full circle.

$$360° \times \frac{1}{10} = 36°$$

17. **$6\pi + 36$:** The perimeter of a sector equals the arc length plus two radii.

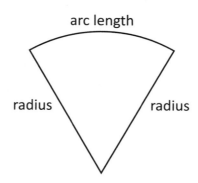

arc length

radius   radius

Begin by finding the circumference and the fraction of the total circle that the central angle represents.

$$C = 2\pi r$$
$$C = 2\pi(18) = 36\pi$$

$$\frac{\text{Central angle}}{\text{Total angle}} = \frac{60}{360} = \frac{1}{6}$$

The arc length is $\frac{1}{6}$ of the circumference: $36\pi \times \frac{1}{6} = 6\pi$.

The perimeter is $6\pi + 18 + 18 = 6\pi + 36$.

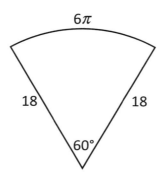

6π

18   18

$60°$

18. **2π:** First, determine what fraction of the circle the sector represents. Use the area of the sector to find the area of the whole circle.

$$A = \pi r^2$$
$$A = \pi(8)^2$$
$$A = 64\pi$$

$$\frac{\text{Area of sector}}{\text{Area of circle}} = \frac{8\pi}{64\pi} = \frac{1}{8}$$

The sector represents $\frac{1}{8}$ of the circle. To find the arc length, find the circumference of the whole circle.

$$C = 2\pi r$$
$$C = 16\pi$$

$$\text{Arc length} = 16\pi \times \frac{1}{8} = 2\pi$$

19. **2:** Begin by determining what portion of the circle the sector represents.

$$\frac{\text{Central angle}}{\text{Total angle}} = \frac{45}{360} = \frac{1}{8}$$

The sector represents $\frac{1}{8}$ of the circle. If the arc length of the sector is $\frac{\pi}{2}$, then the circumference of the whole circle is $\frac{\pi}{2} \times 8 = 4\pi$.

$$C = 2\pi r$$
$$4\pi = 2\pi r$$
$$2 = r$$

20. **Sector of circle B:** First, calculate what you need for sector A. Determine the fraction that the sector represents, the area of the entire circle, and finally the area of sector A.

$$\frac{\text{Central angle}}{\text{Total angle}} = \frac{90}{360} = \frac{1}{4}$$

$$A_A = \pi r_A{}^2$$
$$A_A = \pi(4)^2$$
$$A_A = 16\pi$$

Area of sector A $= \frac{1}{4} \times 16\pi = 4\pi$

Calculate the equivalent information for sector B.

$$\frac{\text{Central angle}}{\text{Total angle}} = \frac{45}{360} = \frac{1}{8}$$

$$A_B = \pi r_B{}^2$$
$$A_B = \pi(6)^2$$
$$A_B = 36\pi$$

Area of sector B $= \frac{1}{8} \times 36\pi = \frac{9}{2}\pi = 4.5\pi$

Because $4.5\pi > 4\pi$, the area of circle B's sector is greater than the area of circle A's sector.

Drill 5

21. **$6 <$ third side < 16:** The sum of the lengths of any two sides of a triangle must be greater than the length of the third side. Therefore, the third side must be less than $5 + 11 = 16$. The third side must also be greater than the difference of the lengths of the other two sides: $11 - 5 = 6$. Therefore, the third side must be between 6 and 16.

22. **8:** This is a Pythagorean triple, a 6–8–10 triangle. When you think you may have a triple, make sure that the longest measure corresponds to the hypotenuse. If you were told that the two *legs* were 6 and 10, then the hypotenuse could not be 8, since the hypotenuse is the longest side of the triangle.

Alternatively, use the Pythagorean theorem to solve.

$$a^2 + b^2 = c^2$$
$$6^2 + b^2 = 10^2$$
$$b^2 = 100 - 36$$

$$b^2 = 64$$
$$b = 8$$

23. **24:** Side *DF* can act as the base, and the line dropping straight down from point *E* to touch side *DF* at a right angle can act as the height. The base is 8 and the height is 6, so find the area:

$$A = \frac{1}{2}bh$$
$$A = \frac{1}{2}(8)(6)$$
$$A = 24$$

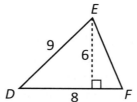

24. **GH:** Although *GI* looks like the longest side, don't trust how the picture looks when the question states that the picture is not drawn to scale. In any triangle, the longest side is opposite the greatest angle. Angle *HIG* is the greatest angle in the triangle, so side *GH* is the longest side.

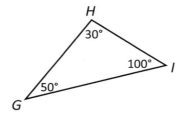

10

25. **80°:** In triangle *ABC*, sides *AB* and *BC* are equal, so their opposite angles are also equal. Therefore, angle *ACB* is also 50°.

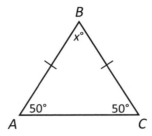

Now find angle *x*:

$$50 + 50 + x = 180$$
$$100 + x = 180$$
$$x = 80$$

Drill 6

26. **(C) 6 and (D) 8:** The sum of the lengths of any two sides of a triangle must be greater than the length of the third side. Therefore, the third side must be less than $4 + 8 = 12$. In addition, the third side must be greater than the difference of the other two sides: $8 - 4 = 4$. Only values between 4 and 12 are possible values for the third side of the triangle. (Note that 4 itself is not a possible value. The third side must be greater than 4.)

27. **65°:** First, solve for angle *BCA*. A straight line equals 180°, so angle *BCA* equals $180 - 125 = 55$.

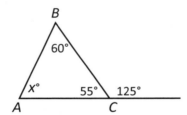

Now solve for *x*.

$$60 + 55 + x = 180$$
$$x = 65$$

28. **9:** This is a tricky one. If the triangle is isosceles, then two sides have equal length, but is the unknown third side 3 or 9? If the third side were 3, then the lengths of two of the sides would not sum to greater than the length of the third side, because $3 + 3$ is not greater than 9.

    The third side must be greater than the difference of the other two sides but smaller than the sum of the other two sides. The third side, then, has to be between $9 - 3 = 6$ and $9 + 3 = 12$.

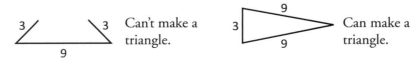

    Because 3 is not a valid possibility, the length of the third side must be 9.

29. **(B) 10:** Start with the rule about the third side of any triangle. The third side must be between $7 - 4 = 3$ and $7 + 4 = 11$. That knocks out answer (C), 14.

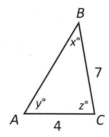

$$3 < \text{third side} < 11$$

Both 6 and 10 are still possibilities according to the rule about the third side of any triangle, though. Did the problem provide any other information that you haven't used yet?

Yes: angle x < angle y < angle z. If this is the case, then side AC < side BC < side AB, or $4 < 7 <$ side AB. Side AB cannot be 6, but it could be 10.

30. **30:** Find the base and the height. Because this is a right triangle, the base can be 12 and the height can be the leg AB. The triangle is a Pythagorean triple, so side $AB = 5$.

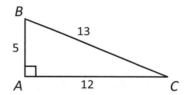

Now calculate the area.

$$A = \frac{1}{2}bh$$
$$A = \frac{1}{2}(12)(5)$$
$$A = 30$$

Drill 7

31. **11:** The perimeter is the sum of the three sides of the triangle. This triangle is not a Pythagorean triple, nor is it a right triangle, so there must be some other way to figure out the unknown third side.

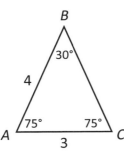

Let x equal the unknown third angle:

$$30 + 75 + x = 180$$
$$x = 75$$

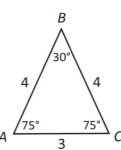

Angle BAC and angle ACB are both 75, so the triangle sides opposite those angles are also equal. Side AB has a length of 4, so side BC also has a length of 4.

The perimeter is $4 + 4 + 3 = 11$.

32. $\sqrt{52}$ **or** $2\sqrt{13}$**:** The area of a right triangle can be calculated using the two legs as the base and height, so use the given area to find the length of the unknown leg.

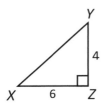

$$A = \frac{1}{2}bh$$
$$12 = \frac{1}{2}(6)h$$
$$12 = 3h$$
$$4 = h$$

Now use the Pythagorean theorem to find the length of the hypotenuse.

$$a^2 + b^2 = c^2$$
$$4^2 + 6^2 = c^2$$
$$16 + 36 = c^2$$
$$52 = c^2$$
$$\sqrt{52} = c$$

You can also write $\sqrt{52}$ as $\sqrt{52} = \sqrt{4 \times 13} = 2\sqrt{13}$.

33. **10:** First, calculate the value of the unknown third angle. In the second step of the math, rearrange the numbers to make the addition easier to perform.

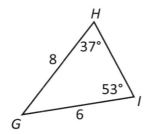

Let x = the third angle.

$$37 + 53 + x = 180$$
$$40 + 50 + x = 180$$
$$90 + x = 180$$
$$x = 90$$

It turns out that the third angle is a right angle! The directions indicated that figures are not necessarily drawn to scale. It may help to redraw the triangle in a way that more closely approximates the real dimensions.

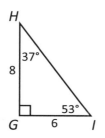

Now you have a chance to recognize that this is a Pythagorean triple in disguise. It's a 6–8–10 triangle, so the hypotenuse equals 10.

34. **Triangle *DEF*:** Start with triangle *ABC*. Two of the angles equal 60°. The third angle must also be 60°, since $60 + 60 + 60 = 180$. Because the triangle is equilateral (all of the angles are the same), all of the sides are the same length, 9. The perimeter is $3(9) = 27$.

Triangle *DEF* is a right triangle and a Pythagorean triple. The hypotenuse equals 13, and the perimeter is $5 + 12 + 13 = 30$.

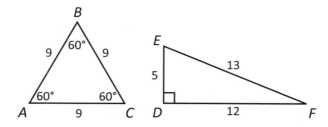

Because $30 > 27$, triangle *DEF* has the greater perimeter.

35. **12:** Redraw the figure and fill in everything you know about triangle *XYZ*.

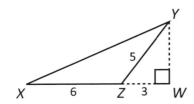

If *XZ* is the base, then *YW* can act as the height. Triangle *ZYW* is a right triangle and Pythagorean triple, a 3–4–5 triangle. Side *YW* = 4.

Now calculate the area of the triangle.

$$A = \frac{1}{2}bh$$

$$A = \frac{1}{2}(6)(4)$$

$$A = 12$$

Drill 8

36. **30:** Opposite sides of a parallelogram are equal, so side *CD* has a length of 7 and side *AD* has a length of 8. The perimeter is $2(7 + 8) = 2(15) = 30$.

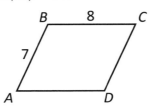

37. **40:** The area of a parallelogram is base × height. In this parallelogram, the base is 10 and the height is 4 (the base and height need to be perpendicular). The area is $10 \times 4 = 40$.

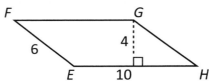

38. **7:** Start with the rectangle, since it provides enough information to calculate the perimeter.

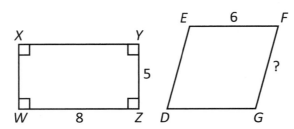

The opposite sides of a rectangle are the same length, so the perimeter of the rectangle is $2(5 + 8) = 2(13) = 26$.

The problem indicates that the perimeters of the two shapes are the same, so use this information to find the length of the unknown side of the parallelogram, FG.

$$26 = 2(6 + FG)$$
$$26 = 12 + 2FG$$
$$14 = 2FG$$
$$7 = FG$$

39. **12:** When you split any parallelogram by its diagonal, you create two identical triangles. In this case, since triangle ABC has an area of 12, triangle ACD must also have an area of 12.

40. **6:** Start by finding the area of rectangle $WXYZ$. The area of a rectangle $= lw = (3)(4) = 12$.

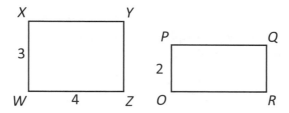

Since the two rectangles have equal areas, rectangle $OPQR$ also has an area of 12. Use this information to find the length of side PQ.

$$A = lw$$
$$12 = l(2)$$
$$6 = l$$

Drill 9

41. **12:** The area of a rectangle equals the length times the width. The length, 4, is already given. Find the width, AD (or BC).

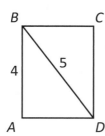

The triangle ABD is a Pythagorean triple—a 3–4–5 triangle. The length of side AD is therefore 3.

$$A = lw$$
$$A = (4)(3)$$
$$A = 12$$

42. **13:** First, use the area of the triangle to find the length of side *BC*.

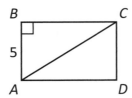

$$A = \frac{1}{2}bh$$

$$30 = \frac{1}{2}(5)h$$

$$\frac{2}{\cancel{5}}\left(\cancel{30}^{6}\right) = h$$

$$12 = h$$

The triangle is a 5–12–13 Pythagorean triple, so the length of the hypotenuse is 13. The hypotenuse, *AC*, is also the diagonal of the rectangle.

43. **18:** First, use the information given for rectangle *ABCD* to find its area.

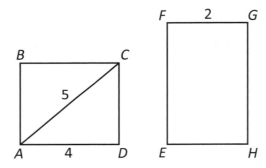

Triangle *ACD* is a 3–4–5 Pythagorean triple, so the side of length *CD* equals 3. The area of rectangle *ABCD* = *lw* = (3)(4) = 12.

According to the question stem, the area of rectangle *EFGH* is three times the area of rectangle *ABCD*, so the area of rectangle *EFGH* equals (3)(12) = 36. Use this information to find the length of side *GH*.

$$A = lw$$

$$36 = l(2)$$

$$18 = l$$

44. **11 and 2:** The question stem allows you to write two equations.

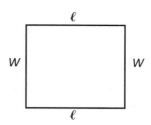

The area of the rectangle is 22, and the perimeter of the rectangle is 26.

$$A = lw = 22$$
$$P = 2(l + w) = 26$$

The second equation can be simplified to $l + w = 13$.

Substitute to solve for the values of the variables. In the second equation, isolate w (or l, your choice).

$$l = 13 - w$$

Substitute $(13 - w)$ for l in the first equation.

$$(13 - w)w = 22$$
$$13w - w^2 = 22 \qquad \text{This is a quadratic, so move everything to one side to set it equal to 0.}$$

$$0 = w^2 - 13w + 22$$
$$0 = (w - 11)(w - 2) \qquad \text{Factor the equation and solve.}$$
$$w = 11 \ \text{ or } \ w = 2$$

Which is it, 11 or 2? It doesn't matter: If $w = 2$, then $l = 13 - 2 = 11$, and if $w = 11$, then $l = 13 - 11 = 2$. These two solutions are the two dimensions of the rectangle.

45. **30:** Start with triangle ABC, since more information is provided for that shape.

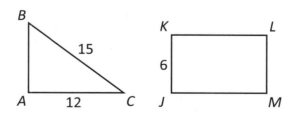

You can use the Pythagorean theorem to find side AB…but examine that triangle for a second. Does anything look familiar?

A 9–12–15 triangle is a 3–4–5 triangle in disguise; all of the measurements are tripled. Therefore, side $AB = 9$. Use that information to find the area of the triangle.

10

$$A = \frac{1}{2}bh$$
$$A = \frac{1}{2}(12)(9)$$
$$A = 54$$

Since the areas of the two shapes are the same, rectangle *JKLM* also has an area of 54. Use this information to find the length of the rectangle.

$$A = lw$$
$$54 = l(6)$$
$$9 = l$$

Now calculate the perimeter: $2(l + w) = 2(9 + 6) = 2(15) = 30$.

Drill 10

46. **20:** A square has four equal sides, so the area of a square is the length of one side squared.

$$A = s^2$$
$$25 = s^2$$
$$5 = s$$

The perimeter of a square is equal to four times the length of one side.

$$P = 4s$$
$$P = 4(5) = 20$$

47. **16:** Begin by drawing the shapes described in the problem.

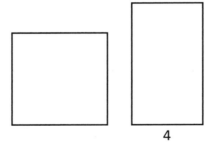

4

A square has four equal sides, so the perimeter is four times the length of one side. Let s = length of a side of the square. Solve for s, then find the area of the square.

$$4s = 32$$
$$s = 8$$
$$A = s^2 = (8)^2 = 64$$

Since the two shapes have the same area, the area of the rectangle is also 64. The width of the rectangle is 4, so solve for the length.

$$A = lw$$
$$64 = l(4)$$
$$16 = l$$

48. **32:** Whenever you're given a figure that combines shapes, find a common link between the two shapes. When a circle is inscribed in a square, the diameter of the circle is the same as the length of one side of the square.

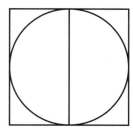

You can use the area of the circle to find the radius and then the diameter.

$$A = \pi r^2$$
$$16\pi = \pi r^2$$
$$16 = r^2$$
$$4 = r$$

If the radius is 4, then the diameter is 8. The diameter also equals one side of the square; find the perimeter.

$$P = 4s$$
$$P = 4(8) = 32$$

49. $\sqrt{72}$ **or** $6\sqrt{2}$**:** Use the area of the square to find the length of one side.

$$A = s^2$$
$$36 = s^2$$
$$6 = s$$

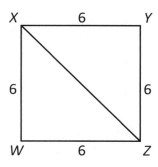

Now use the Pythagorean theorem to find the length of the diagonal *XZ*.

$$a^2 + b^2 = c^2$$
$$6^2 + 6^2 = c^2$$
$$36 + 36 = c^2$$
$$72 = c^2$$

$\sqrt{72}$ can also be written as $\sqrt{72} = \sqrt{36 \times 2} = 6\sqrt{2}$.

50. **4:** Start with the triangle. It's a 3–4–5 Pythagorean triple, so the hypotenuse is 5.

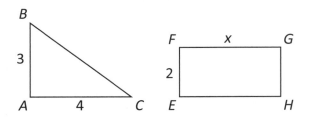

The perimeter of triangle *ABC* is $3 + 4 + 5 = 12$.

Since the two shapes have the same perimeter, the perimeter of rectangle *EFGH* is also 12. Use this information to find the value of *x*.

$$P = 2(l + w)$$
$$12 = 2(2 + x)$$
$$12 = 4 + 2x$$
$$8 = 2x$$
$$4 = x$$

Drill 11

51.

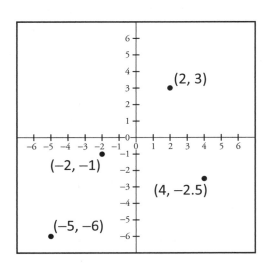

52.   A. (3, 0)

B. (−3, 2)

C. (1, −5)

D. (0, −3)

53.   **−4:** Find $x = 3$ on the x–axis, then go straight down to the line and draw a point. Next, move straight left to the y–axis to find the corresponding y–value. The point is (3, −4).

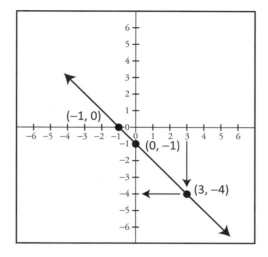

54.   **−2:** Find $y = −4$ on the y–axis, then go straight left to the line and draw a point. Next, move straight up to the x–axis to find the corresponding x–value. The point is (−2, −4).

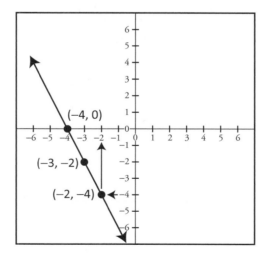

55. **Yes:** Plug the point $(-3, 5)$ into the equation $y = -2x - 1$. If the math works, then this point does lie on this line. If the math does not work, then this point does not lie on this line.

$$y = -2x - 1$$
$$(5) = -2(-3) - 1$$
$$5 = 6 - 1$$
$$5 = 5 \text{ True!}$$

The math is true, so this point does lie on this line.

Drill 12

56. **2:** To find the y-coordinate, plug in 2 for x and solve for y.

$$y = 3(2) - 4$$
$$y = 6 - 4 = 2$$

The y-coordinate is 2.

57. **−7:** The equation of the line $y = -2x - 7$ is already in $y = mx + b$ form. In this form, b stands for the y-intercept, so look at the equation to find the y-intercept. The y-intercept is $b = -7$. The point is $(0, -7)$. (For the y-intercept, the x value is always 0.)

58. The equation is in $y = mx + b$ form. The y-intercept, b, is 3, so place a point at $(0, 3)$. The slope, m, is $\frac{1}{2}$. Count up 1 and to the right 2 to get to the point $(2, 4)$. Draw a line to connect the two points.

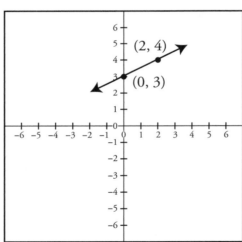

10

59. First, put the equation into $y = mx + b$ form. Multiply both sides by 2.

$$y = -x + 2$$

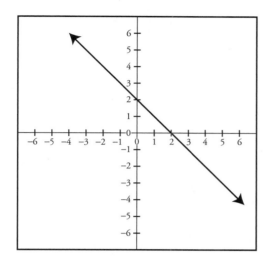

The y-intercept is 2, so place a point at $(0, 2)$.

The slope (m) is -1, so the line drops to the right. Count down 1 and right 1 and place another dot.

Draw a line to connect the two dots.

10

GLOSSARY

absolute value: The distance from 0 on the number line for a particular term. For example, the absolute value of -7 is 7, or $|-7| = 7$.

arc length: A section of a circle's circumference.

area: The space enclosed by a given closed shape on a plane; the formula depends on the specific shape. For example, the area of a rectangle equals *length × width*.

axis: One of the two number lines (*x*-axis or *y*-axis) used to indicate position on a coordinate plane.

base: 1) In the expression b^n, the variable b represents the base. The base is multiplied by itself n times. For example, $3^4 = 3 \times 3 \times 3 \times 3$, where 3 is the base. 2) Can refer to a side of a triangle when calculating the area of the triangle.

center (circle): The point from which any point on a circle's circumference is equidistant. The distance from the center to any point on the circumference is called the *radius*.

central angle (circle): The angle created by any two radii (lines drawn from the center of the circle to any point on the circumference).

circle: A set of points in a plane that are equidistant from a fixed center point.

circumference: The measure of the perimeter of a circle. The circumference of a circle can be found with the formula $C = 2\pi r$, where C is the circumference, r is the radius, and π is a constant (which you can approximate as 3.14).

coefficient: A number that is multiplied by a variable. In the equation $y = 2x + 5$, the coefficient of x is 2.

common denominator: When adding or subtracting fractions, first find a common denominator, generally the smallest common multiple of both numbers. (Note: You can also use the double-cross technique to add fractions.)

Example:

Given $\frac{3}{5} + \frac{1}{10}$, the two denominators are 5 and 10. The smallest multiple of both numbers is 10. Use 10 as the common denominator.

constant: In an equation or expression, a number that doesn't change. For example, in the equation $y = 3x + 2$, the number 2 is a constant. The x and y, on the other hand, are *variables*; they can have many possible values. Note: The number 3 is also constant in the sense that it doesn't change, but because it is attached to the variable x, its official name is a *coefficient*.

coordinate plane: Consists of a horizontal axis (typically labeled "*x*") and a vertical axis (typically labeled "*y*"), crossing at the number 0 on both axes.

decimal: Most commonly refers to numbers that fall in between integers, such as 1.5. An integer can be expressed in decimal form, though, such as 2.00 or 3.4×10^2. A decimal can express a part-to-whole relationship, just as a percent or fraction can.

Example:

1.2 is a decimal. The integers 1 and 2 are not decimals. An integer written as 1.0, however, is considered a decimal. The decimal 0.2 is equivalent to 20% or to $\frac{2}{10} = \frac{1}{5}$.

denominator: The bottom of a fraction. In the fraction $\frac{7}{2}$, 2 is the denominator.

diameter: A line segment that passes through the center of a circle and whose endpoints lie on the circle's circumference.

difference: When one number is subtracted from another, the difference is what is left over. The difference of 7 and 5 is 2, because $7 - 5 = 2$.

digit: The ten numbers 0, 1, 2, 3, 4, 5, 6, 7, 8, and 9. Use digits in combination to represent other numbers (e.g., 12 uses the digits 1 and 2, or 4.38 uses the digits 4, 3, and 8).

distribute: To multiply a term across other terms. For example, given the expression $3(x + 2)$, to distribute the 3, multiply each term in the parentheses by 3: $3(x + 2) = 3x + 6$. For the reverse math, see: *factor (verb)*.

distributed form: Presenting an expression as a sum or difference. In distributed form, terms are added or subtracted. $x^2 - 1$ is in distributed form, as is $x^2 + 2x + 1$. In contrast, $(x + 1)(x - 1)$ is not in distributed form; it is in *factored form*.

divisible: If an integer x divided by another integer y yields an integer, then x is said to be divisible by y.

Example:

> Because 12 divided by 3 yields the integer 4, 12 is divisible by 3. On the other hand, 12 divided by 5 does not yield an integer. Therefore, 12 is not divisible by 5.

divisor: The part of a division operation that comes after the division sign. In the operation $22 \div 4$ (or $\frac{22}{4}$), 4 is the divisor. *Divisor* is also a synonym for *factor (noun)*.

equation: A combination of mathematical expressions and symbols that contains an equals sign. For example, $3 + 7 = 10$ is an equation, as is $x + y = 3$. An equation makes a statement: left side equals right side.

equilateral triangle: A triangle in which all three angles are equal (and all three are 60°); furthermore, all three sides are of equal length.

even: An integer is even if it is divisible by 2. The integer 14 is even because $\frac{14}{2} = 7$, which is an integer.

exponent: In the expression b^n, the variable n represents the exponent. The exponent indicates how many times to multiply the base, b, by itself. For example, $4^3 = 4 \times 4 \times 4$, where 3 is the exponent.

expression: A combination of numbers and mathematical symbols that does not contain an equals sign. For example, xy is an expression, as is $x + 3$. An expression represents a quantity.

factor (noun): If integer a is divisible by integer b, then integer b is said to be a factor of integer a. For instance, 12 is divisible by 6 (because $12 \div 6 = 2$, an integer), so 6 is a factor of 12. On the GMAT, problems about factors are typically limited to positive integers; in this case, factors are equal to or smaller than the integer in question. The factors of 12 are 1, 2, 3, 4, 6, and 12. Note: For any given integer, 1 and the number itself are always factors of that number.

factor (verb): To pull a common factor out of multiple terms. For example, given the expression $3x + 6$, factor out the common term of 3: $3x + 6 = 3(x + 2)$. For the reverse math, see: *distribute*.

factor foundation rule: If a is a factor of b, and b is a factor of c, then a is also a factor of c. For example, 2 is a factor of 10, and 10 is a factor of 60. Therefore, 2 is also a factor of 60.

factor tree: Use the "factor tree" to break any number down into its prime factors (see: *prime number*). For example:

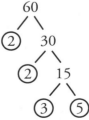

factored form: Presenting an expression as a product. In factored form, expressions are multiplied together. The expression $(x + 1)(x - 1)$ is in factored form: $(x + 1)$ and $(x - 1)$ are the factors. In contrast, $x^2 - 1$ is not in factored form; it is in *distributed form*.

FOIL: **F**irst, **O**utside, **I**nside, **L**ast; an acronym to remember the method for converting from factored to distributed form in a quadratic equation or expression. For example, $(x + 2)(x - 3)$ is a quadratic expression in factored form. Multiply the First, Outside, Inside, and Last terms to get the distributed form.

Example:

Factored form: $(y + 2)(y - 3)$

First: $y \times y = y^2$

Outside: $y \times (-3) = -3y$

Inside: $y \times 2 = 2y$

Last: $2 \times (-3) = -6$

Distributed form: $y^2 - 3y + 2y - 6 = y^2 - y - 6$

fraction: A way to express numbers that fall in between integers (though integers can also be expressed in fractional form). A fraction expresses a part-to-whole relationship in terms of a numerator (the part) and a denominator (the whole). For example, $\frac{3}{4}$ is a fraction, as is $\frac{6}{5}$. (The latter fraction has a special name. See: *improper fraction*.)

hypotenuse: The longest side of a right triangle. The hypotenuse is opposite the right angle.

improper fraction: Fractions that are greater than 1 or less than −1, such as $\frac{7}{2}$ or $-\frac{4}{3}$. Improper fractions can also be written as *mixed numbers*. For example, $\frac{7}{2} = 3\frac{1}{2}$.

inequality: A comparison of quantities that have different values. There are four ways to express inequalities: less than ($<$), less than or equal to ($\leq$), greater than ($>$), or greater than or equal to ($\geq$). Inequalities can be manipulated in the same way as equations with one exception: when multiplying or dividing by a negative number, the inequality sign must be flipped to the other direction.

integers: Numbers, such as −1, 0, 1, 2, and 3, that, in their most reduced form, have no fractional part. Integers include the counting numbers (1, 2, 3, …), their negative counterparts (−1, −2, −3, …), and 0.

interior angles: The angles that appear in the interior of a closed shape. For example, a triangle has three interior angles.

isosceles triangle: A triangle in which two of the three angles are equal; furthermore, the sides opposite the two equal angles are equal in length.

line: A straight one-dimensional figure having no thickness or curvature and extending infinitely in both directions. On the GMAT, lines are by definition perfectly straight.

line segment: A continuous, finite section of a line. The sides of a triangle or of a rectangle are line segments.

linear equation: An equation that does not contain exponents or multiple variables multiplied together. For example, $x + y = 3$ is a linear equation, but $xy = 3$ and $y = x^2$ are not. When plotted on a coordinate plane, linear equations create lines.

mixed number: An integer combined with a proper fraction. For example, $3\frac{1}{2}$ is a mixed number. Mixed numbers can also be written as *improper fractions*: $3\frac{1}{2} = \frac{7}{2}$.

multiple: Multiples are integers formed by multiplying some integer by any other integer. For example, 12 is a multiple of 12 (12×1), as are 24 ($= 12 \times 2$), 36 ($= 12 \times 3$), 48 ($= 12 \times 4$), and 60 ($= 12 \times 5$). (Negative multiples are possible in mathematics, but you typically don't have to use them to get GMAT questions correct. You can usually just stick with positive multiples.)

negative: Any number to the left of 0 on a number line; can be integer or non-integer.

negative exponent: Any exponent less than 0. To find a value for a term with a negative exponent, put the term containing the exponent in the denominator of a fraction and make the exponent positive, such as $4^{-2} = \frac{1}{4^2}$. Alternatively, if the term with the negative exponent is already in the denominator of a fraction, flip the fraction and make the exponent positive: $\frac{1}{3^{-2}} = 3^2 = 9$.

number line: A straight line that represents all of the numbers from negative infinity to positive infinity.

numerator: The top of a fraction. In the fraction $\frac{7}{2}$, the numerator is 7.

odd: An integer that is not divisible by 2 (i.e., division by 2 results in a non-integer value). The integer 15 is odd because $\frac{15}{2} = 7.5$, which is not an integer.

order of operations: The order in which mathematical operations must be carried out in order to simplify an expression. See: *PEMDAS*.

origin: The coordinate pair (0, 0) represents the origin of a coordinate plane.

parallelogram: A four-sided closed shape composed of straight lines in which the sides opposite one another are equal and the angles opposite one another are equal.

PEMDAS: An acronym that stands for Parentheses, Exponents, Multiplication, Division, Addition, Subtraction; used to remember the order of operations.

percent: Literally, "per 100"; expresses a special part-to-whole relationship between a number (the part) and 100 (the whole). A special type of fraction or decimal that involves the number 100 (e.g., 50% = 50 out of 100).

perfect cube: The cube of an integer.

perfect square: The square of an integer.

perimeter: In a polygon, the sum of the lengths of the sides.

perpendicular: Lines that intersect at a 90° angle.

place value: The "position" of a specific digit in a number. In the number 32.4, the digit 3 is in the tens place. The digit 2 is in the units place. The digit 4 is in the tenths place.

plane: A flat two-dimensional surface that extends infinitely in every direction.

point: An object that exists in a single location on the coordinate plane. Each point has a unique x-coordinate and y-coordinate that together describe its location. For example, $(1, -2)$ is a point.

polygon: A two-dimensional closed shape made exclusively of line segments. For example, a triangle is a polygon, as is a rectangle. A circle is a closed shape, but it is not a polygon because it does not contain line segments.

positive: Any number to the right of 0 on a number line; can be an integer or a non-integer.

prime factorization: A number expressed as a product of its *prime numbers*. For example, the prime factorization of 60 is $2 \times 2 \times 3 \times 5$. Every number has a unique prime factorization; that is, no two numbers have the same prime factorization.

prime number: A positive integer with *exactly* two different factors: 1 and itself. The number 1 does not qualify as prime because it has only one factor, not two. The number 2 is the smallest prime number; it is also the only even prime number. The numbers 2, 3, 5, 7, 11, and 13 are prime, for example.

product: The end result when two numbers are multiplied together. For example, 20 is the product of 4 and 5.

Pythagorean theorem: A formula used to calculate the sides of a right triangle: $a^2 + b^2 = c^2$, where a and b are the two sides (legs) that create the 90° angle, and c is the length of the side (hypotenuse) across from the 90° angle.

Pythagorean triple: A set of three numbers that describes the lengths of the three sides of a right triangle in which all three sides have integer lengths. Common Pythagorean triples are 3–4–5, 6–8–10, and 5–12–13.

quadrant: One-quarter of the coordinate plane. Bounded on two sides by the x- and y-axes.

quadratic expression: An expression including a variable raised to the second power (and no higher powers). Commonly of the form $ax^2 + bx + c$, where a, b, and c are constants. (A **quadratic equation** contains both a quadratic expression and an equals sign.)

quotient: The result of dividing one number by another. The quotient of $10 \div 5$ is 2.

radius: A line segment that connects the center of a circle with any point on that circle's circumference. Plural: *radii*.

reciprocal: The product of a number and its reciprocal is always 1. To get the reciprocal of an integer, put that integer in the denominator of a fraction with numerator 1. The reciprocal of 3 is $\frac{1}{3}$. To get the reciprocal of a fraction, switch the numerator and the denominator. The reciprocal of $\frac{2}{3}$ is $\frac{3}{2}$.

rectangle: A four-sided closed shape in which all of the angles equal 90° and in which the sides opposite one another are equal. Rectangles are also parallelograms.

right triangle: A triangle that includes a 90°, or right, angle.

root: The opposite of an exponent (in a sense). The square root of 16 (written $\sqrt{16}$) is the number (or numbers) that, when multiplied by itself, will yield 16. In this case, both 4 and -4 equal 16 when squared. However, when the GMAT provides the root sign for an even root, such as a square root, the only accepted answer is the positive root. That is, $\sqrt{16} = 4$, *not* $+4$ or -4. In contrast, the equation $x^2 = 16$ has *two* solutions, $+4$ and -4.

sector: A "wedge" or "slice of pie" of a circle, bounded by two radii and the arc connecting those two radii.

simplify (in general): Change the form of given math to a simpler form. For example, add two numbers in parentheses, combine like terms, or divide every term in an equation by the same number.

simplify (fractions): Reducing numerators and denominators to their smallest form by taking out common factors. Dividing the numerator and denominator by the same number does not change the value of the fraction.

Example:

Simplify $\frac{21}{6}$. Divide both the numerator and the denominator by a common factor, in this case, 3: $\frac{21}{6} = \frac{7}{2}$.

slope: In a coordinate plane, "rise over run," or the distance the line runs vertically over the distance the line runs horizontally. The slope of any given line is constant over the length of that line.

special products: The set of three commonly used quadratic forms found on the GMAT:

square of a sum: $(x + y)^2 = x^2 + 2xy + y^2$

square of a difference: $(x - y)^2 = x^2 - 2xy + y^2$

difference of squares: $(x + y)(x - y) = x^2 - y^2$

square: A four-sided closed shape in which all of the angles equal 90° and all of the sides are equal. Squares are also rectangles and parallelograms.

square root: See *root*.

sum: The result when two numbers are added together. The sum of 4 and 7 is 11.

term: Parts within an expression or equation that are separated by either a plus sign or a minus sign. For example, in the expression $x + 3$, "x" and "3" are each separate terms.

triangle: A three-sided closed shape composed of straight lines; the interior angles sum to 180°.

variable: Letter used as a substitute for an unknown value, or number. Common letters for variables are x, y, z, and t. In contrast to a constant, a variable is a value that can change (hence the term *variable*). In the equation $y = 3x + 2$, both y and x are variables.

whole number: See *integer*.

***x*-axis:** A horizontal number line that indicates left–right position on a coordinate plane.

***x*-coordinate:** The number that indicates where a point lies along the *x*-axis. The *x*-coordinate is always written first in the parentheses. The *x*-coordinate of $(2, -1)$ is 2.

x-intercept: The point where a line crosses the x-axis (that is, when $y = 0$).

y-axis: A vertical number line that indicates up–down position on a coordinate plane.

y-coordinate: The number that indicates where a point lies along the y-axis. The y-coordinate is always written second in the parentheses. The y-coordinate of $(2, -1)$ is -1.

y-intercept: The point where a line crosses the y-axis (i.e., when $x = 0$). In the equation of a line $y = mx + b$, the y-intercept equals b. The coordinates of the y-intercept are $(0, b)$.

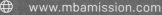

Go beyond books.
Try us for free.

In Person

Find a GMAT course near you
and attend the first session
free, no strings attached.

**Find your city at
manhattanprep.com/gmat/classes**

Online

Enjoy the flexibility of prepping
from home or the office with
our online course.

**See the full schedule at
manhattanprep.com/gmat/classes**

On Demand

Prep where you are, when you
want with GMAT Interact™—
our on-demand course.

**Try 5 full lessons for free at
manhattanprep.com/gmat/interact**

Not sure which is right for you? Try all three!
Or, give us a call, and we'll help you figure out
which program fits you best.

Prep made personal.

Whether you want quick coaching in a particular GMAT subject area or a comprehensive study plan developed around your goals, we've got you covered. Our expert GMAT instructors can help you hit your top score.

CHECK OUT THESE REVIEWS FROM MANHATTAN PREP STUDENTS.

Contact us at 800-576-4628 or gmat@manhattanprep.com for more information about your GMAT study options.